Nanocomposite Materials for Biomedical and Energy Storage Applications

Edited by Ashutosh Sharma

Published in London, United Kingdom

Nanocomposite Materials for Biomedical and Energy Storage Applications
http://dx.doi.org/10.5772/intechopen.95130
Edited by Ashutosh Sharma

Contributors
Samadhan R. Waghmode, Amol A. Dudhane, Vaibhav P. Mhaindarkar, Tahira Mahmood, Rahmat Ali, Abdul Naeem, Murtaza Sayed, Mohammed A. Sharaf, Andrzej Kloczkowski, Muhammad Hafeez, Yousef Faraj, Ruzhen Xie, Mahdi Razm-Pa, Farzin Emami, Honghao Hou, Shuai Liu, Rurong Lin, Jie Zhang, Chunyi Pu, Jianxing Huang, Rosa-Alejandra Hernadez-Esquivel, Elvia Zarate-Hernández, Gabriela Navarro-Tovar, Patricia Aguirre-Bañuelos, Lakshmana Reddy Nagappagari, Santosh S. Patil, Ganesh Shamrao Kamble, Diksha E. Shinde, Kiyoung Lee, Abid Ullah, Monika Wysocka-Żołopa, Emilia Grądzka, Krzysztof Winkler, Mourad Chemek, Mabrouk Ali, Ben Braieck Mourad, Ventirini Jany Wérry, Alimi Kamel, Masoud Safari, Valiollah Nobakht, Anielle Christine Almeida Silva, Weslley Rick Viana Sampaio, Petteson Linniker Carvalho Serra, Rômulo Ríbeiro Magalhães de Sousa, Noelio Oliveira Dantas

First published in London, United Kingdom, 2022 by IntechOpen
IntechOpen is the global imprint of INTECHOPEN LIMITED, registered in England and Wales, registration number: 11086078, 5 Princes Gate Court, London, SW7 2QJ, United Kingdom

British Library Cataloguing-in-Publication Data
A catalogue record for this book is available from the British Library

Additional hard and PDF copies can be obtained from orders@intechopen.com

Nanocomposite Materials for Biomedical and Energy Storage Applications
Edited by Ashutosh Sharma
p. cm.
Print ISBN 978-1-80355-618-5
Online ISBN 978-1-80355-619-2
eBook (PDF) ISBN 978-1-80355-620-8

For EU product safety concerns:
IN TECH d.o.o., Prolaz Marije Krucifikse Kozulić 3, 51000 Rijeka, Croatia,
info@intechopen.com or visit our website at intechopen.com.

Meet the editor

Ashutosh Sharma is an assistant professor in the Department of Materials Science and Engineering, Ajou University, Suwon, South Korea. He earned his Ph.D. in Metallurgical and Materials Engineering from the Indian Institute of Technology (IIT) Kharagpur. His research interests include electrochemical deposition, lead-free soldering and brazing, additive manufacturing, high entropy alloys, gas sensors, and biomaterials. Dr. Sharma is a life member of various scientific and professional bodies. He has more than 120 international journals, 17 patents, 3 book chapters, and 1 book to his credit. He received an Extraction and Processing Division (EPD) award from The Minerals, Metals and Materials Society (TMS) for his outstanding contribution to nonferrous materials processing.

Contents

Preface **IX**

Section 1
Nanocomposite Materials - Science and Technology **1**

Chapter 1 **3**
Recent Progress and Overview of Nanocomposites
by Muhammad Hafeez

Chapter 2 **33**
Investigating the Role of Auger Recombination on the Performance of a Self-Assembled Quantum Dot Laser
by Mahdi Razm-Pa and Farzin Emami

Chapter 3 **53**
Improved Nanocomposite Materials and Their Applications
by Tahira Mahmood, Abid Ullah and Rahmat Ali

Chapter 4 **77**
Nanocomposites Thin Films: Manufacturing and Applications
by Weslley Rick Viana Sampaio, Petteson Linniker Carvalho Serra, Noelio Oliveira Dantas, Rômulo Ríbeiro Magalhães de Sousa and Anielle Christine Almeida Silva

Section 2
Nanocomposite Materials for Biomedical Applications **91**

Chapter 5 **93**
Nanocomposite Biomaterials for Tissue Engineering and Regenerative Medicine Applications
by Shuai Liu, Rurong Lin, Chunyi Pu, Jianxing Huang, Jie Zhang and Honghao Hou

Chapter 6 **123**
Syzygium cumini Mediated Green Synthesis of Silver Nanoparticles for Reduction of 4-Nitrophenol and Assessment of Its Antibacterial Activity
by Samadhan R. Waghmode, Amol A. Dudhane and Vaibhav P. Mhaindarkar

Chapter 7 135
Solid Lipid Nanoparticles (SLN)
by Rosa-Alejandra Hernández-Esquivel, Gabriela Navarro-Tovar, Elvia Zárate-Hernández and Patricia Aguirre-Bañuelos

Section 3
Advanced Nanocomposites for Energy Storage and Conversion **161**

Chapter 8 163
Graphene Related Materials and Composites: Strategies and Their Photocatalytic Applications in Environmental Remediation
by Santosh S. Patil, Lakshmana Reddy Nagappagari, Ganesh Kamble, Diksha E. Shinde and Kiyoung Lee

Chapter 9 193
Processing of Graphene/Elastomer Nanocomposites: A Minireview
by Mohammed A. Sharaf and Andrzej Kloczkowski

Chapter 10 209
Perovskite-Based Nanomaterials and Nanocomposites for Photocatalytic Decontamination of Water
by Yousef Faraj and Ruzhen Xie

Chapter 11 237
Encapsulation of Metal Nanoparticles (MNPs) as Catalyst
by Masoud Safari and Valiollah Nobakht

Section 4
Polymer Nanocomposites for Diverse Engineering Applications **265**

Chapter 12 267
Conducting Polymer 1-D Composites: Formation, Structure and Application
by Monika Wysocka-Żołopa, Emilia Grądzka and Krzysztof Winkler

Chapter 13 291
Synthesis, Experimental and Theoretical Investigations on the Optical and Electronic Properties of New Organic Active Layer for a New Generation of Organic Light-Emitting Diode
by Mourad Chemek, Ali Mabrouk, Mourad Ben Braieck, Jany Wérry Ventirini and Alimi Kamel

Chapter 14 307
Design and Fabrication of Microencapsulated Phase Change Materials for Energy/Thermal Energy Storage and Other Versatile Applications
by Tahira Mahmood, Rahmat Ali, Abdul Naeem and Murtaza Syed

Preface

Nanocomposite materials are the backbone of various fundamental and engineering applications requiring optimized sets of properties that are lightweight and fulfill energy demands. Nanocomposites such as carbon nanomaterials have attracted the attention of researchers in both industry and academia.

This book presents and discusses the development and applications of nanocomposites in fields such as biomedicine, health care, microelectronics, energy storage devices, and the environment. It is divided into four sections. Chapters discuss nanocomposite materials and their applications, nanocomposite materials for biomedicine and energy storage and conversion, graphene and quantum dots, organic light-emitting diodes, biological nanoparticles (solid lipid nanoparticles), nanocomposite materials for environmental remediation, and more.

This book is a useful resource for undergraduate and postgraduate students, industry and academic personnel, interdisciplinary materials scientists, nanocomposite manufacturers, and other interested readers.

I take this opportunity to thank all the contributing authors, co-authors, and publication staff from IntechOpen for their efforts in preparing this book. We hope that the new experiments and research presented herein will be the foundation for the development of new composite materials and technologies for humankind. I hope that readers enjoy this book and that it will serve as an aid to create new exciting materials with unique properties.

Dr. Ashutosh Sharma
Assistant Professor,
Department of Materials Science and Engineering,
Ajou University,
Suwon, South Korea

Section 1

Nanocomposite Materials - Science and Technology

Chapter 1

Recent Progress and Overview of Nanocomposites

Muhammad Hafeez

Abstract

Nanocomposites are versatile materials because of possessing superior properties as compared to their parent materials. Due to their improved electrical, mechanical, thermomechanical, electronic, optoelectronic, thermal, and magnetic properties, these materials are receiving much attention from researchers all over the world. In every field, the focus of the research is to develop such materials which have low weight, superior strength, and enhanced performance as well as cost competitiveness in comparison to existing materials. The nanocomposite materials have been used in the fields of avionics, biomedical, auto industry, sports industry, oil/gas, construction, food industry, agriculture industry, and information technology. This chapter addresses the synthesis, unique properties, and diverse applications of nanocomposites in different fields.

Keywords: nanocomposites, synthesis, properties, applications, ceramic, metal, polymer

1. Introduction

Composites consist of two materials; one of the materials is called the reinforcing phase (fibers, sheets, or particles), and is embedded in the other material called the matrix phase. The composite materials have more strength as compared to constituent materials. Wood is a common example of natural composites. Snails and oysters shells are also examples of natural composite materials. The history of the usage of nanocomposites is quite old. Egyptian have mixed straws with clay to form bricks. Mongols have used the composites in warfare and even in recent times, during World War II, the composite-based materials were used in military appliances, and in the modern era, a large number of composites are used in different fields [1].

The nanocomposite is a solid compound made up of several layers where at least one of the layers has one, two, or three dimensions with a nanometer size [2]. In nanocomposites, the atoms of the materials are arranged in the form of clusters or small grains. A solid multiphase having one of the 1, 2, or 3 phases less than 100 nm is called nanocomposites [3]. Nanoparticles, nanoclays, and nanofibers are examples of nanocomposites. Nanocomposites find their applications in the fields of medical and engineering.

Currently, scientists are facing the problem of data loss due to the overheating of imaging spectrographs in the Hubble Space Telescope. A probable solution is the use of a carbon nanotube that could dissipate the excess heat. The nanocomposites are lighter in weight and it is estimated that in near future the steel and aluminum-based products used in different engineering applications might be replaced with nanocomposites.

A Nanocomposite (NC) consists of several stages where, at least one, two, or three dimensions are in the nanometric range. Taking the size of objects up to the nanometer level creates optical connectors that are very important in the development of building material structures. Nanocomposites (NCs) provide opportunities on a completely new scale to solve hurdles ranging from the medical, pharma-industry to food packaging and from the electronics to energy-producing industries. Nanocomposites can be divided into three categories, which are as follows:

- Metal matrix nanocomposites (MMNCs).
- Ceramic matrix nanocomposites (CMNCs).
- Polymer matrix nanocomposites (PMNCs).

The aim of the development of nanocomposites is to designate which raw materials and processes are best suited to produce specific nanomaterials by studying their uses, benefits, and drawbacks. Transformation factor where the size of the material is made less than a certain level is known as "Sensitive size." Taking the size of the material down to the nanometer level fabricates the interactions with phase interfaces that become necessary in the development of building material structures. The ratio within the surface area to the volume of a reinforced material is used in synthesizing nanocomposites has directly been related to the structure-property relationship. A material nanocomposite (10^{-9} m diameter) that is made from non-metallic, metallic, or polymeric constituents by a certain process offers added benefits to retain their primary features and overcome defects by expressing some new characteristics. Such materials present a multiphase crossover in the matrix as well as in reinforcing materials. Reinforced material is a dispersed surface in the form of composite materials, such as fiberglass and organic fibers, whereas, matrix material consists of a uniform state and contains metallic, non-metallic, and polymer-based materials [4]. Nanocomposites represent a new way to surpass the limitations posed by microcomposites and the monoliths which have become the objects of the future. NCs have the following advantages over other materials:

- The ratio between surface area and volume is high that allows a small fill size.
- Superior mechanical characteristics that are minimum wastage of power and initial resistance and possess high ductility.
- Having improved visual features (the particle size anchors the light transmission).

High robustness and effective impact are associated with nanoparticles along with the incorporation of composite matrix. But, insufficient understanding of structural materials and material to structural relationships, the need for easy particle breakdown and dispersion are the major hurdles in nanocomposite synthesis (see **Figure 1**).

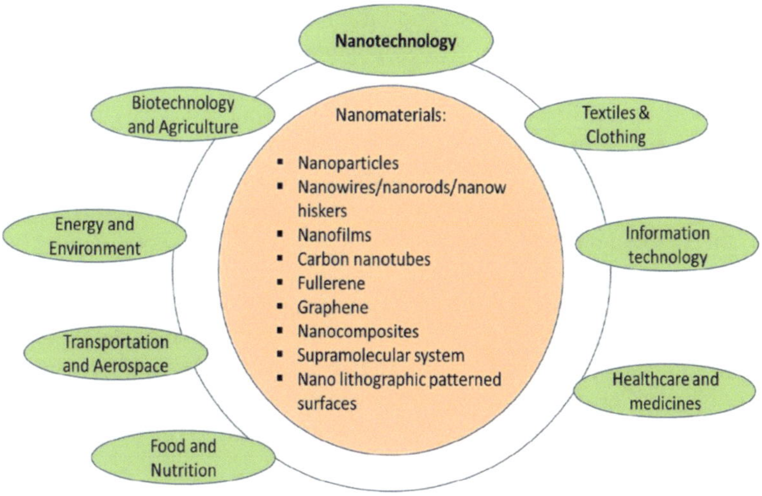

Figure 1.
Broader classification and applications of nanotechnology.

2. Classification of nanocomposites

2.1 Metal matrix nanocomposites (MMCs)

Since the late nineteenth century, composite materials were widely applicable in many systems with greater efficacy [5]. MMCs are one of the largest groups of compounds that are often strengthened with clay. The combination of metals and ceramic structures offers a variety of applications. There have been different ways of making MMCs but, powdered metallurgy is considered a unique process. Indeed, powdered metallurgy (PM) has been considered an effective method that has transformed the material industry by allowing them to build complex structural elements that control the precise strength, flexibility of composite design, and produce highly soluble materials with desirable compositions [6]. It seems that there are two main factors that determine the mechanisms of action in MMCs, that is the desired features of a material and technical constraints. If manufacturing a special type of alloy, for example, it is imperative of establishing a hot phase that allows a phase transformation into a metallic wire form; however, the scientific possibility still exists to design it from a large billet but, technological hurdles would have not allowed it to do so. We would otherwise need to apply high strain and thermal modulus, which can eventually damage the required structure. Therefore, multistep processing channels are used to improve features and overcome technical barriers that are the backbone of the process. To elaborate further, two-step sintering (TSS) and multistep isothermal forging (MIF), and isothermal rolling (IR) as a good plastic deformation (SPD) process have been used.

The sorting of building materials to meet the desired mechanical requirements followed by fine grain sizes and high density is a challenge for MMCs, as it requires very high temperature and often requires to be aligned with mechanical stress in a standard environment, that is very costly for samples having large sizes. The process offered a constricted geometry of the sample matrix [7]. Such a challenge can be remedied by using pressure-less two-step sintering (TSS) where the exfoliated sintering has resulted in fine heat resistance-granular material that is highly pure and more stable. TSS is a robust and productive method for obtaining a good microstructure with theoretical density. The mechanism includes heating of the composite until it reaches a high temperature and gains 70–90% of theoretical density then returns to a lower temperature to obtain higher density with controlled grain growth. As the grain boundary responsible for grain growth, it has a higher enthalpy compared to the grain boundary distribution at a certain temperature, the latter accomplishes during the second-step sintering in mild thermal conditions, suppresses grain growth, and cause pores annihilation and full densification. In addition, the TSS can overcome the problems associated with performing cold compression/filtration and/or compaction of the water phase leading to the metal agglomeration, respectively. Many other studies have shown that MMCs obtained through different processes provide a higher initial temperature for many clay materials, such as zinc oxide, magnesium-niobium-doped yttrium oxide, barium titinate, and silicon carbide, as well as other compounds. Ceramic matrix materials include compounds that are titanium boride 40% by weight, titanium nitride, and aluminum oxide 10% by volume.

2.2 Ceramic matrix nanocomposites (CMNCs)

Ceramic matrix nanocomposites (CMNCs) are added with solitary or multiple layers of ceramics to strengthen the crack resistance, heat absorption, and chemical resistance. Whereas, the main flaws of ceramics are their stiffness and less durability that keep them away from being used for industrial references. The limitation has been overcome by the production of ceramic-matrix (CMNCs) nanocomposites. The CMNC model incorporates a matrix in which energy dispelling components (fiber, platelets, or particulates) are added to CMNCs to reduce stiffness and increase crack resistance [2]. Raw materials for CMNC matrixes include alumina, SiC, SiN, etc. Generally, all reinforcements of the nanocomposites are of nanometric sizes. Iron and other metal powders: TiO_2, silica, clay are used for amorphous reinforcements. The most common reinforcements are clays and silicates, due to their low particle sizes and well-studied chemical interactions. The addition of clays and silica layers even in small amounts modify matrix structures. Many different approaches are designed for CMNCs integrations. Recently modified techniques are single source-precursor technology that is based on melt spinning of mixed raw materials followed by pyrolysis of the nanofibers. Some established mechanisms are PM; polymeric monomer method, spray-pyrolysis, and vapor methodologies (CVD and PVD) [8]. Chemical methods are the sol-gel process, colloidal method, and rain synthesis [9].

Mixtures of metal amalgamation and mechanical milling are widely used to process a promising program but these methods require an accurate measurement of powdered concentrations to produce a system in a metastable state and then there are a few steps that make strong semifinal products. Various combinations of metallic reactants are another method of directly producing the metallic bulk, for example, the Mg system is a hybrid matrix that includes fusion welding and composite casting requiring metal in the form of a liquid or roll cladding/bonding of solid-state welding. In all of these ways, metals present a diffusion-bound under relatively moderate pressure at a higher temperature for a long time; so, it is not possible to produce a good microstructures. In the process, the metal disk is pressed in high pressure environment with simultaneous torsion straining and processing, which is usually carried out at room temperature (RT), the process is equally effective even on hard as well as on amorphous substances, such as Mg alloys. Also, processed metals usually show improvements in physical and mechanical properties through the use of critical grain refinement and deep introduction of point and line disorders (see **Tables 1**–**3**).

Class	Matrix/ Reinforcements	Properties	Reference
Ceramic/Matrix	$Al_2O_3/NdAlO_3$	Improved Photoluminescence	[10, 11]
	Al_2O_3/Mo	Improved strength and toughness	[12, 13]
	Al_2O_3/W	—	[11]
	Al_2O_3/ZrO_2	—	[14]
	Al_2O_3/SiC	—	[15]
	Si_3N_4/SiC	—	[16, 17]
	$MoSi_2/ZrO_2$	—	[18]
	$B4C/TiB_2$	—	[19]
	MgO/SiC	—	[20]
	Al_2O_3/TiO_2		[13]
	Al_2O_3/CNT		[18, 21]

Table 1.
Examples of ceramic-based nanocomposites.

Matrix/reinforcement	Ag/Au	Improved catalytic activity	[22]
	Ag-Pd	Improved catalytic activity	[22]
	Au-Pd	Improved catalytic activity	[22]
	Cu/Nb	Improved microhardness	[23]
	Al/AlN	Higher compression resistance	[24]
	Al/SiC	Improved hardness	[25]
	CNT/Sb	Improvement in Li intercalation properties	[26]
	Cu/Al2O3	Improved microhardness	[27]
	CNT/Fe3O4	Improved electrical conductivity	[28]
	Fe-Cr/Al2O3	—	[21]; [29]
	Co/Cr	—	[12]; [19]
	Fe/MgO	—	[30]
	Mg/CNT	—	[31]
	Al/CNT	—	[32]

Table 2.
Examples of metal-based nanocomposites.

When fine-grained ceramic or other solid particles are embedded in a "soft" metal matrix to form metal matrix compounds (MMCs), the elements of the matrix materials can be greatly improved and strengthened. The strengthening of the mechanism for MMCs has been tested by many researchers. It has been thought that desired characteristics of composite metal structures with nano-sized ceramic particles (1.0–100 nm), called MMNCs, can be greatly improved even in these lowest volume conditions. Currently, mechanical mixing (e.g., high-power ball milling) for metal

Class	Matrix/Reinforcements	Properties	Reference
Polymers/metal oxide	PANI/WO_3		[33]
	PPANI/CeO_2	Thermally stable material	[34]
	PANI/ Sm_2O_3, La_2O_3	—	[35]
	PANI/La-Nd	Electromagnetic interference	[36]
	PANI/ Ce-TiO_2	sensor	[37]
	PANI/Nd_2O_3:Al_2O_3	Dielectric constant	
	PPY/Sm_2O_3	Supercapacitor	[38, 39]
	PPY/Y_2O_3	Semiconductor	[30]
	PPY/Nb_2O_5	—	[38]
	PPY/CeO_2	Sensor	[40]
	PPY/RuO_2	Supercapacitor	[41]
	PPY/RuO_2	Supercapacitor	[42]
	Polyindole (PIN) TiO_2	Semiconductor	[43]
	PEO/La_2O_3	Semiconductor and Solid Polymer Electrolyte (SPE)	[44]
	PVA/Ho/Gd	Optical display	[45]
	PVDF/La_2O_3	Thermally stable material	[46]
Polymers/NPs	Poly (styrene sulfonate) Co_3O_4 nanoparticles	Humidity Sensing	[47]
	Polypyrrole (PPy) Titanium dioxide (TiO_2)	Gas sensor Applications	[48]
	Polyaniline (PANI) Halloysite	Supercapacitor Applications	[49]
	Poly (allylamine hydrochloride) Graphene Oxide	Enhanced Mechanical Properties	[50]
	poly-L-lysine Graphene Oxide	Bio-scaffold coating	[50]
	Gelatin Graphene	Cellular imaging / Drug Delivery	[47]
	Sodium alginate Graphene	Oxide Tissue Engineering	[51]
	poly(lactic-co-glycolic acid) Graphene Oxide	Enhance Mechanical and Thermal properties	[52]
	Polyurethane (PU) Graphene Oxide	Enhance Mechanical and Thermal properties	[53]
	Poly(propylene fumarate) Graphene Oxide	Tissue Engineering	[54]
	polyvinylidene difluoride (PVDF) Graphene/ZnO	Temperature Sensing Applications	[55]

PANI = polyanialine, PPY = polypyrrole, PVA = Polyvinyl Alcohol, PEO = Poly(ethylene oxide), PVDF = Polyvinylidene fluoride.

Table 3.
Examples of polymer-based nanocomposites.

and ceramic powders is generally used to study the characteristics of MMNCs. Mixing ceramic particles with nanosize is energy as well as a time-consuming and costly procedure. Exfoliation, like the liquid-phase process, is best known for its ability to produce products with complex shapes. It will be desirable to synthesize MMNCs parts that are not as heavy as cast with the distribution of good reinforcement and integrity of the structure. However, there are ceramic particles with nanosize that put forth several problems that is very difficult to disperse the same is true for liquid metals because of their unwetting nature, the metal matrix having large surface to volume ratios, which facilitates agglomeration and cluster formation. Powerful ultrasonic waves have been proven very helpful in the context that they produce important indirect effects of liquids, namely transient cavitation as well as acoustic radiation. Acoustic cavitation covers the formation, growth, folding, and collapse of small object bubbles, which produce momentary (in microseconds) small "hot spots" that can attain temperatures (5000°C), pressures (1000 atm), and temperature rise and drops of 10^{10} K/s. The combination of impact with higher temperatures can also create improvements in the wetting between liquids and particles, thus facilitating the preparation of diffused compounds with effective microparticles.

It is thought that strong cavitation of the microscale transient, as well as macroscopic dispersion, may effectively disperse nanoparticles into soluble alloys and improve their wetting, thus making them more productive in performance as highly castable, light-weighted MMNCs.

Most CMNCs have low fracture resistance and are brittle. In addition to the discovery of ceramic coated CMNCs and silicon carbide (SiC), the modern focus is on the construction of ceramic-based nanocomposites with improved properties. Carbon nanotubes agglomeration increases the material's toughness by energy quenching through elastic modulus in the deformation stage. However, the design complexities have put a limit on the syntheses of these nanomaterials. The main drawback has been the nonuniformity of carbon nanotubes (CNTs) in the matrix suspension. The deformation of CMNCs has often been associated with high thermal and reactive environments that occur during the production of CMNCs. Nevertheless, there is sizeable progress in the field of nanocomposites but still, these are just preliminary steps to develop nanocomposites, a significant amount of exploration and effort is further required to ultrafine these manufacturing techniques. For example, a team from the University of California, Davis, has developed alumina ceramic by combining a single wall of carbon nanotubes (SWCNTs) with Al_2O_3 nanopowders using PM method. The resulting nanocomposite had advanced thermal, electronic, and mechanical characteristics. The highly potent anisotropic nanocomposite has a thermal ratio of 3:1 in an aligned plane. Electrical conductivity was far better than pure alumina matrix. Most importantly, the fracture strength was thrice higher than alumina with the crack resistance, heat absorption, and shock resistance capacity. Recently, at Tohoku University, a research group has synthesized a sophisticated CMNC on alumina ceramic through multi-walled carbon nanotubes. This process has reduced the phase separation that has resulted in a nanocompound with more uniformity in its structural phase. The addition of 0.9% acid-contained MWCNTs produced a component with a crack capacity of 5.90–0.27 MPa m1/2, greater than pure alumina NC (3–5 MPa m1/2) and a stronger bending capacity of 27%. A Chinese group of Qingdao University of Science and Technology has reported the MWCNT/zircona CMNCs produced by spark-sintering process had 18% higher fracture strength as compared to pure zircona. Another study by US Nano Labs has prepared a high-density boron carbide (B_4C) containing CMNCs. This composite was produced by the hot pressure-sintering process. However, none of these techniques have produced significant fracture toughness and heat dispelling properties, such as those in SiC-fiber-reinforced composite.

2.3 Polymer nanocomposites (PMNCs)

While nanotechnology still presents a picture of the future, nanocomposites set an example for realistic and rapidly booming applications. For instance, Geoff Ogilvy won the 2006 US Open golf tournament by using a nanomaterial-reinforced polymer-based club. Nanocomposites include materials that is CNTs, mineral materials, metals, and other fillers that can greatly improve composite structures. They attract a lot of awareness and some have been commercially available, having abilities to offer all kinds of uniqueness. Polymer-based products are the best-selling categories of NCs and covered global revenue of approximately, 223 million US dollars, in the year 2009. The nanomaterial's inclusions to the polymeric materials can enhance polymer characteristics that is robustness and strength, Young's modulus, impact endurance and scratch proofing, heat absorption, chemical defiance with electrical insulation and thermal adherence, stability toward the thermal shocks. Currently, minerals compounds and CNTs-based materials are more widely used than NPs. One of the premier commercial systems for PMNCs was used by Toyota, which has used nanoclay with nylon-6 PMNC [56] in their engine component showed an excellent result. In the late 1980s, Toyota Central Research Labs partnered with Ube Industries, a Japanese supplier of fossils, to cement a new 6-nylon composite coated with layers of montmorillonite (naturally occurring silicate clay). The component of this clay has enhanced Toyota's new model's performance which subsequently found its uses in a time belt cover, benefiting from improved temperature adherence and size stability. Since then, few car manufacturers have used nanocomposites of clay material in auto parts, such as rocker box coverings, body panels; the latter is 60% lighter and is more fracture-proof than regular automotive parts. The cargo bed for the 2005 GM model Hummer used approximately 3-kilogram of molded parts of nanoclay/polypropylene nanocomposite in its trim, mid-bridge, canvas panel, and box protectors. Polymer barrier technology was also benefited from these material NCs. Nylon/nanoclay composite is also applied for beverage bottles and in the food packaging industries. The addition of clay can significantly reduce gas/vapor infiltration, as clay platelets and thus prevent mobility, leading to significant improvements in shelf life. CNT-based nanocomposites are gaining increased industrial use from sports and leisure to technology, automotive, and defense motives. CNTs are attractive because of their excellent physical properties that often surpass many highly advanced materials and are now embedded in many polymeric NCs. Many automotive systems are sprayed with electrostatic paints. Plastic body panels need to be carefully processed for the paint to work properly. CNT is being applied as an alternative to carbon black, an expensive primer. The extra edge is being low CNT loading is needed to acquire the required conduction for the polymer to retain half of its actual length than 3–4% length saved when using carbon black. Importantly it is ensured that a panel must maintain its strength at a critical decrease of temperatures and never breaks. In addition, CNTs are so minute and used for such a low load that the higher end of class "A" is available in obtained NCs. The high-power output of CNT-nanocomposites is also utilized in the electronic industry, mainly to reduce the chances of damage caused by electrostatic accumulation or emissions. The PMNCs have found their applications in integrated circuits (IC). Joint Electronic Engineering Council trays wafer carriers and IC test that burn sockets because of high potential differences, combined with these materials having superior thermophysical properties to avoid the disaster. An example of a substance used in the industry is the Plasticyl range of CNTs/thermoplastic and nanocomposites, produced by the Nanocyl component as a precursor. Other benefits of PMNCs are seawater-cooled intercoolers on large diesel engines and in the power stations, where PMNCs will offer a robust substitution to copper-containing alloy, thermal rescue systems from fire hydrants and flue gases, working under 3008 celsius, whereas commercially used MMNCs systems lose their robustness in the chemical management as well as in processing industries where fissile environments prevail [56]. Demonstrating their strength and toughness, these materials have found applications from being used in baseball bats, bicycle frames, and power boats to military boats and aircraft. The leading company in nanocomposite technology is Applied Nanotech Holdings.

However, adding PMNCs, especially CNTs, in a resin or other matrices is not an easy task. Problems, such as segmentation, merging, poor disintegration, and poor adherence to host, should be overcome during integration. Some companies have developed methods that are specific for certain NP. For example, Zyvex uses new technologies based on solid composite polymers, in which large interactions within the polymer core and the nanotube surface occur with noncovalent ("aromatic") interactions. Although these interactions are much weaker with fragile bonds than covalent interactions, their total impact strengthens the composite leading to stable systems. Similarly, dispersing nanoparticles of clay onto polymers requires special techniques, most commonly involving solution, *in situ* polymerization, and intercalation using clay with appropriate treatment. Some manufacturers now produce CNTs based on chemical incorporation on compounds while others offer PMNCs masterbatches, which usually contain 10–20% polymer composite by weight. A variety of polymers, including acrylonitrile, poly styrene, butadiene, polybutylene, polycarbonate, polystyrene, terephthalate, and polyamide. Similarly, masterbatches containing scattered nanoclays are available commercially. As CNTs have excellent properties overall that often they overshadow many highly advanced material compounds. Therefore, the chain resources are easily existing now to produce composite materials having more adapted and refined characteristics.

3. Synthesis of nanocomposite materials

3.1 Synthesis through melt intercalation and template synthesis methods (sol-gel technology)

Melt-intercalative polymerization has been attributed as a process in which *in situ* sheets polymerization occurs. It is based on the release of embedded silicate inside an aqueous precursor in which polymerization is initiated thermally, by irradiation, by diffusion of an appropriate initiator, or by natural raw material. Initially, nylon-6 montmorillonite nanocomposite was synthesized by this technique, then its production facilitated the synthesis of other thermosetplastics and nanocomposites. *In situ* polymerization is the simplest form of thermosetclay NCs. Disadvantages of the method are; fast reaction rates and reliability on clay incorporation, swelling of clay, measurement of the dispersed layer monomer of clay, and oligomer is formed when polymerization is yet to be completed. The method is free from the solvent presence and does not require a matrix and contains layered silica in a fused state. Conventionally, used methodologies include injection molding or extrusion in which thermoplastic NCs are mechanically mixed with organophilic clay at high temperatures. In this step, the polymer gets exfoliated to form the desired nanocomposite. This method is very simple to prepare thermoplastic nanocomposites. In the event of failure of *in situ* polymerization or formation of improper polymers, this method has been used.

3.2 Template synthesis (sol-gel technology)

An aqueous solution or a gel containing material and silicate block has been utilized during this process. During nucleation processing, the inorganic host crystals grow and are adsorbed within the layered surface. The sol-gel method can enhance the elimination between silicate layers through the single-step process in the absence of important oniumions, with some disadvantages. First, in the composition of clay compounds, the clay mineral requires high amounts of heat energy, which decomposes polymers matrix. Also, the negative tendency has been found while merging during silicate growth. Sol-gel process has commonly been employed for the generation of dual-layer nanocomposites, but with very little variation in concentrated silicates [2]. The natural features of the matrix structure have allowed it to be the most widely used synthetic material.

The main advantages of the method are as follows:

- Limitations of spray pyrolysis and efficacy in producing ultrafine grains.
- More uniform and well-oriented nanopowders with the larger surface area are used in multicomponent systems.
- Production of large-scale uniform, nano-sized particles.
- The sol-gel process is quite versatile, processing at low temperature, better homogeneity, rigorous stoichiometrically controlled, and renders pure products.
- The composite is porous having a low wear-resistivity and is weakly bonded.
- Include rapid solidification process (RSP), effective, reliable, and simple.
- High strength, high specific modulus, a combination of high performance, and low density of reinforced fibers making a low modulus.
- In heavily denser fiber glasses, the specific elastic modulus of the fiber-glass resins is slightly lower than MNCs.
- High shock absorbance, unlike classic materials, where fracture propagations cause breakages, nanocomposites have low matrix toughness with interfacial de-bonding and fiber splitting.
- There are only a few numbers of nanofibers, which are observed with fracture-effect on load transfer through the matrix, on intact nanofibers. When nanocomposite material is loaded for short time bearing capacity has not been affected, even encompassing defects in the desired nanocomposite.

The drawback with metal-metal nanocomposites is agglomeration and nonhomogeneous composition. The preparation of high-quality polymer nanocomposite materials using appropriate processing methods is essential to achieve high NC performance. Unique processing methods have been designed for the preparation of polymer nanocomposites. One universal method of preparation for all nanocomposites is not possible due to the structural and chemical differences of NCs and the different types of materials used. Each process requires specific processing conditions depending on the synthesis method, the type of nanoscale filler, and the required structures. In general, processing different technologies does not produce the same results.

3.3 Layered filled/nanocomposites (LFNCs)

Since silicate clay is hydrophilic, it is not suitable for mixing and blending with many compounds. In addition, the electrostatic forces cause the solid accumulation of platelets of clay. The neighboring platelets can share counter-ions, resulting in stacked platelets. It does not work with untreated clay to form nanocomposites because most of the clay matter is trapped internally and shows an interaction with layered nanocomposites. So, the clay must be processed well, before it can be utilized to prepare a nanocomposite material. The ion-exchange method is commonly used to obtain molded clays which make it more compatible with organic nanocomposites. After that, the clay can be mixed with different materials to get the desired product. Toyota has begun extensive research on nanocomposites molding and done a lot of work on loaded nanocomposites. There are four main processes that are used for the synthesis of polymer composites. These processes are as follows:

1. Intercalation of polymer or pre-polymer from solution.

2. *In situ* intercalative polymerization.

3. Direct melt intercalation.

4. Template synthesis for layered silicate/polymer nanocomposites.

The silicate layers are hardened and the polymer is melted to further processing stages. The concentrated silicate is swollen in a solvent, for example, chloroform, toluene, or water. Thereafter, the surface silicate and polymer solutions are mixed, the polymer chains bind and the solution inside the silicate layers evaporates. The composite structure remains the same during solvent removal, resulting in nanocomposite being deposited between the moving layers. Because of their excellent properties, that often used in materials, that are embedded in many nanocomposites. For example, amino acids convert montmorillonite (MMT) which is degraded by caprolactam monomer at 100 degrees centigrade and initiate its ring opening to detect MMT/nylon-6 nanocomposites. Ammonium cation of amino acids prefers the separation of caprolactam. The number of carbon atoms in the amino acid moiety greatly affects the flammability, which indicates that the concentration of caprolactam monomer is higher.

In the second process, embedded silicate begins to swell in an aqueous monomer mixture to form a polymer solution between the coated clay layers. Although the methods of interlamellar polymerization are best known for using concentrated silicates. Polymer nanocomposites are receiving a lot of attention due to the nanocomposite activity of MMT/nylon-6. In addition, two-step *in situ* polymerization was used to prepare MMT/polymer nanocomposite. These two steps include the preparation of the treated MMT solution and the mixing of polymers, respectively.

In the third process, melt intercalation occurs directly and composite silicates are combined with molten state NPs without the requirement of solvent. The polymer mixture is drawn by cutting over the softening area of the polymer suspension. The expanded chains of the polymer penetrate the intermediate layers of silicate from the melting of the polymer mass during the shrinkage. Fourthly, the process enforces as polymer suspension behaves as a template to form layered clay material. Silicon-based polymer materials are made from *in situ* hydrothermal crystallization, where a colloidal matrix of polymer gel and silicon-based NCs are being synthesized. This technique is being radially used to assemble nanocomposites with a double layer where silicates are formed in a solid solution consisting of building blocks of silicate and polymer precursor. This process is most suitable for water-soluble polymers, such as hydroxypropylmethylcellulose (HPMC), poly (dimethyldi-allylammonium) (PDDA), poly (vinyl-pyrrolidone) (PVPyr), and poly (aniline) (PANI)).

Polymerization techniques are well known for using concentrated silicates. Polymer nanocomposites are receiving high admiration due to nanocomposite induction. The *in situ* polymerization process could also be used to prepare nanocomposites loaded with graphene oxide. Natural graphite flakes are efficient and wearable structures with a carbon axis at the normal lattice. Because there are no active ionic groups of natural graphite flakes present, it is difficult to combine monomers onto graphite components to form graphite/polymer NCs with ion-exchange interactions. However, the dispersion of graphite has many holes with a diameter of 2–10 μm. Firstly, the graphitic dispersion starts interacting with a polymer solution with the help of the sonication technique and then polymers get embedded in the dispersed graphite holes and the solvent is released, thereafter. Graphene and composites are then obtained through heat transfer or immersion process. In addition, an electrostatic bonding process has been reported for the preparation of graphene/polymer nanocomposites [29]. First, polystyrene (PS) latex was synthesized using hexadecyl trimethyl ammonium bromide (cationic surfactant), creating favorable conditions in the areas of PS micelles. In addition, CNT/polymer resin (such as epoxy) nanocomposites could be prepared by means of thermal compression. The thin layer of the nanotube network is first detected by multiple nanotube dispersing steps and suspension filters, the large CNT sheets are then processed as a permeable resin. These large sheets are assembled to form solid nanocomposites for thermal mechanics.

3.4 Synthesis of chitosan nanocomposites (Ch-NCs)

To date, there have been several reported studies on the integration of Ch-NCs using a variety of integrated approaches. Researchers have developed several novel methods for the synthesis of Ch-NCs. Such methods include emulsion droplet coalescence, micellar modification, ionic gelation, precipitation, sieving, and spray drying. These methods have been used in the integration of chitosan-based materials which are used for drug delivery and other biomedical applications. However, the use of nanocomposites for agro-applications is still very limited. This can only happen if nanocomposite sources are economical and consistent. To ascertain the desirable characteristics, chitosan has been used for nanocomposite synthesis. As per the literature, ionic gelation methods and spray suspension methods have been considered as the most suitable synthetic methods for the production of large Ch-NCs.

The mechanism of ionic gelation has been discussed. In this system, well-charged amino groups are combined with the less well-gelled tripolyphosphate (TPP). TPP is an anionic cross-linker that binds to the chitosan molecule and converts it into nanoparticles. TPP is nontoxic, so it is used in the production of chitosan-based nanomaterials (ChNMs). The plant response to nanocomposites used depends on a number of factors, including particle size, size distribution index, higher zeta capacity, and component nature. Nanocomposites and their activities with naturally occurring materials have introduced the environmentally friendly pollution-free method to deal with many challenges. Manufactured nanocomposites can be used as a foliar application, seed growth, and in soil mixing.

Chitosan-based nanomaterials are very extensively tested on plants to identify various factors, such as antimicrobial, adhesive, antioxidant. Chitosan can be used as a single ingredient or combined with other substances, such as copper (Cu), zinc (Zn), and silver (Ag), to synthesize the material of interest. Chitosan exhibits a strong metal bonding due to the availability of free amine groups throughout the polymeric spine of chitosan. The Zn^{+2} and Cu^{+2} have an important role in plant growth and germination; therefore, researchers have focused more on these two metal ions, by combining them with chitosan substrate.

3.5 Chitosan-Zn nanocomposites (Ch-ZnNCs)

The researchers have incorporated a variety of plant micronutrients onto NCs, including zinc. Zinc (Zn) was named as an integral part of the plant micronutrient in 1869. The addition of Zn to plants was intended to ensure its continued availability and increased efficiency. Zinc also protects plants from different environmental hazards (sun, water, etc.). Interactions between Zn-chitosan molecules have been demonstrated by using analytical methods, such as FT-IR and X-ray diffraction. The amino moiety in chitosan has shown two different styles, which are as follows:

1. There was only a solitary amino moiety that showed any type of bonding with chitosan in a pendant pattern.

2. Bridging pattern was found in metal ions when two or more amino groups got embedded in a metal chunk.

In 2018, Ch-Zn nanocomposite was prepared by the incorporation of low-molecularized chitosan molecules by iron-containing organosol. In a standard test, Zn granules (0.5 g), toluene (120 ml), and chitosan (4 g) were used. The synthesized NC was also tested for the physicochemical parameters by using different analytical methods, such as; SEM, TEM, and XRF. The nanocomposite, when combined with iron, has shown excellent antifungal activity against *Rhizoctonia solani*. Du, Niu, in 2009, loaded Zn^{2+} granules into chitosan solution to improve antibacterial activity [48]. In summary, $ZnSO_4$ solution was obtained by adding 0.3%, w/v Zn-granules to chitosan-solution (dissolved in 1% (v/v) acetic acid and TPP (1% w/v). The results of the study showed that the increased concentration of Zn^{2+} significantly improved the potential zeta of nanocomposite which led to an increase in the combined antibacterial activity. Chitosan zinc oxide nanocomposites (Ch-ZnO) were also tested for antifungal activity against Fusarium wilt (created by Fusarium oxysporum F.sp. Ciceri in chickpea). In addition to reducing recorded disease ~40% after using Ch-ZnO; and contributed to the increased growth of chickpeas.

3.6 Chitosan-Cu nanocomposites (Ch-CuNCs)

Copper (Cu) is one of the most important nutrients in plants. Although excessive use of Cu is harmful to all plants, Cu is allowed for organic farming. There are several reported copper-based fungicides. Various formulas are designed for the successful absorption of Cu by plants. It acts as an elicitor in plant cells to accelerate enzymatic activity. In other compounds, chitosan-copper nanocomposites (Ch-CuNCs) have also been tested for their antifungal activity in tomato inhibitors; Alternaria solani and Fusarium oxysporum. TEM, SEM-EDS, AAS-TEM, and SEM micrographs successfully demonstrated the inclusion of Cu in the chitosan matrix. Ch-CuNCs inhibits 70.5% and 73.5% mycelia growth and 61.5 and 83.0% algae growth rate in Alternaria solicit and Fusarium oxysporum, respectively. Plant lesion control was demonstrated when a significant decrease was observed in nano formulation-treated plants. The percentage efficacy of disease control (PEDC) success rate for Ch-Cu was recorded as 87.7%. In another report, Ch-Cu nanocomposite has shown significant antifungal activity against Sclerotium rolfsii and Rhizoctonia solani. The synthesized compound shows a means diameter of around 2–3 nm and is shown to be evenly distributed in nanocomposite uniforms. The results showed excellent results for the prevention of tested fungal diseases. Nanoparticles synthesized in acetone, produced a much higher degree of inhibition compared to those inferred by using toluene solvent. Jaiswal et al. synthesized Ch-CuNCs by adding copper sulfate to a chitosan solution followed by the incorporation of NaOH. The size of the copper particles produced was recorded as 700–750 nm. The solution is applied to plants referred to as a fungicide. The results revealed an important protective effect built against fungal pathogens. Chitosan-Cu nanocomposite has also been shown to be an important growth promoter in a variety of plants that performs Ch-Cu nano formulation and is combined with maize seedlings. Nano formulation has shown promising effects on plant growth by reducing the activity of α-amylase and protease enzyme and increasing the amount of protein content in seed germination (see **Figure 2**).

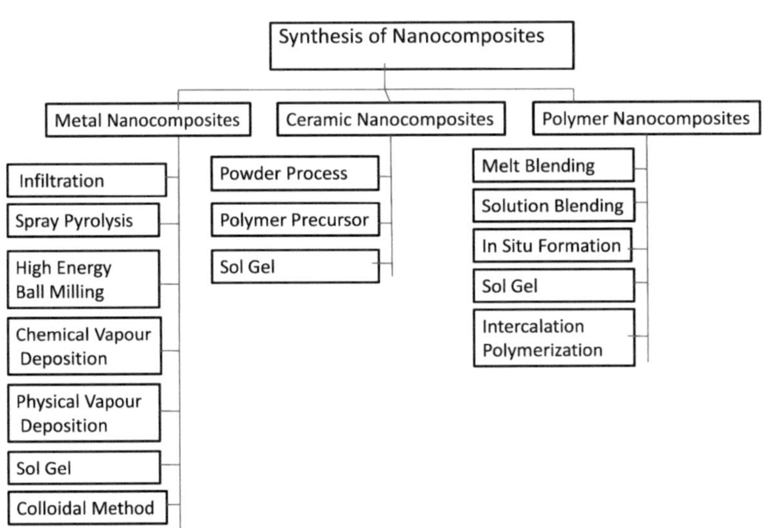

Figure 2.
Flow chart diagram for the synthesis of nanocomposites.

4. Properties

4.1 Electrical properties

The addition of CNTs to composite materials has had a significant impact on improving the conductive properties of nanocomposites. It is reported that the addition of CNTs improves the mechanistic and thermal characteristics of nanocomposites. Multiscale strengthening with NPs greatly improves enthalpy and electronic efficiency in related NCs. Conventional filters, such as carbon and glass filters are a viable solution in developing a combination of multiple functions. The doping of carbon black and nanotubes has led to the improved electronic operations of polymer film and organic sheets. Electrical conductivity depends on the concentration or the amount of filling material applied to nanopolymers. Semicrystalline polyamino compounds have shown better electrical performance than noncrystalline polycarbonate. When polymer films are applied to an organic sheet in nanofibers agglomeration, it causes an increase in electrical activity. The insulator material such as polycarbonate can be made conductive by adding it to nanocomposite material of varying compositions. The cheapest plastic known is made from nanocomposite material has both mechanical and optical features and is gaining future use. By assembling the right amount of CNTs in plastic, it becomes an electrical device. Cheap plastic is used to make optical discs, used in high-performance air-conditioning products and to shield these from electrical transformations and pulsation that cause the failure of the product. By changing the number of CNTs in polycarbonate, the performance of nanocomposite is also improved. The mixing of conductive particles to the polymer has a reasonable impact on the dielectric properties of composites. By the advent of electrical devices (capacitors, resistors, and others) on printed circuit boards, advanced compounds of nanocomposites have prevailing properties, such as:

- The integrated capacitors can be functionalized to produce large capacitance.
- The greater compatibility was found by industrial PCB fabrication with other composite materials.
- Robust process.
- Abandoning of leaded materials.
- The higher number of life cycles.
- Greater reliability and when it is required to increase dielectric properties of NCs then it has been added with ferroelectric ceramics material that has more permeation to work with that is $BaTiO_3$.

4.2 Organic conducting material nanocomposites (OMNCs)

They have a wide variety of properties with some important features, such as ease of processing, recycling, cost-effectiveness, and sustainability. The nonstandard semiconductors possess better electronic features, such as high thermal conductivity, high throughput, and high electrical conductivity. The inorganic NPs semiconductors materials have better luminescent and image processing properties. By combining both the polymers NPs and inorganic semiconductors form a hybrid nanocomposite that became the dominant candidate for photovoltaic cells. A variety of materials are used to form NCs. High performance has been achieved by mixing the CNTs to titina and P_3HT with the ability of power conservation. The addition of CNTs improves the thermomechanical characteristics of nanocomposites. Multistage strengthening with nanoparticles enhances electronic efficiency. The carbo-glass amalgamation is a viable solution for developing a composite having multiple functions. The addition of carbon allotropes has led to the improved electrical conduction of polymer and organic sheets. There are many factors that affect a hybrid system and require research to improve these systems. Overall, the nanotubes array and the nanorod-based hybrid system work better. Second, the interaction between organic and inorganic elements determines the effectiveness of cost sharing as well. Additionally, the alignment of power levels in the interface is one of the most important aspects of hybrid systems. Therefore, greater caution is required during the selection of these genotypes. Some guidelines should be followed to improve these systems. The correct combination of inorganics and metal semiconductors should also be taken into account. So,

- In the process of charge separation, a nanostructure with a large interface should be used.
- Good contact between inorganic and biological elements should occur.

The nanostructure network greatly assists in hybrid systems, TiO_2 has been employed as among the most widely used nanomaterial in our daily lives.

4.3 Multifunctional properties

Multifunctional materials, such as nanocomposites, are mostly applied as active sensors that perform multiple functions. Au (Pt) doped with α Fe_2O_3, nanospindles work in many ways to combine co-oxidation, and gas sensing devices. Catalytic activity is measured in a stainless steel bed reactor, while a gas measuring device performs gas sensor testing. It was found that the activity of various NPs led to higher performance in both functions compared to α Fe_2O_3. The reason for the improved effect is due to the active Au-NPs that act as a catalyst for sensitive reactions and also exhibit high efficiency in low-temperature co-oxidation. In 2010, thick and dense oxidized nanorod was produced in a row to form a strong fabric with good resistance to washing and pressure cycles [3]. Polymeric materials attract a lot of attention because of their advanced properties and functional performance in various industries. In structural features, thermoset polymers are very important in fields, such as automotive and aviation. In addition, the high strength of the thermoset polymers makes them compatible with their metal counterparts and creates a multi-layered environment. Recently, advances in nanotechnology introduced many innovative features in NCs. These benefits include an increase in strength, lightness, and durability. Nanocomposites are receiving a lot of attention because of the advanced mechanistic properties that have improved their stability.

There are different materials that are used in the production of thermoset nanocomposites. One of the most extensively used material components is carbon nanoparticles (CNPs) and nanoclays. One of the major hurdles is the dispersing of NPs on the matter substrate during the synthesis process. Nanocomposites contain

10–12% nanoclays with greater strength and durability than nanocalcium products. In nanoclay components, nanoparticles are extracted and synthesized. This improves the mechanical and physical performance of the filler and matrix optical connector which is very helpful in eliminating stress by improving the mechanical properties of nanocomposites. High-pressure mixing is better than direct mixing that can induce clay breakage. Titanate conversion is used for better spreading of nanoparticles. Due to the excellent mechanical properties, the requirement for low filling load, reinforcement strength, less weight, and corrosive environment of nanoparticles are found in some materials, such as cellulose, which is an ideal ingredient for the development of enviro-friendly polymers. Many researchers have focused on high-quality performance, extraction, and mechanical performance of filling polymer matrix in varying proportions. There are some challenges to the formation of nanocomposites, such as low dispersion of natural solvents, agglomeration, inclination, and hydrophilic nature. Due to growing environmental concerns, regulations have placed a great deal of interest in developing enviro-safe materials. Natural fibers have many advantages over synthetic due to their eco-friendly nature, but working with natural fibers, we cannot find the same strength that we can get from synthetic fibers.

Cellulose nanocrystals have been used in systematic and geometrically modeled cellulose fillers in a variety of useful products. It has been found that it will improve the mechanistic and thermal range of nanocomposites. Structure toughness with NPs affects the heat resistivity and electronic mobility in related NCs. As compared to traditional fillers, woody cellulose offers multidimensional combinations of variable functions. Microcrystalline cellulose as a colloidal matrix found in water with high solid concentrations, such as Celish (Trade name from Daicel Corporation) which provides 10% cellulose slurry and nanofibers. Solid-liquid crystals are used in a variety of optical applications. Researchers have successfully developed optically transparent wood cellulose nanocomposites with a small young's modulus and low thermal increase. In addition, they have successfully applied an electroluminescent to flexible transparent cellulose nanocomposites resulting in a low coefficient of thermal expansion. To prevent the scattering of ionic diffusivity, cellulose whiskers (less than 10% concentrate) can be used in low-density electrolyte polymers having applications in lithium batteries. Low-density loudspeaker membranes with high Young's modulus can be made from melamine-formaldehyde and micro-fibrillated cellulose. Electrospun cellulose nanofibers are used as an affective membrane that allows the purification and penetration of molecules based on physical or chemical characteristics instead of the weight or size of a molecule. Cellulose nanofibers are an excellent part of biological systems due to their load-bearing properties, low toxicity, excellent mechanical properties, biodegradability decay, and biocompatibility. Cellulose nanocomposites can be obtained from soft pulp from wood by mechanical fibrillation process. Mixing of the mixture can also be done using nanowhiskers and semiconducting polymers. NCs are very useful in producing stable materials with improved performance and mechanical properties. Scientists are trying to modify thermoset NCs to use Polyoles of vegetable oil-based chemicals instead of bio-based resins to stabilize and reduce dependence on petroleum-based resins. According to a recent study, nanocomposites can also be made from environmentally friendly vegetable oils.

4.4 Thermomechanical properties

Previously, thermoplastic materials were used with nanocellulose materials, showing the advantage of high crack resistance and recycling. The strength and durability of nanocomposites are greatly enhanced by the use of nanocellulose on a thermoplastic composite particulate or composite-based dispersion with the benefits of nanocellulose in resin interaction and the limited surface area of cellulose fibers. Fiber impacts on the mechanical/thermal properties of biocomposites based on carbon nanofibers (CNFs). It has been found that up to 40% of fiber content laminar increase in fiber modulus was observed using phenolic resins. With the inclusion of CNF in epoxy resins, a significant increase in the glass transition temperature was found. With 5% epoxy film of CNFs at 30°C, modulus showed an increase from 2.6 GPa to 3.1 GPa. In addition to the changing temperature of the glass, a significant increase has been reported. The mechanical properties have been drastically improved by adding up to 2% CNFs by weight while continuous addition of CNFs reduces mechanical and thermal conversion features due to agglomeration. Increased reinforcement of CNFs around bamboo fibers in the poly-lactic acid (PLA) matrix has been found to bind CNFs and improves fracture resistance that prevents fracture growth. But when, the CNFs did not gain weight of 1%, the cause of the fractured impact was increased by 200%. It has been investigated that the processing of CNFs could lead to safer, lightweight compounds with different properties, such as barriers and open spaces with multiple applications in electronics, sensors, energy storage, packaging, medicine, and automotive production. Nanocellulose films are also used to induce the barrier properties to the resulting composites. In addition, high-porosity aerogels, ease out the gas outflows and due to their hydrophobic properties, moisture absorbance has also been aided. Nanocellulose integration offers a wide range of applications that include weapon systems and flexible display devices. Highly effective NCs are possible using CNFs.

Many efforts have been made to prepare metal-reinforced material nanocomposites which have structural significance and greater toughness compared to their contemporary counterparts. But they still exhibit larger differences while analyzing their physicomechanical properties. The strengthening process includes metal oxide scattering, stiffening, preventing premature solidification, load transfer, and difference in coefficient of thermal expansion in MMNCs. MMNCs combine metal components with ceramic precursors to having enhanced mechanical properties and toughness.

4.5 Biological properties

Nanocomposites can be a part of the living organisms present in this diverse biosphere. The materials used must be structurally, biologically, physically, chemically, and mechanically compatible with the surrounding tissue. Since the mechanical properties are mainly affected by the elastic modulus, the transfer of load, durability, and higher strengths are of particular interest. Metals, polymers, and ceramic composites are approved for the synthesis of the required materials. Some examples of polymer filling compounds are given below:

- The bone fractures can be repaired by using these fillers as external fixators with the help of epoxide carbon fiber composite.
- Bone fixing screws and as replacement of bone plates in different body parts.
- The bone joints can be replaced by using these fillers, for example., carbon fibers (PEEK) are applied as total hip replacement material.

4.6 Optical properties

The impact of particle treatment on nanocomposite substances is very important. The *in situ* sensitization of alumina nano-sized particles dispersed in Methyl methacrylate resin and successive polymerization incorporates a better level of optical transmission at near-infrared exposure than untreated Al_2O_3. Experiments have found that the gold nanoparticles (AuNPs) when incorporated on polyethylene oxide/polyvinyl pyrrolidone (PEO/PVP) components show an increasing trend of AuNPs to combinations of visual parameters tha increases urbach strength, and optical power gap is also raised.

4.7 Magnetic properties

There are two classes of NCs compounds exhibiting magnetic properties, one containing metal NPs and the other ferrite NPs. Basically, lack of hysteresis, shows 18 increased superparamagnetic activity in ferrite NPs. Nanocomposite containing 2.8% concentration of ferrites was found to have no hysteresis at room temperature and was clearly visible. They also found that nanocomposites containing γ-Fe_2O_3 NPs in the electromagnetic polymer matrix were also free of hysteresis. Nanoparticles from nickel oxide synthesis in Polyvinyl cinnamate also show magnetic properties. They found a ferromagnetic state in nickel nanocomposites. Additionally, weight gain, magnetic response, and hysteresis values are obtained with the incorporation of nickel oxide NPs on the nanocomposite material.

5. Applications

5.1 Applications as structural materials

Thermoset polymers composites and nanocomposites are very important in today's ultratech world. Items, such as seafood, automobiles, aircraft, are examples of nanocomposites applications. Most importantly, the improved structural modification, NCs of special strength make them compatible with metal materials incorporated in various locations. These materials are easy to process and therefore, have a broad range of applications. With the advent of nanotechnology, NCs offer numerous benefits as synthetic nanomaterials, such as stability, lightweight, and sustainability. Nanocomposites are receiving a lot of attention because of their advanced mechanical properties that have increased the reliability of these materials. Different materials are used in the production of different types of nanocomposites. The most widely used compounds are carbon nanoparticles loaded on nanoclays. Dispersion of NPs is one of the major hurdles faced by researchers recently. Nanocomposites contain 10–12% of nanoclays that are more potent and stronger than nanocalcium compounds.

The nanoscale size significantly improves physicochemical and chemical interactions. The morphology found in nanocomposites can change the phase, important for the development of various structures. Mixing and aero treatment are two important factors that improve the performance of a given NC. The variety of combinations between matrix, synthetic additives, and nanofillers allows for a wide range of materials used in fire reactions, electronic structures, optical performance, mechanical and thermal properties. Improvement of filling quality greatly improves the distribution of such nanocomposites through multiple applications.

The impact of nanoparticles on the mechanical properties of polymer composites is an important factor to consider. The addition of nanofiller significantly reduced the resistance coefficient compared to pristine epoxy. Besides, 1% by weight of nanofillers showed better results than 3% by weight of nanofillers, which was unexpected. It may be due to agglomeration particles leading to poor dispersion in the epoxy matrix.

5.2 Applications of nanoclays

Nanoclays improve the mechanical and physical characteristics of the filler and matrix optical interface, which is very helpful in eliminating pressures by improving the mechanical properties of nanocomposites. Due to the impressive mechanical properties, requirements, such as low filling, stiffness, low weight, and biodegradability of nanoparticle materials, make them ideal for enviro-compliant material development. Many researchers have focused on high-quality optimization, extraction, and machine performance by filling the matrix polymer with varying degrees. There are some challenges to the formation of NCs namely, severe dispersion of organo-solvents, agglomeration, and the presence of a hydrophilic environment.

5.3 Applications of nanocrystal of naturally occurring materials

Due to the growing environmental concerns, it has indeed been emphasized that natural fibers have many advantages over synthetics because of their eco-friendly nature. However, when we use natural fibers, we cannot get more stiffness than synthetic fibers. Cellulose nanocrystals have been used as structural and geometric models of "cellulose" in various functional products. In addition, the active incorporation of electroluminescent compounds in naturally found nanocomposites results in a lower coefficient of thermal expansion. To prevent ionic diffusivity, cellulose whiskers (less than 10%) can be applied to low-density electrolyte polymers that are used in lithium-ion batteries.

5.4 Applications in coating materials

Organic and inorganic composites based on alkoxy-lanes and alko-oxides have great uses in hard-coated eye glass lenses. The addition of NPs to epoxy silanes acts as a linker between organic and inorganic moieties that greatly enhance abrasion control without affecting the transparency of the glass material. Nanocomposites have also been developed for low surface free energy coatings. To add up in the mechanical properties, nanotubular materials are extensively used that gave strength to NCs where light weight and hardness are required for the resultant NCs. The use of nonlinear optics, including optical sensor protection from high-intensity laser beams, flat panel displays, electromechanical actuators, light-emitting diodes, field-effect transistors, supercapacitors, and optical limiters are some applications of CNTs. Nanocomposites have provided significant advances over conventional NCs in tropical, mechanical, electrical, and barrier properties. Furthermore, it maintains transparency and reduces the flammability of the polymer matrix.

5.5 Applications in industries

The industrial applications include:

- **Automotives:** Use of nanocomposite materials in gas tanks, bumpers, interior, and exterior panels of automotive.

- **Construction industry:** Use of composites in building materials and sectional or structural panels.

- **Aerospace:** Highly efficient and heat resistant material panels of NPs are operating in the aerospace industries.

- **Electrical and electronics:** The uses of NCs in electrical switchboards and electronic components of various devices are increasing every day.

- **Food Packaging:** NCs are also used in containers and wrapping films during the food-packaging process.

- **Oils and gas pipelines:** Rusting on these pipelines is a major problem for steel products that reduce the life of infrastructure that is implanted. There is a need for more attention to find a solution to this problem. Often, corrosion and failure are not considered in the construction and installation of plumbing systems because one cannot measure the damage of corrosion due to unforeseen environmental events. The multiple issues of physical damage, corrosion, and natural phenomenon cause sudden failures. The use of NCs has major importance in the production and repairing of damaged pipeline structures. Nanocomposites provide excellent benefits in the production of pipes and overlay toward repairing rusty steel pipelines. The cost and performance of fiber-reinforced nanocomposite as an alternative to steel pipes is recently gaining importance. Therefore, nanocomposite structures are designed according to an engineering perspective. Each component provides a certain function and interaction that aids the structure and creates a significant difference in their performance.

The reinforced materials add up some important advantages to the plumbing and pipeline systems are as follows:

- The nanocomposites exhibit anisotropic properties that provide strength to extraordinary collapse and burst due to pressure increase.

- These nano-derived pipes have increased load-carrying ability, high compressibility, and have greater tensile strength.

- The use of welding or joining these nano pipes to a long distance is not required.

- Very few of these materials are required to be replaced. Therefore, the replacement ratio is very less than metallic pipes.

- They are highly corrosion resistant.
- They fulfill all the standards set for gas and oil pipelines.

5.5.1 Applications in automobile industry and auto parts

Nanocomposites are well-suited for automotive parts due to their tolerance to breakage during harsh damages. Nanocomposites loaded with supports are useful to give strength to the different parts of a material where high efficacy is required. Because of the high pollution concerns, car manufacturers are working on the development of such technologies that can effectively control the problem. Therefore, nanocomposites are used to deal with this problem. Nanocomposites have improved barriers, heat resistance, impact adherence, and mechanical properties than conventional compounds. Therefore, the development of structural compounds having properties, such as biodegradability and recycling is a challenging goal. Such compounds are extensively used in automotive body parts. Industries are deeply concerned about the following factors using NCs:

- Aesthetical perspectives.
- Ability to recycle the used raw materials.
- Reduction of heavily weighted products.
- High performance with greater precision.

Nanotechnology is a driving force that has brought many changes in different industries at the level of component development, material selection, and system execution.

Nanocomposites are also used in:

- Development of sophisticated computer chips and capacitors with the help of thin films.
- It is used as a polymer electrolyte in different battery systems.
- Structural NCs are used in fuel tanks.
- Nanocomposites are used in the automotive industry.
- The blades and high-performance impellers are derived by using NCs.
- They are used in gas-oxygen barrier systems, thereby reducing the delamination/cracking of composite materials.
- Fiber-reinforced nanocomposites have excellent mechanical advantages, but the problem of delamination appears due to different mismatches.

5.6 Biomedical applications

Carbon nanotubes (CNTs) are well investigated for their importance in health-related features, especially those which are used in medicine. CNTs provide strength for metal matrix compounds, composite matrix compounds, and ceramic matrix compounds. They also improve mechanical properties, *in vitro* cell testing, and biocompatibility (*In vivo*). Composites having flexible mechanical properties could be used in tissue engineering, genetic engineering, medicine, scaffolding, implants, and as fillers of various compounds to expand their mechanical properties. The modified materials, such as plocaprolactum, are used in tissue engineering because of their robustness and adaptability in living tissues, it is the material of choice in the field of biomedical engineering. Strengthened compounds of CNTs not only serve as a bone marrow transplant, but their use in medical systems undoubtedly gave opportunities for advancement for the next generation in the fields of, biologically designed organisms, such as tissue engineering, cell therapy, drug delivery, and diagnostic equipment production. Magnetic sponges based on local, flexible, and readily available NC units are excellent for medical usage.

5.7 The application in insulative materials

The addition of nanocomposite to inorganic compounds improves their physical properties and gave the number of applications based on inorganic composites within the material. Nanocomposites are composed of carbon nanotubes as filler material, are used in the electronics industry. Their electrical insulation range and thermal properties have made them suitable to be used in an area where insulative electrical characteristics are required. The challenges of agglomerations are one of the biggest problems when adding nanomaterials to composite materials. Due to agglomerations, nanomaterials cannot be dispersed evenly, which is a major obstacle to the commercial application of nanocomposites. When agglomeration occurs during mass production then the uniform structures are more difficult to achieve. Some other applications are as follows:

- Thermoelectric devices with quite flexible properties.
- When combined with organonano they are used for dying.
- Inorganic NCs show photoelectric characters and are used in photocells.
- Polyethylene glycol NCs are used for effective drug delivery systems.
- Polyethers Ag-based nanocomposites are used in antimicrobial and biomedical remedies.
- Tissue engineering technique includes inorganic nanocomposites of calcium phosphate and poly lactic acid.
- MMNCs, such as cobalt oxide nanocomposites, are applied for humidity sensing in meteorology.
- Reduced graphene oxide NCs have applications in energy-harvesting devices.
- The doped alumina-zinc nanocomposites are used to increase the dielectric constant and also increase the conductivity of the composite.

- PMMA doped iron oxide composites are used for electromagnetic uses and also have shielding capacity.

- Cellulose loaded with copper and ZnO have shown significant UV resistance, therefore, they are used as UV-protection devices.

5.8 Agro-applications

The use of agro-biomass is a promising and continuous process of producing naturally occurring NCs, where agro-biomass is used simultaneously. In this regard, lignin-derived agro-biomass is an economical resource for the production of functional biomaterials that are compatible and sustainable. In the case of metal oxide nanocomposites (MONCs), the use of lignin-based MONCs should be extended to dynamic fields, for example, ultraviolet (UV) protection, photocatalysis, and antimicrobial agents. The development of lignocellulosic biomass as raw material should be a viable option for the development of UV protective materials from an industrial point of view.

Today different integration methods for the production of nanocomposites (CMNCs, MMNCs, and PMNCs) are available, but limitations and barriers also exist, which require the exploration of new techniques and engineered methodologies. These NCs offer improved performance in addition to their monolithic and micro-composite counterparts and are well-adapted partners to overpower the constraints of many existing materials and devices. Today the use of nanocomposites is taking place rapidly, but still, their full potential has not been attained. Making such highly stress-tolerant equipment, low-load reinforcements, gas barrier, and flame-related adherence create potential applications and marketplaces for these items. So, these nanocomposites discussed have the potential to create a new material age in the future.

6. Conclusion

Nanocomposites are versatile materials and are the current focus of research across the world. Different methods, for example, sol-gel, electrospinning, precipitation, melt mixing, solution mixing, *in situ* polymerization, *in situ* interactive polymerization, intercalation, melt intercalation, and template synthesis, are used for the synthesis of nanocomposites. The nanocomposites have unique electrical, thermal, and mechanical properties and hence are materials of choice in a variety of fields, for example, automotive industry, aerospace, semiconductor, electronics, and biomedical applications. Due to their light weight and broader mechanical strength, in the future, these materials will replace the metals and their alloys in different fields. Progress is going on for silicon-carbon nanocomposites for the manufacturing of lithium-ion batteries with greater performance output, nanotube-polymer composites for faster bone healing, polymers-based windmill materials for better performance, development of lightweight and efficient sensors, and development of magneto-fluorescent materials for better tumor imaging. Polymer nanocomposites are the material of choice for packaging companies in near future. The future challenge is to synthesize such materials that bear better material properties and price compatibility in comparison to existing materials.

DOI: http://dx.doi.org/10.5772/intechopen.102469

Author details

Muhammad Hafeez
Department of Chemistry, University of Azad Jammu and Kashmir, Muzaffarabad, Pakistan

*Address all correspondence to: muhammadhafeezchemist@gmail.com

References

[1] Gholampour A, Ozbakkaloglu T. A review of natural fiber composites: Properties, modification and processing techniques, characterization, applications. Journal of Materials Science. 2019;**55**(3):829-892

[2] Omanović-Mikličanin E, Badnjević A, Kazlagić A, Hajlovac M. Nanocomposites: A brief review. Health and Technology. 2020;**10**(1):51-59

[3] Hassan T, Salam A, Khan A, Khan Saif U, Khanzada H, Wasim M, et al. Functional nanocomposites and their potential applications: A review. Journal of Polymer Research. 2021;**28**(2):1-22

[4] Banerjee R, Ray SS. Foamability and special applications of microcellular thermoplastic polymers: A review on recent advances and future direction. Macromolecular Materials and Engineering. 2020;**305**(10):2000366

[5] Chen H, Du X, Sun J, Wu R, Wang Y, Xu C. Template-free synthesis of novel Co3O4 micro-bundles assembled with flakes for high-performance hybrid supercapacitors. Ceramics International. 2021;**47**(1):716-724

[6] Alem SAA, Latifi R, Angizi S, Mohamadbeigi N, Rajabi M, Ghasali E, et al. Development of metal matrix composites and nanocomposites via double-pressing double-sintering (DPDS) method. Materials Today Communications. 2020;**25**:101245

[7] Hao L, Tang D, Sun T, Xiong W, Feng Z, Evans KE, et al. Direct ink writing of mineral materials: A review. International Journal of Precision Engineering and Manufacturing-Green Technology. 2021;**8**:665-685

[8] Kusdianto K, Sari TD, Laksono MA, Madhania S, Winardi S. Fabrication and Application of ZnO-Ag Nanocomposite Materials Prepared by Gas-Phase Methods, Feb 2021.

[9] Ghimire PP, Jaroniec M. Renaissance of Stöber method for synthesis of colloidal particles: New developments and opportunities. Journal of Colloid and Interface Science. 2021;**584**:838-865

[10] Mathur S, Veith M, Shen H, Hufner S, Jilavi MH. Structural and optical properties of NdAlO3 nanocrystals embedded in an Al2O3 matrix. Chemistry of Materials. 2002;**14**: 568-582

[11] Babar S, Mane Anil U, Angel Y-G, Elham M, Haasch RT, Elam JW. W:Al2O3 nanocomposite thin films with tunable optical properties prepared by atomic layer deposition. The Journal of Physical Chemistry C. 2016;**120**(27):14681-14689

[12] Miller KJ, Colletti A, Papi PJ, McHenry ME. Fe–Co–Cr nanocomposites for application in self-regulated rf heating. Journal of Applied Physics. 2010;**8**:2521

[13] Luo T, Wei X, Zhao H, Cai G, Zheng X. Tribology properties of Al2O3/TiO2 nanocomposites as lubricant additives. Ceramics International. 2014;**40**(7):10103-10109

[14] Sarkar D, Adak S, Mitra NK. Preparation and characterization of an Al2O3–ZrO2 nanocomposite, part I: Powder synthesis and transformation behavior during fracture. Composites Part A: Applied Science and Manufacturing. 2007;**38**(1):124-131

[15] Xiao B, Byrne PJ, Wheatley PS, Wragg DS, Zhao X, Fletcher AJ, et al.

Chemically blockable transformation and ultraselective low-pressure gas adsorption in a non-porous metal organic framework. Nature Chemistry. 2009;**1**(4):289-294

[16] Qin M, Yanqing Y, Mokuang K, Qunji X. Microstructures and mechanical properties of hot-pressed MoSi2±matrix composites reinforced with SiC and ZrO2 particles. Composites Science and Technology. 2001;**7**;963-971

[17] Hnatko M, Galusek D, Šajgalík P. Low-cost preparation of Si3N4–SiC micro/nano composites by in-situ carbothermal reduction of silica in silicon nitride matrix. Journal of the European Ceramic Society. 2004;**24**(2):189-195

[18] Wang Z, Huang J, Mao J, Guo Q, Chen Z, Lai Y. Metal–organic frameworks and their derivatives with graphene composites: Preparation and applications in electrocatalysis and photocatalysis. Journal of Materials Chemistry A. 2020;**8**(6):2934-2961

[19] Saeedi Heydari M, Baharvandi HR. Comparing the effects of different sintering methods for ceramics on the physical and mechanical properties of B4C–TiB2 nanocomposites. International Journal of Refractory Metals and Hard Materials. 2015;**51**:224-232

[20] Surani F, Abdizadeh H, Baharvandi Hamid R, Nemati ALI. Effect of MgO nano particles on sintering behavior of Al2O3-SiC-MgO nano composites. International Journal of Modern Physics: Conference Series. 2012;**05**:568-573

[21] Laurent C, Demai JJ, Rousset A. Fe-Cr/Al2O3 metal-ceramic composites: Nature and size of the metal particles formed during hydrogen reduction. Materials Research Society. 1994;**1**:229

[22] Vokhidova NR, et al. Synthesis and stabilization of cobalt and copper nanoparticles by using Bombyx morichitosan. Journal of the Korean Physical Society. 2016;**69.8**:1295-1300

[23] Ha JM, Hur SH, Pathak A, Jeong J-E, Woo HY. Recent advances in organic luminescent materials with narrowband emission. NPG Asia Materials. 2021;**13**(1):1-36

[24] Amosov AP, Titova YV, Timoshkin IY, Kuzina AA. Fabrication of Al-AlN nanocomposites. Key Engineering Materials. 2016;**684**:302-309

[25] Khdair AI, Fathy A. Enhanced strength and ductility of Al-SiC nanocomposites synthesized by accumulative roll bonding. Journal of Materials Research and Technology. 2020;**9**(1):478-489

[26] Chen WX, Lee JY, Liu Z. The nanocomposites of carbon nanotube with Sb and SnSb0.5 as Li-ion battery anodes. Carbon. 2003;**41**(5):959-966

[27] Sadoun AM, Mohammed MM, Fathy A, El-Kady OA. Effect of Al2O3 addition on hardness and wear behavior of Cu–Al2O3 electro-less coated Ag nanocomposite. Journal of Materials Research and Technology. 2020;**9**(3):5024-5033

[28] Rahmawati R, et al. The synthesis of Fe_3O_4/MWCNT nanocomposites from local iron sands for electrochemical sensors. In: AIP Conference Proceedings. Vol. 1958. No. 1. AIP Publishing LLC; 2018

[29] Liu Y, Xuan W, Cui Y. Engineering homochiral metal-organic frameworks for heterogeneous asymmetric catalysis and enantioselective separation. Advanced Materials. 2010;**22**(37):4112-4135

[30] Ge L, Wang W, Peng Z, Tan F, Wang X, Chen J, et al. Facile fabrication of Fe@MgO magnetic nanocomposites for efficient removal of heavy metal ion and dye from water. Powder Technology. 2018;**326**:393-401

[31] Abazari S, Shamsipur A, Bakhsheshi-Rad HR, Ismail AF, Sharif S, Razzaghi M, et al. Carbon nanotubes (CNTs)-reinforced magnesium-based matrix composites: A comprehensive review. Materials (Basel). 2020;**13**(19): 4421

[32] Sharma A, Gupta G, Paul J. A comprehensive review on the dispersion and survivability issues of carbon nanotubes in Al/CNT nanocomposites fabricated via friction stir processing. Carbon Letters. 2021;**31**(3):339-370

[33] Deng T, Yoon CK, Jin Q, Li M, Liu Z, Gracias David H. Self-folding graphene-polymer bilayers. Applied Physics Letters. 2015;**106**(20):203-108

[34] Loomis J, King B, Panchapakesan B. Layer dependent mechanical responses of graphene composites to near-infrared light. Applied Physics Letters. 2012;**100**(7):073108

[35] Loomis J, King B, Burkhead T, Xu P, Bessler N, Terentjev E, et al. Graphene-nanoplatelet-based photomechanical actuators. Nanotechnology. 2012;**23**(4):045501

[36] Rafiqi FA, Majid K. Synthesis, characterization, luminescence properties and thermal studies of polyaniline and polythiophene composites with rare earth terbium(III) complex. Synthetic Metals. 2015;**202**: 147-156

[37] Ansari S, Rahima C, Muralidharan MN. Photomechanical characteristics of thermally reduced graphene oxide – Polydimethylsiloxane nanocomposites. Polymer-Plastics Technology and Engineering. 2013; **52**(15):1604-1610

[38] Liu P, Wang Y, Wang X, Yang C, Yanfeng Y. Polypyrrole-coated samarium oxide nanobelts: Fabrication, characterization, and application in supercapacitors. Journal of Nanoparticle Research. 2012;**14**(11):1-11

[39] Bo S, Wu Z, Zhu Y, Moon K-s, Wong CP. Three-dimensional graphene-based composite for flexible electronic applications. Electronic Components & Technology Conference. 2015;**4**;1889-1898

[40] Du Y, Dong N, Zhang M, Zhu K, Na R, Zhang S, et al. Covalent functionalization of graphene oxide with porphyrin and porphyrin incorporated polymers for optical limiting. Physical Chemistry Chemical Physics. 2017;**19**(3):2252-2260

[41] Zang J, Bao S-J, Li CM, Bian H, Cui X, Bao Q, et al. Well-aligned cone-shaped nanostructure of Polypyrrole/ RuO2 and its electrochemical supercapacitor. Journal of Physical Chemistry C. 2008;**112**:14843

[42] Owens DR, Zinman B, Bolli G. Alternative routes of insulin delivery. DM. 2003;**20**(11):886-898

[43] Tan L, Gan L, Hu J, Zhu Y, Han J. Functional shape memory composite nanofibers with graphene oxide filler. Composites Part A: Applied Science and Manufacturing. 2015;**76**:115-123

[44] Ramesh Kumara P, Venkateswerlub M, Satyanarayanaa N. Synthesis and electrical conductivity studies of PEO + LiClO4 + TiO2 + MoO3 nanocomposite solid polymer electrolyte. NSTI-Nanotech. 2012;**6**:598-601

[45] Thompson BC, Moulton SE, Ding J, Richardson R, Cameron A, O'Leary S, et al. Optimising the incorporation and release of a neurotrophic factor using conducting polypyrrole. Journal of Controlled Release. 2006;**116**(3):285-294

[46] Herrasti P, Kulak AN, Bavykin DV, de Léon C, Ponce ZJ, Walsh FC. Electrodeposition of polypyrrole–titanate nanotube composites coatings and their corrosion resistance. Electrochimica Acta. 2011;**56**(3):1323-1328

[47] Khan J, Ilyas S, Akram B, Ahmad K, Hafeez M, Siddiq M, et al. Zno/NiO coated multi-walled carbon nanotubes for textile dyes degradation. Arabian Journal of Chemistry. 2018;**11**(6):880-896

[48] Zhang D, Jiang C, Sun Y'e, Zhou Q. Layer-by-layer self-assembly of tricobalt tetroxide-polymer nanocomposite toward high-performance humidity-sensing. Journal of Alloys and Compounds. 2017;**711**:652-658

[49] Huang H, Yao J, Chen H, Zeng X, Chen C, She X, et al. Facile preparation of halloysite/polyaniline nanocomposites via in situ polymerization and layer-by-layer assembly with good supercapacitor performance. Journal of Materials Science. 2016;**51**(8):4047-4054

[50] Qi W, Xue Z, Yuan W, Wang H. Layer-by-layer assembled graphene oxide composite films for enhanced mechanical properties and fibroblast cell affinity. Journal of Materials Chemistry B. 2014;**2**(3):325-331

[51] Yang J-H, Lin S-H, Lee Y-D. Preparation and characterization of poly(l-lactide)–graphene composites using the in situ ring-opening polymerization of PLLA with graphene as the initiator. Journal of Materials Chemistry. 2012;**22**(21):10805-10815

[52] Parka JJ, Yua EJ, Leeb W-K, Haa C-S. Mechanical properties and degradation studies of poly(D,L-lactide-co-glycolide) 50:50/graphene oxide nanocomposite films. Polymers Advanced Technologies. 2013;**25**:48-54

[53] Khan J, Siddiq M, Akram B, Ashraf Muhammad A. In-situ synthesis of CuO nanoparticles in P(NIPAM-co-AAA) microgel, structural characterization, catalytic and biological applications. Arabian Journal of Chemistry. 2018;**11**(6):897-909

[54] Sitharaman B, Shi X, Walboomers XF, Liao H, Cuijpers V, Wilson LJ, et al. In vivo biocompatibility of ultra-short single-walled carbon nanotube/biodegradable polymer nanocomposites for bone tissue engineering. Bone. 2008;**43**(2):362-370

[55] Lalwani G, Henslee AM, Farshid B, Lin L, Kasper FK, Qin YX, et al. Two-dimensional nanostructure-reinforced biodegradable polymeric nanocomposites for bone tissue engineering. Biomacromolecules. 2013;**14**(3):900-909

[56] Singh H, Sharma A, Bhardwaj Sanjeev K, Arya Shailendra K, Bhardwaj N, Khatri M. Recent advances in the applications of nano-agrochemicals for sustainable agricultural development. Environmental Science: Processes & Impacts. 2021;**23**(2):213-239

Chapter 2

Investigating the Role of Auger Recombination on the Performance of a Self-Assembled Quantum Dot Laser

Mahdi Razm-Pa and Farzin Emami

Abstract

We first examine the relaxation dynamics inside quantum dot structures. After presenting the rate equations model, we investigate the effect of some parameters introduced in the rate equation on the performance of quantum dot lasers. The effects of QDs coverage factor, inhomogeneous broadening, which its physical source is the size fluctuation of quantum dot in forming self-assembled quantum dots, as well as cavity length, on SAQD laser have been analyzed. Then, based on the rate equations, a circuit model will be introduced. Finally the effect of phonon bottleneck and Auger recombination on the performance of quantum dot lasing, are examined. It is shown that, there is more output power and quantum efficiency, and higher modulation bandwidths when Auger recombination is considered for these lasers.

Keywords: self-assembled quantum-dot laser (SAQDL), inhomogeneous broadening, phonon bottleneck, coverage factor, relaxation dynamic

1. Introduction

The reduction in dimensionality caused by confining electrons (or holes) to a thin semiconductor layer leads to a dramatic change in their behavior. This principle can be developed by further reducing the dimensionality of the electron's environment from a two-dimensional quantum well to a one-dimensional quantum wire and eventually to a zero-dimensional quantum dot. In this context, however, dimensionality refers to the degree of electron momentum freedom. In fact, within a quantum wire, unlike the quantum well where the electron is confined in just one dimension, it is confined in two dimensions and thus, the freedom degree is reduced to one. In a quantum dot, the electron is confined in all three-dimensions, hence reducing the degree of freedom to zero. Under certain growth conditions, when a thin semiconductor layer grows on a substrate having a completely different lattice constant, the thin layer is spontaneously arranged or changes into quantum dots through

IntechOpen

self-assembles while attempting to minimize the total strain energy between the bonds. Microscopy can show quantum dots that are in the shapes of pyramids, square based, and tetrahedron [1]. The performance expected from quantum dot lasers is often due to the density of their quasi-atomic states. Using the quantum structures confined in some dimensions will reduce the momentum freedom of the carrier in a certain direction. Ideally, carriers are completely enclosed in quantum dots. Therefore, the density of quantum dot states, that is, the number of states per volume unit and per energy unit, is expressed by the delta function. The gain spectrum amplitude is determined only by homogeneous broadening due to intraband relaxation at the quantum dot. These gain properties are the basis of the features that give the quantum dot laser some advantages over conventional lasers [2, 3]. The effect of carrier dynamics on the performance of quantum dot laser and the possibility of bi-exciton lasing have been studied. Bi-excitons are achieved if the increase in ground state of the quantum dot reaches the laser threshold, and if the carrier relaxation is rapidly below 100 ps, the laser will be observed [4, 5]. Carrier relaxation in quantum dots (QDs) is studied widely when applications of these devices are reported for optical communications [6]. The problem of *phonon bottlenecks* in quantum dots has sparked heated debate over whether carrier relaxation in a discrete ground state is significantly slow due to the lack of phonons required to meet the conservation rule. Because of *phonon bottleneck* [7] problem in these types of lasers, it is found that some important parameters, such as threshold current density, quantum efficiency and modulation response, are deteriorated [8]. To overcome these problems, one can manage the level spacing of a QD to decrease the carrier relaxations which inherently may decreases the laser performance, since these long relaxation times are comparable to the semiconductor radiative and non-radiative lifetimes. To have fast relaxation, overcome the phonon bottleneck problem and to improve the threshold current, external quantum efficiency, and modulation response, Auger recombination effects may be considered [9]. The most useful and well-known method to study the statics and dynamics of the carrier and photon numbers in these lasers, is to solving the rate equations for them [10]. By using the rate equations in this article and taking the Auger effect into account, a new circuit model for InGaAs-GaAs self-assembled QD (SAQD) laser will be suggested. Indeed, the Auger effect on QD laser performance are considered.

2. Theoretical background

2.1 Carrier capture and relaxation dynamics into a quantum dots

The carrier relaxation process within quantum dots is actually two steps, as shown in the **Figure 1**.

One is the carrier relaxation from continuous energy levels within the discrete levels of the quantum dot (A, Red Color). Another is the relaxation between discrete levels within dots (B, Blue Color). In many light experiments, as well as in quantum dot lasers, carriers go through these two steps unless they are brought directly into discrete levels by excitation resonant or tunneling. Since the principle of energy conservation must be satisfied for the carrier to relax, relaxing carriers transfer the corresponding energy to other particles (such as phonons) and to other carriers in the bulk. Therefore, "the relaxation rate strongly depends on the density of final levels and on the number of particles other than the transition matrix elements" [4].

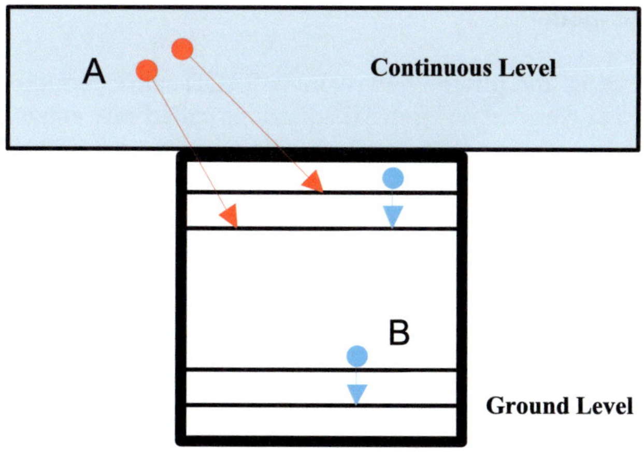

Figure 1.
Carrier relaxation in a quantum dot [1]. (A, Red Color) The relaxation from continuous levels. (B, Blue Color) The relaxation between discrete levels.

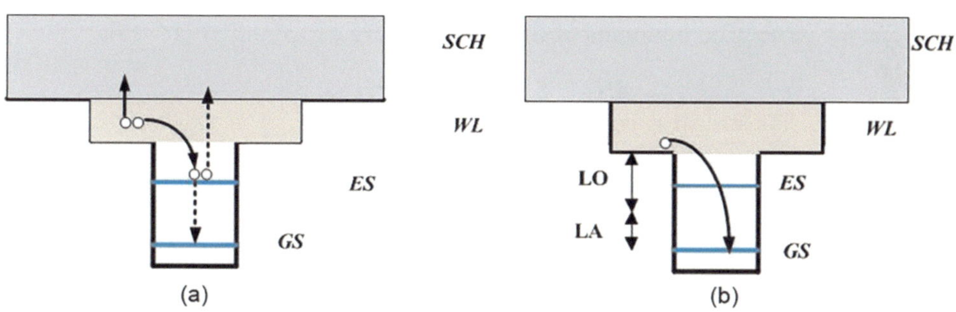

Figure 2.
Carrier capture and relaxation processes [4, 12]. (a) LO- and LA-phonons. (b) Auger processes.

Introducing the LO- and LA-phonons, it is possible to satisfy the energy conservation rule [11], is shown in **Figure 2a** [4, 12]. Such a two-phonon process decreases the lifetimes severely [4], but it cannot be adequate to relax back the carriers inside the dots deeply. Carrier trapping into the dot energy levels and hence energy conservation rule, are satisfied whenever the number of carriers outside the dots are increased. So, Auger process can be proper phenomenon due to increase of the captured carriers that occupy the QD energy levels [13]. Therefore, with more injected currents or equivalently more injected carriers, the relaxation lifetime increases and Auger process can be effective due to carrier relaxation into deep lower levels by a step-like energy decrement. Illustration of the carrier relaxation processes, considering the Auger effect, is shown in **Figure 2b**. Auger scattering is more important when the density of excited carriers is high. As illustrated in **Figure 2b**, one of the two electrons in the wetting layer (WL) is captured by the Auger mechanism in the excited state (ES) of QD, while the other electron is emitted upward in the WL or separate confinement heterostructure (SCH). Then, the electron which is captured in ES transfers its energy to the third electron of the barrier or even more probably to another electron in the excited state (ES) of QD and relaxes downward in the QD ground state (GS).

2.2 Rate equation model

Usually, the carrier and photon behaviors in SAQD semiconductor lasers are expressed by a set of coupled differential equations called rate equations [14].

In **Figure 3**, there is a simple energy band diagram to explain different levels, ground state (GS), excited state (ES), and wetting layer (WL) state, in these structures [14]. In this model, there are some injected carriers into a SCH barrier with a rate of I/e where I is the injected current and e is the electron charge.

They can either relax with a time of τ_s in the WL state or come back from the barrier with a time of τ_{qe}. These carriers can be captured by different dot sizes from the WL state. Assume that each dot has only two distinct energy states; GS and excited state (ES). In these levels, the captured carriers have a time of τ_c from WL to ES and relaxed carriers have a time of τ_d from ES to GS. They can come back in the reverse path at the times of τ_{eES} an τ_{eGS}, respectively. There are several radiative or non-radiative recombination times for carriers. τ_{Sr} is carrier recombination in the SCH region and τ_{qr} is carrier recombination in the wetting layer (WL). τ_r is recombination in the QD. Assume that an excited emission is only due to an electron hole recombination in ES and GS. The photons emitted from the laser cavity have a rate of S/τ_p, where S is the number of photons and τ_p is the photon lifetime. The governing rate equations for various components of these carriers are explained as [14–16]:

$$\frac{dN_s}{dt} = \frac{I}{e} - \frac{N_s}{\tau_s} - \frac{N_s}{\tau_{sr}} + \frac{N_q}{\tau_{qe}}, \tag{1}$$

$$\frac{dN_q}{dt} = \frac{N_s}{\tau_s} + \frac{N_{ES}}{\tau_{e_{ES}}} - \frac{N_q}{\tau_{qr}} - \frac{N_q}{\tau_{qe}} - \frac{N_q}{\tau_c}, \tag{2}$$

$$\frac{dN_{ES}}{dt} = \frac{N_q}{\tau_c} + \frac{N_{GS}(1-P_{ES})}{\tau_{e_{GS}}} - \frac{N_{ES}}{\tau_r} - \frac{N_{ES}}{\tau_{e_{ES}}} - \frac{N_{ES}}{\tau_d} - \frac{\left({}^{c}\!/\!{}_{n_r}\right) g_{mES}^{(1)} \Gamma}{1 + \varepsilon_{mES} \Gamma {}^{S}\!/\!{}_{V_a}} S, \tag{3}$$

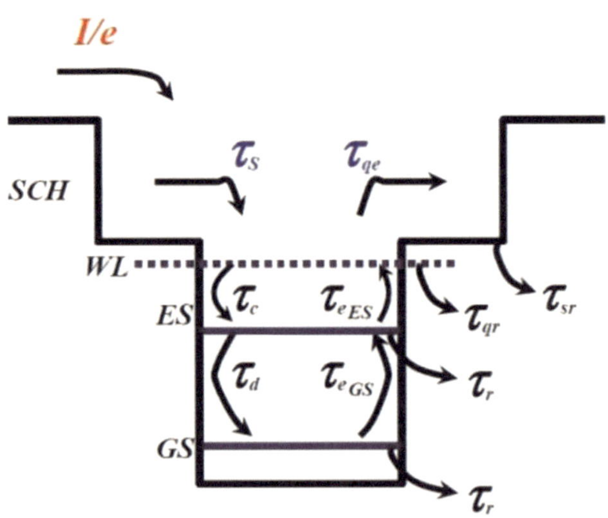

Figure 3.
A sample energy band diagram for a SADQ laser considering the excited state.

$$\frac{dN_{GS}}{dt} = \frac{N_{ES}}{\tau_d} - \frac{N_{GS}}{\tau_r} - \frac{N_{GS}(1 - P_{ES})}{\tau_{e_{GS}}} - \frac{\left(c/n_r\right) g_{mGS}^{(1)} \Gamma}{1 + \varepsilon_{mGS} \Gamma S / V_a} S, \tag{4}$$

$$\frac{dS}{dt} = \frac{\left(c/n_r\right) g_{mES}^{(1)} \Gamma}{1 + \varepsilon_{mES} \Gamma S / V_a} S + \frac{\left(c/n_r\right) g_{mGS}^{(1)} \Gamma}{1 + \varepsilon_{mGS} \Gamma S / V_a} S - \frac{S}{\tau_p} + \frac{\beta N_{GS}}{\tau_r} + \frac{\beta N_{ES}}{\tau_r} \tag{5}$$

With the following parameters: N_s: number of carriers in the SCH layer, N_q: number of carriers in the WL layer, N_{ES}: number of carriers in the ES layer, and N_{GS}: number of carriers in the GS layer.

In the above equations, Γ is the optical confinement factor, Γ_0 is the inhomogeneous broadening of the optical gain, c is the speed of light, β is the spontaneous coupling efficiency, and the photon lifetime τ_p can be found from [17]:

$$\tau_p^{-1} = \left(\frac{c}{n_r}\right) \left\{ \alpha_i + \frac{\ln\left[1/(R_1 R_2)\right]}{2L} \right\} \tag{6}$$

Where R_1 and R_2 are the facet reflectivity of the laser cavity with a length of L and the internal loss of α_i, and n_r is the cavity refractive index. Where the nonlinear gain coefficient (ε_{mGS} and ε_{mES}) are defined as [4, 16, 18]:

$$\varepsilon_{mGS} = \frac{e^2}{2 n_r^2 \varepsilon_0 m_0^2} \frac{\left|P_{c,v}^{\sigma}\right|^2}{E_{GS}} \frac{1}{\Gamma_{cv} \Gamma_{\|}} \tag{7}$$

$$\varepsilon_{mES} = \frac{e^2}{2 n_r^2 \varepsilon_0 m_0^2} \frac{\left|P_{c,v}^{\sigma}\right|^2}{E_{ES}} \frac{1}{\Gamma_{cv} \Gamma_{\|}} \tag{8}$$

and $\Gamma_{\|} \equiv 1/\tau_p$ is the longitudinal relaxation constant and Γ_{cv} is the scattering or polarization de-phasing rate. Based on the density matrix theory the linear optical gain of the active region, with a dot density of N_D, can be found as [10, 16, 19]:

$$g_{mGS}^{(1)} = \frac{2.35\sqrt{2\pi} e^2 \hbar}{c n_r \varepsilon_0 m_0^2} \frac{\left|P_{cv}^{\sigma}\right|^2}{E_{GS}} \frac{(2P_{GS} - 1)}{\Gamma_0} N_D D_{GS} \tag{9}$$

$$g_{mES}^{(1)} = \frac{2.35\sqrt{2\pi} e^2 \hbar}{c n_r \varepsilon_0 m_0^2} \frac{\left|P_{cv}^{\sigma}\right|^2}{E_{ES}} \frac{(2P_{ES} - 1)}{\Gamma_0} N_D D_{ES} \tag{10}$$

Parameters D_{GS} and D_{ES} in Eqs. (5) and (6), impose the GS and ES degeneracy which are considered to be 2 and 4, respectively [15]. $P^{\sigma}{}_{c,v}$ is a matrix element, related to the overlap integral $I^{\sigma}{}_{c,v}$ between the envelope functions of an electron and hole, defined as [19]:

$$\left|P_{c,v}^{\sigma}\right|^2 = \left|I_{c,v}\right|^2 M^2 \tag{11}$$

Where M^2 is the first-order K-P interaction between conduction and valence bands, with a separation of E_g and is equal to:

$$M^2 = \frac{m_0^2}{12m_e^*}\frac{E_g(E_g+\Delta)}{E_g+\frac{2\Delta}{3}} \tag{12}$$

In this relation, E_g is band gap, m^*_e is the effective mass for electrons and Δ is defined as the spin-orbit interaction energy for a QD material. Based on the Pauli's Exclusion Principle, the occupation probabilities at the ES and GS states of the QD are defined as [19]:

$$P_{GS} = \frac{N_{GS}}{2N_D V_a D_{GS}} \tag{13}$$

$$P_{ES} = \frac{N_{ES}}{2N_D V_a D_{ES}} \tag{14}$$

Where $V_a = LWd$ is the cavity volume with length L, width W and thickness d. Effects of the non-uniform dot size can be included in the relaxation, capture, and escape times with the following definitions [14–16, 18]:

$$\tau_c = \frac{\tau_{c0}}{(1-P_{ES})} \tag{15}$$

$$\tau_d = \frac{\tau_{d0}}{(1-P_{GS})} \tag{16}$$

Where τ_{c0} is the capture time from WL to ES and τ_{d0} is the relaxation time from EL to GS with the assumption of an empty final state. Without the stimulated emission and at room temperature, the system must converge to a quasi-thermal equilibrium based on a Fermi distribution. To have this condition, the carrier capture time, τ_{c0}, and relaxation time, τ_{d0}, and the carrier escape times, τ_{eGS}, τ_{eES}, should be satisfied by the following relations [14, 15]:

$$\tau_{e_{GS}} = \tau_{d0}\left(\frac{D_{GS}}{D_{ES}}\right)e^{\frac{E_{ES}-E_{GS}}{kT}} \tag{17}$$

$$\tau_{e_{ES}} = \tau_{c0}\left(\frac{D_{ES}N_d}{\rho_{WL_{eff}}}\right)e^{\frac{E_{WL}-E_{ES}}{kT}} \tag{18}$$

$$\rho_{WL_{eff}} \equiv \left(\frac{m_{e_{WL}}kT}{\pi\hbar^2}\right) \tag{19}$$

Where ρ_{WLeff} is the effective density of states per unit area of the WL and N_d is the QD density per unit area. Both phonon- and Auger-assisted capture and relaxation are taken into account phenomenologically through the relation [16, 20, 21]:

$$\tau_{c0}^{-1} = \tau_{c01}^{-1} + C_W N_q \tag{20}$$

$$\tau_{d0}^{-1} = \tau_{d01}^{-1} + C_E N_q \tag{21}$$

Where $\tau^{-1}{}_{c01}$, $\tau^{-1}{}_{d01}$ are characteristic rates of phonon-assisted capture and interlevel relaxation processes, respectively, and C_W and C_E are the coefficients for

Auger-assisted relaxation related to the WL and the ES, respectively. Assume that all of the carriers are injected into the WL layer or equivalently $\tau_{qe} = \tau_{sr} \rightarrow \infty$.

2.3 Effect of parameter variations

Section 2.3 covers investigation of the effects of some parameters expressed in the rate equations on the performance of quantum dot laser. This section is based on the results obtained the reference [22].

2.3.1 Inhomogeneous broadening Γ_0

Figure 4 shows the results of our simulations on the *L-I* curve of SAQD laser considering the effects of inhomogeneous broadening. Note that, the physical origin of this parameter is in the random size of the QDs [4]. As shown in the figure, for higher inhomogeneous broadening factor, more threshold current or equivalently more injection currents are needed but in this case, the external quantum efficiency does not have dominant changes. In other words, for higher Γ_0 the output power of the laser decreases. Physically speaking, for increased inhomogeneous broadening factors there is a higher occupation probabilities in the lasing action of the structure, so that the relaxation times increase and the output power decreases.

This effect can be studied in the dynamic response of a SAQD laser, too. The results of such variations for the frequency responses are plotted in **Figure 5**. The simulation results show that for higher inhomogeneous broadening factors the frequency responses of the laser deteriorates.

2.3.2 Carrier recombination time τ_{qr} in a WL

The effect of carrier recombination in WL, τ_{qr} which equals WL crystal quality, has been shown on *L-I* feature in **Figure 6**. As τ_{qr} degrades, the carriers would find more opportunities to recombine through non-radiative process out of the quantum dot which results in degradation of external quantum efficiency. The findings indicate that τ_{qr} degradation also increases the threshold current.

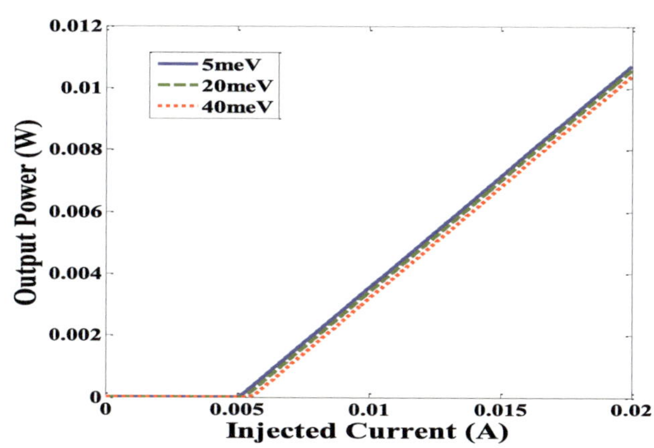

Figure 4.
Effects of the inhomogeneous broadening variations on the L-I *curve of SAQD laser considering the excited state with different values of the broadening parameter Γ_o = 5, 20, and 40 meV.*

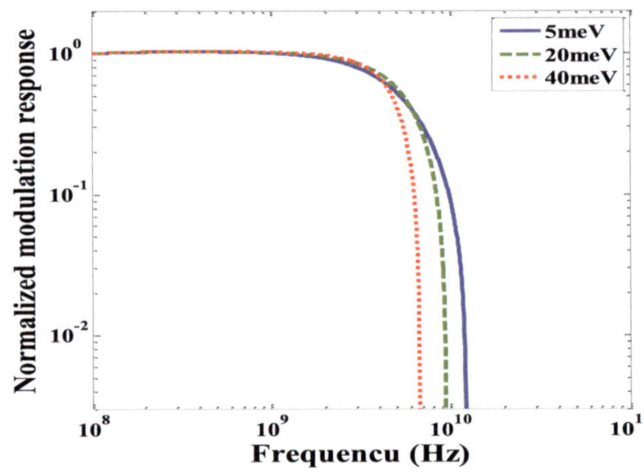

Figure 5.
Effects of the inhomogeneous broadening variations on the frequency response of SAQD laser with different values of Γ_o *= 5, 20, and 40 meV.*

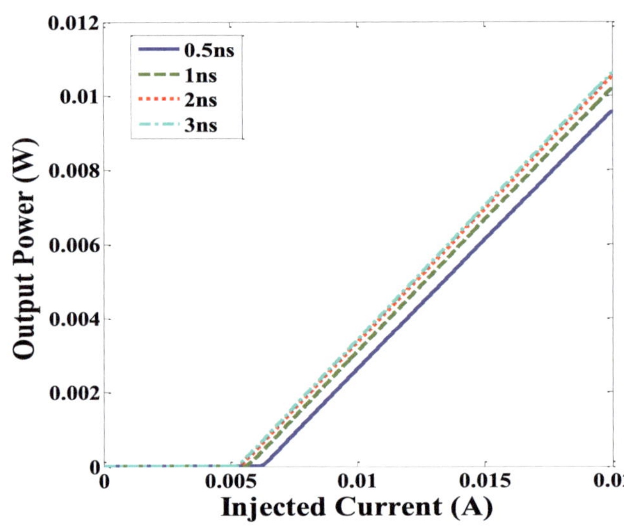

Figure 6.
The L-I *curve of a SAQD laser for different values of carrier recombinations in WL.*

The effect of carrier recombination in the WL state on the small-signal frequency response of laser is shown in **Figure 7**. As shown, the carrier recombination in the WL state has no considerable effect on modulation response.

2.3.3 Carrier recombination inside quantum dot, τ_r

Figure 8 shows the effect of carrier recombination inside quantum dot, τ_r, on the *L-I* curve. While the effect of τ_r on threshold current is significant, it does not have a considerable effect on external quantum efficiency.

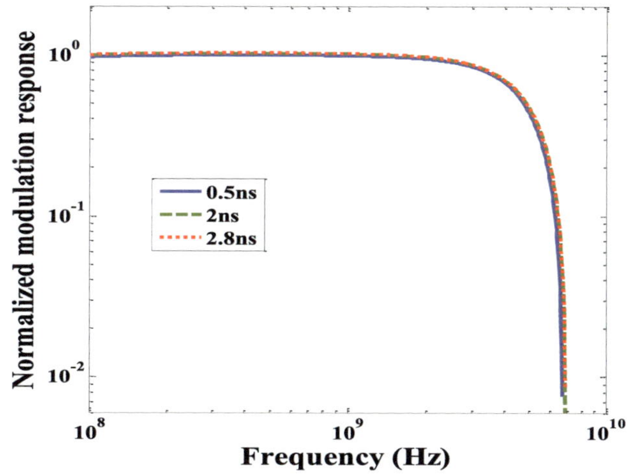

Figure 7.
The modulation response of a SAQD laser for different values of carrier recombination in WL.

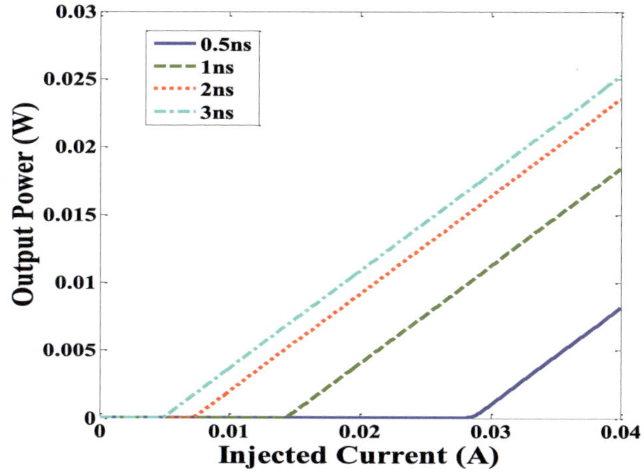

Figure 8.
The L-I *curve of a SAQD laser for different values of carrier recombination inside quantum dot.*

The effect of carrier recombination inside quantum dot on frequency response has been shown in **Figure 9**. What is important here is that, as τ_r decreases from 2.8 to 0.5 ns, the frequency response degrades. Therefore, to prevent the effect of phonon bottleneck on frequency response, the recombination lifetime within quantum dots τ_r must be much longer than the carrier relaxation time (the carrier relaxation time is about a few pico-seconds).

2.3.4 Carrier escape time from ground state to excited τ_{eGS} state and from excited state to WL state τ_{eES}

The effect of carrier escape time from ground state to excited state and from excited state to WL state on *L-I* feature has been shown in **Figure 10**. As the carrier

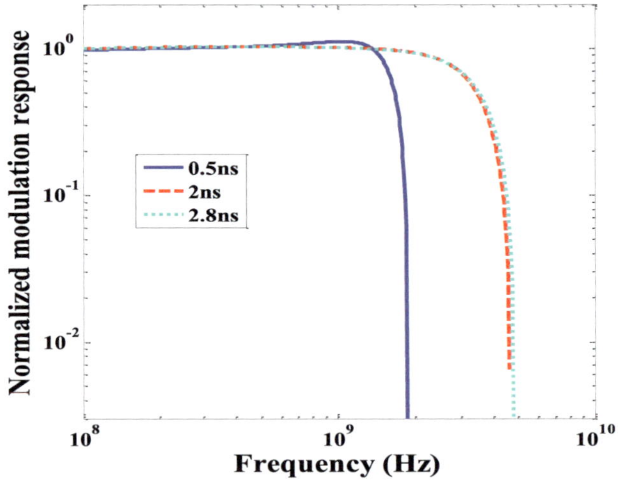

Figure 9.
The modulation response of a SAQD laser for different values of carrier recombination inside quantum dot.

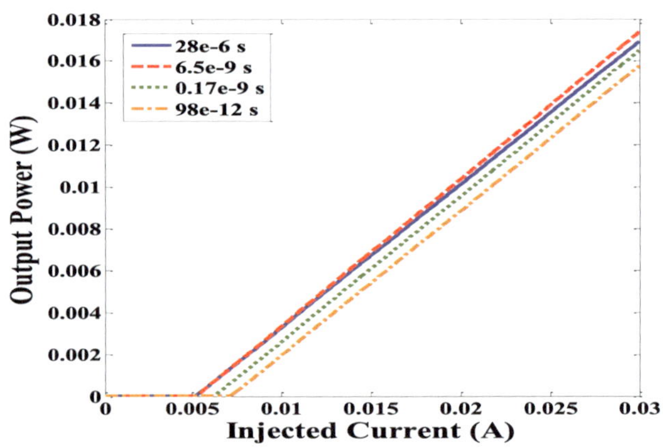

Figure 10.
The L-I *curve of a SAQD laser for different values of carrier escape time from ground state to excited state and from excited state to wetting layer.*

escape time degrades, the number of carriers in WL states (N_q) increases. These carriers are generally used due to carrier non-radiative recombination (with τ_{qr} lifetime) and this leads to the increase of the threshold current. Generally, for more decrement in τ_{qr}, there is more increment in the threshold current.

Figure 11 also shows that, as the carrier escape time degrades from ground state to excited state and from excited state to wetting layer, the frequency response degrades as well.

2.3.5 Coverage factor ξ

Figure 12 show the frequency response for different amounts of QDs coverage factor ξ = 0.1, 0.2, 0.4, relaxation time 100 ps and inhomogeneous broadening of Γ_0 = 20 meV. As shown in **Figure 12**, the increase in coverage factor due to the

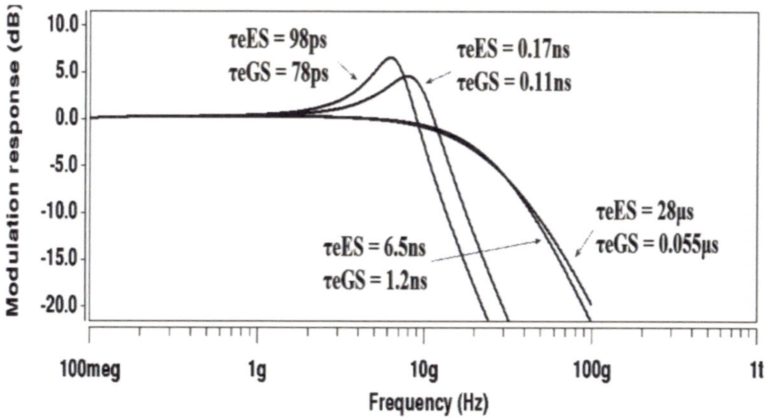

Figure 11.
The modulation response of a SAQD laser for different values of carrier escape time from ground state to excited state and from excited state to wetting layer.

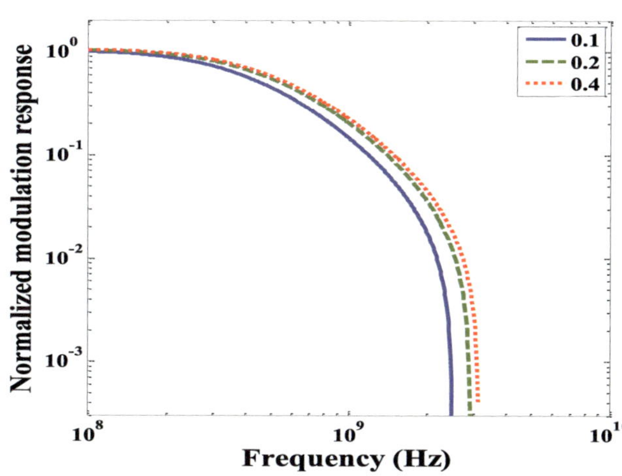

Figure 12.
Modulation response of a SAQD laser at different coverage factors; Γ_o = 20 meV and the relaxation time is 100 ps.

increase in the volumetric density of QDs (N_D) leads to the decrease in P_{GS} and P_{ES}, that is, filling probability of the GS and ES. As filling probability of the GS and ES decreases, the relaxation rate increases for the carrier inside the GS and ES, and this leads to the increase in 3 dB bandwidth.

Figure 13 shows the simulation results for the effects of coverage factor ξ = 0.1, 0.2, 0.4, inhomogeneous broadening Γ_0 = 5 meV and relaxation time 1 ps on the frequency response band width. To achieve a high-speed modulation, higher than 10 GHz, not only the relaxation lifetime should be decreased to about 1 ps but also the QD coverage factor should also be increased and the inhomogeneous broadening should be decreased.

2.3.6 Cavity lengths L

Figure 14 shows the *L-I* characteristics for different cavity lengths. The increase in cavity length leads to loss degradation and output power increase.

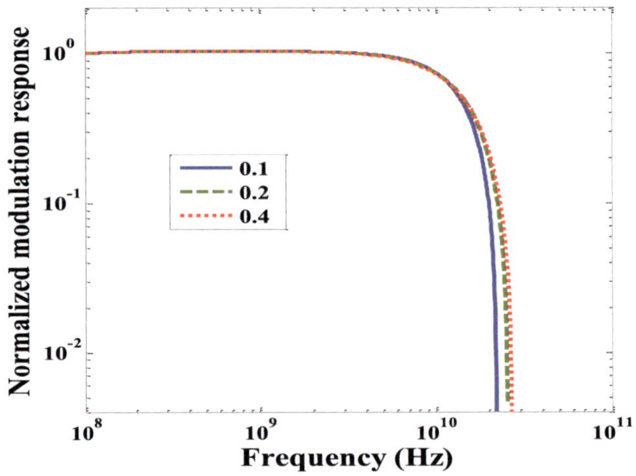

Figure 13.
The modulation response of a SAQD laser for different values of coverage factor and Γ_o = 5 meV and relaxation time 1 ps.

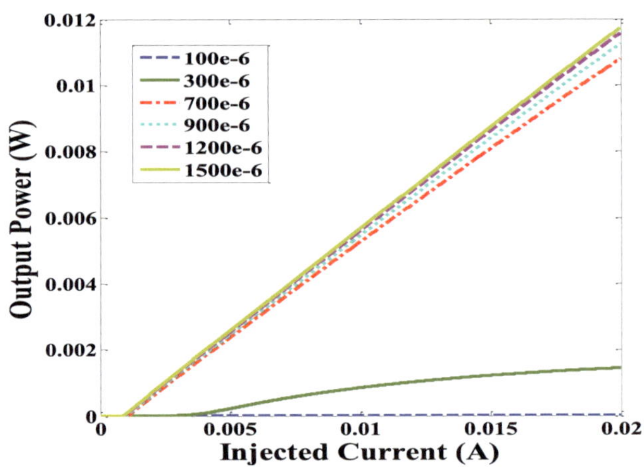

Figure 14.
L-I *curve of a SAQD laser at different cavity lengths.*

2.3.7 QD height

Figure 15 shows the simulation results of modulation response for different quantities of QD height. As the QD height degrades, the modulation band width improves. The reason for modulation band width improvement while the QD height degrades can be caused by increasing carrier confinement within quantum dot in growth direction (*z*-) which leads to increase excited stimulated emission rate, respectively.

2.3.8 Stripe width of the laser cavity

Figure 16 shows the effect of stripe width of the laser cavity on frequency response. As the stripe width of the laser cavity degrades, the modulation band width improves. It is inferred from the figure that degradation of the stripe width of the laser

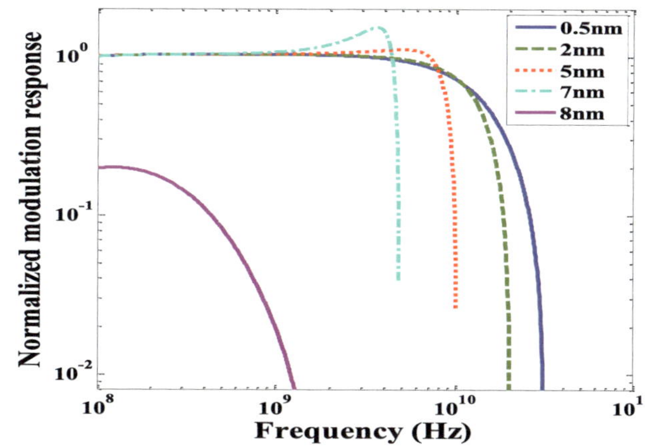

Figure 15.
Modulation response of a SAQD laser at different QD heights.

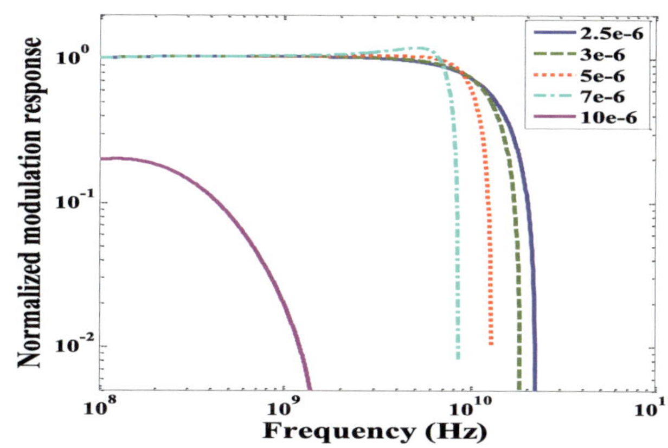

Figure 16.
Modulation response of a SAQD laser at different stripe widths of the laser cavity.

cavity and therefore the degradation of the active region can provide a higher total capture rate. Hence, it results in a greater modulation band width.

3. Circuit model implementation

To solve the rate equations of a SAQD laser, considering the excited state and Auger effect, a conceptual equivalent electrical circuit is proposed is shown in **Figure 17**. The aforementioned equations convert to some simple electrical circuit equations and then, the resulting circuit is simulated by a circuit simulator such as HSPICE [23]. The corresponding parameters for the equivalent circuit model of SAQD lasers are described in detail in Ref. [24].

4. Simulation results

In this simulation, typical parameters, which are shown in **Table 1**, are used.

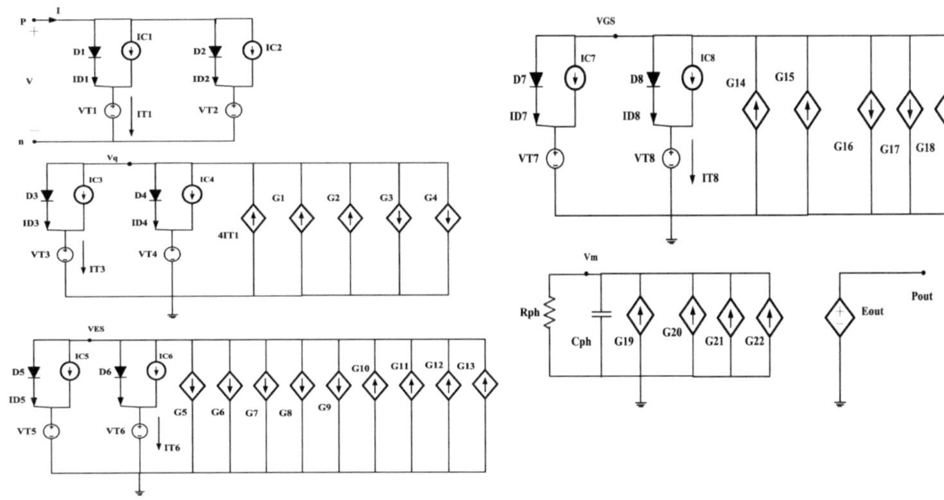

Figure 17.
A modified equivalent circuit model of SAQD lasers, considering the Auger effect.

Quantity	Value
Inhomogeneous broadening, Γ_0	20 meV
Diffusion in SCH, τ_s	6 ns
SCH recombination, τ_{sr}	4.5 ns
WL recombination, τ_{qr}	3 ns
Capture from WL to ES, τ_{c0}	1 ps
Capture from ES to GS, τ_{d0}	7 ps
ES and GS recombination, τ_r	2.8 ns
Energy separation SCH and WL, state	84 meV
Average energy separation WL and ES	100 meV
Average energy separation ES and GS	80 meV
Average recombination energy from GS, EGS	0.96 eV
Average recombination energy from ES, EES	1.04 eV
Spin-orbit interaction energy of QD material, Δ	0.35 eV
Facet reflectivity, R_1, and R_2	30%, 90%
Active region length, L_{ca}	900 μm
SCH thickness, H_b	90 nm
WL thickness, H_{WL}	1 nm
Active region width, w	10 μm
Active region volume, V_a	2.2×10^{-16} m^3
QD density of QD per layer, N_d	5×10^{10} cm^{-2}
QD density per unit volume, N_D	6.3×10^{22} m^{-3}
QD optical confinement factor, Γ	0.06
Intrinsic absorption coefficient, α_i	1 cm^{-1}

Quantity	Value
Spontaneous emission coupling efficiency, β	10^{-4}
Output power coupling coefficient, η_c	0.449
SCH, WL, and QD (GS, ES) diode ideality factors, $n_S = n_q = n_{GS} = n_{ES}$	2

Table 1.
Typical parameters used in the simulation [14, 15, 23].

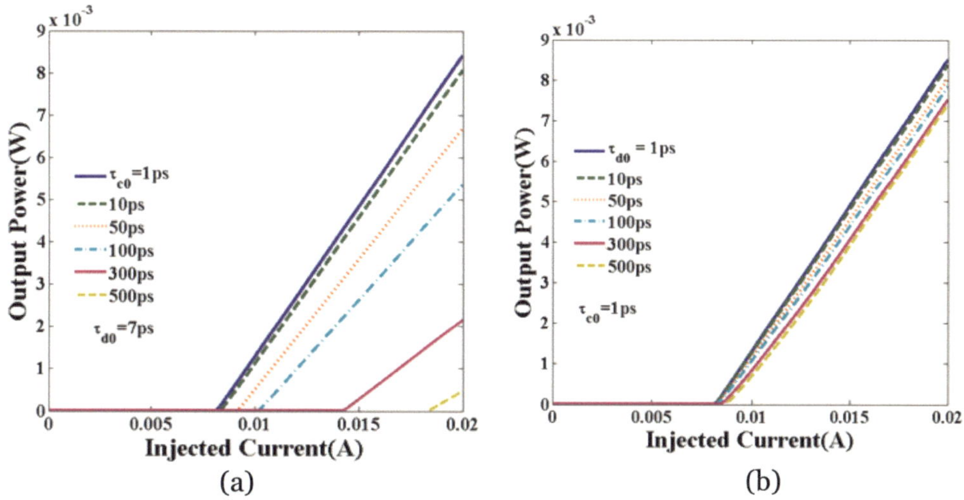

Figure 18.
The L-I curve of a SAQD laser neglecting the Auger effect. (a) Various capture times τ_{co} = 1, 10, 50, 100, 300, and 500 ps at a fixed relaxation time of τ_{do} = 7 ps; (b) various relaxation times τ_{do} = 1, 10, 50, 100, 300, and 500 ps at a fixed capture time of τ_{Co} = 1 ps.

Figure 18 illustrates the output power as a function of injected currents neglecting the Auger effect for $C_w = C_E = 0$, $\tau_r = 2.8$ ns and $\tau_{qr} = 0.5$ ns, when the carrier relaxation times τ_{d0} and capture times τ_{c0}, changes from 1 to 500 ps. As seen, there are more threshold currents for longer lifetimes. This is along with lower slopes for the output power curves which mean lower quantum efficiencies for increased lifetimes. Physically, such effect is due to the phonon bottleneck in QD lasers [4].

This phenomenon can be considered in dynamic response of the laser, too. Application of the proposed model to SAQD lasers considering this effect for dynamic response of the laser are shown in **Figure 19**. With higher relaxation times τ_{d0} and capture times τ_{c0}, the frequency response deteriorates and SAQD laser would have lower bandwidths.

Figure 20(a) and **(b)** illustrate the output power as a function of the injected currents considering the Auger effect for $C_W = 1 \times 10^{-14}$ m^3/s, $C_E = 1 \times 10^{-12}$ m^3/s, and $C_W = C_E = 0$ with $\tau_{C01} = \tau_{d01} = 100$ ps. Inhomogeneous broadening $\Gamma_0 = 20$ meV and recombination lifetimes of $\tau_r = 2.8$ ns, $\tau_{qr} = 3$ ns and $\tau_r = 2.8$ ns, $\tau_{qr} = 0.5$ ns, have been used for **Figure 20(a)** and **(b)**, respectively. It is obvious that the Auger process increases the efficiency, degrades the threshold current, and provides a high output power. This is because, by taking the Auger effect into account, the capture times and carrier relaxation lifetime decreases from WL to ES and from ES to GS. This means the degradation of phonon bottleneck effect. In other words, in QD elements

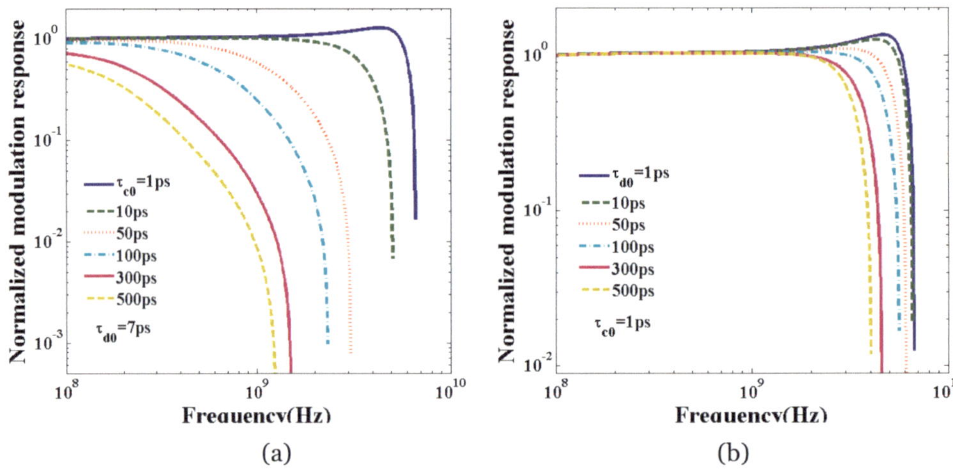

Figure 19.
Modulation response of a SAQD laser neglecting the Auger effect. (a) Various capture times τ_{co} = 1, 10, 50, 100, 300, and 500 ps at a fixed relaxation time of τ_{do} = 7 ps; (b) various relaxation times τ_{do} = 1, 10, 50, 100, 300, and 500 ps at a fixed capture time of τ_{Co} = 1 ps.

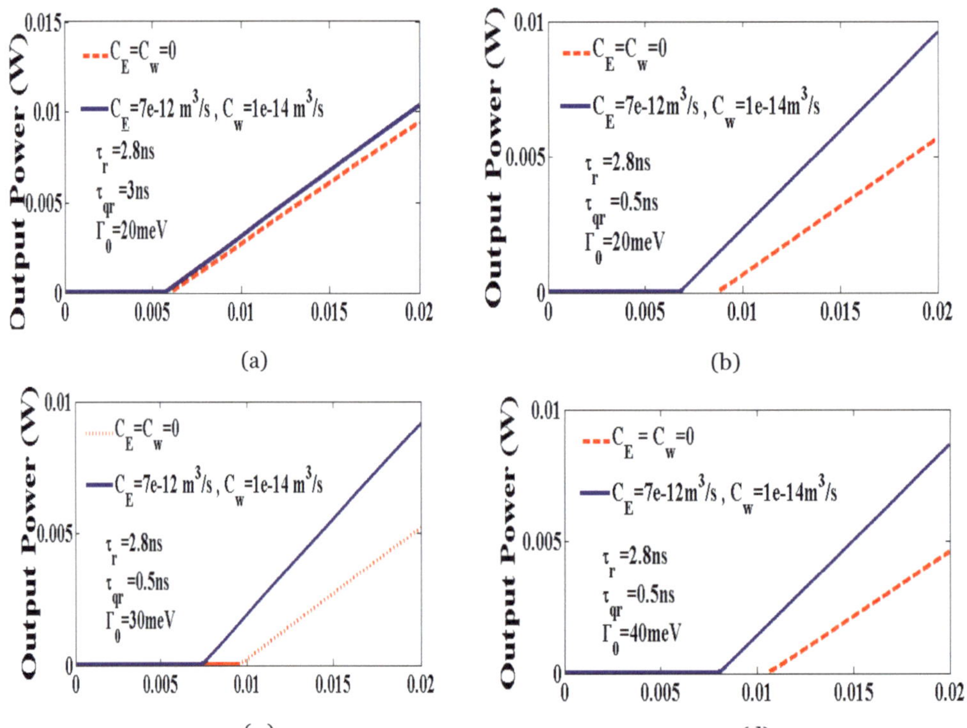

Figure 20.
The L-I curve of a SAQD laser considering the Auger effect. Different values of recombination lifetimes (a) τ_r = 2.8 ns, τ_{qr} = 3 ns, Γ_o = 20 meV; (b) τ_r = 2.8 ns, τ_{qr} = 0.5 ns, Γ_o = 20 meV; and different values of inhomogeneous (c) Γ_o = 30 meV; (d) Γ_o = 40 meV.

with a low density of the carriers in wetting layer (N_q), both relaxation and capture processes are mainly phonon assisted. However, as the QD laser pumping increases, the density of the wetting layer increases, too. This process is first slow but then it occurs in a more pronounced way. This is because, if the pumping increases from zero, most of the carriers get captured into the QDs. On the other hand, based on Pauli Exclusion Principle in QDs, the number of electron state in QD active layer, that is, GS and ES, has been limited. Therefore, if at first the GS and then the ES are filled with electrons, relaxation and capture in these states will get saturated as it is shown by factors (1-P_{ES}) and (1-P_{GS}) in Eqs. (11) and (12), respectively. Thus, the drain of carriers from the WL towards the dots slows down and this leads to the formation of the WL carrier reservoir. The carriers in the reservoir increase the Auger assisted capture speed as shown in Eqs. (16) and (17). On the other hand, by decreasing τ_{qr} from 3 to 0.5 ns, the carriers find more chances for recombining via non-radiative process outside the QDs. This results in the degradation of quantum efficiency and increase of threshold current simultaneously. **Figure 20(c)** and **(d)** illustrate the output power as a function of the injected currents for inhomogeneous broadenings Γ_0 = 30 meV and Γ_0 = 40 meV, respectively. As the inhomogeneous broadening increases, the optical gain decreases [see Eqs. (5) and (6)], and this results in larger occupation probabilities in leasing, and longer relaxation lifetime. Therefore, the quantum efficiency and output power decrease and threshold current increases.

Figure 21 shows the effect of Auger coefficient increment on the small-signal frequency response of the laser for different values of $C_W = 1 \times 10^{-14}$ m^3/s, $C_E = 7 \times 10^{-12}$ m^3/s, and $C_W = C_E = 0$. The inhomogeneous broadening is Γ_0 = 20 meV. The recombination lifetimes of τ_r = 2.8 ns and τ_{qr} = 3 ns have been used. The simulation reveals that when the Auger coefficient increases, the relaxation lifetime decreases. It means that the phonon bottleneck effect is degraded and as a result, the modulation bandwidth increases. On the other hand, as Auger coefficient increases and carrier relaxation times τ_{d0} and capture times τ_{c0} decreases, the 3 dB frequency increases. What is important here is that, as τ_r decreases from 2.8 to 0.5 ns, the frequency response degrades. Therefore, to prevent the effect of phonon bottleneck on frequency response, the recombination lifetime within quantum dots τ_r must be much longer than the carrier relaxation time (the carrier relaxation time is about a few pico-seconds) [4]. Results obtained by using the presented circuit model are in good agreement with the results calculated by solving the rate equations numerically and also with experiments reported so far [4, 7].

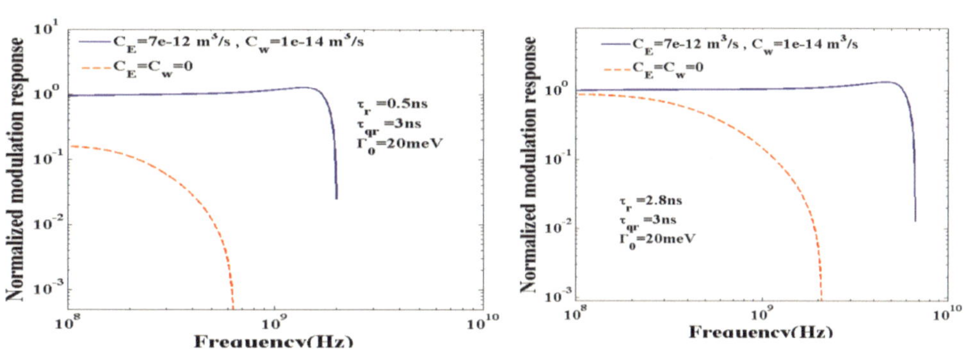

(b) (a)

Figure 21.
Modulation response of a SAQD laser considering the Auger effect for different values of carrier recombination inside quantum dot (a) τ_r *= 2.8 ns and (b)* τ_r *= 0.5 ns.*

5. Conclusions

The Auger effect can be overcome in the phonon bottleneck problem in SAQD lasers propose a modified equivalent circuit for simulation of the rate equations, considering this effect for the first time. The static and dynamic behavior of these lasers were studied. It is found that, if we want to solve the rate equations of a SAQD laser, it is possible to compensate the effect of phonon bottleneck by introducing the Auger effect. This can decrease the threshold current and increase the quantum efficiency of the SAQD laser, so the output power can be raised. It is shown that the Auger effect can increase the modulation bandwidth of a SAQD laser. Quantum dot lasers have been more favored for applications in optical communications, silicon photonics, lightwave telecommunication systems, high power lasers, Q-switched or mode locked lasers for short pulse generation and broad band light sources. The future implications of quantum dot lasers are issues such as quantum dot lasers and integration with Si photonic integrated circuits and colloidal quantum dot lasers.

Author details

Mahdi Razm-Pa[1*] and Farzin Emami[2]

1 Electronic Department of Charkhab Kian Electronic Devices Company, Borazjan, Bushehr, Iran

2 Optoelectronic Research Center, Electronic Department of Shiraz University of Technology, Shiraz, Iran

*Address all correspondence to: razmpa.mahdi@gmail.com

References

[1] Harrison P. Quantum Wells, Wires and Dots: Theoretical and Computational Physics of Semiconductor Nanostructures. New Jersey, USA: Wiley, John Wiley & Sons, Inc; 2010. p. 564. DOI: 10.1002/0470010827

[2] Steiner T. Semiconductor Nanostructures for Optoelectronic Applications. Bingley, United Kingdom: Artech House Publishers; 2004. p. 412. DOI: 10.1108/SR.2004.24.3.320.3

[3] Grillot F, Norman JC, Duan J, Zhang Z, Dong B, Huang H, et al. Physics and applications of quantum dot lasers for silicon photonics. Nanophotonics. 2020;**9**(6):16. DOI: 10.1515/nanoph-2019-0570

[4] Sugawara M. Self-assembled InGaAs/GaAs quantum dots. In: Semi-Conductors and Semimetals. Vol. 60. Washington, USA: Optical Publishing Group; 1999. p. 385. DOI: 10.1364/QO.1999.QMC4

[5] Razm-Pa M, Emami F. Simulation of the effect of biexciton oscillator strength on performance of self-assembled quantum-dot laser. Journal of Nanoelectronics and Optoelectronics. 2018;**13**:493-500. DOI: 10.1166/jno.2018.2091

[6] Morris D, Perret N, Fafard S. Carrier energy relaxation by means of Auger processes in InAs/GaAs self-assembled quantum dots. Applied Physics Letters. 1999;**75**(23):3593-3595. DOI: 10.1063/1.125398

[7] Sugawara M, Mukai K, Shoji H. Effect of phonon bottleneck on quantum-dot laser performance. Applied Physics Letters. 1997;**71**(19):2791-2793. DOI: 10.1063/1.120135

[8] Xin QL, Nakayama H, Arakawa Y. Phonon bottleneck in quantum dots: Role of lifetime of the confined optical phonons. Physical Review B. 1999;**59**(7): 5069-5073. DOI: 10.1103/PhysRevB.59.5069

[9] Magnusdottir I, Bischoff S, Uskov AV, Mork J. Geometry dependence of Auger carrier capture rates into cone-shaped self-assembled quantum dots. Physical Review B. 2003;**67**(20):205326/1-205326/5. DOI: 10.1103/PhysRevB.67.205326

[10] Sugawara M, Mukai K, Nakata Y, Ishikawa H, Sakamoto A. Effect of homogeneous broadening of optical gain on lasing spectra inself-assembled Inx Ga1-x As/GaAs quantum dot lasers. Physical Review B: Condensed Matter. 2000;**61**(11):7595-7603. DOI: 10.1103/PhysRevB.61.7595

[11] Feldmann J, Cundiff ST, Arzberger M, Bohm G, Abstreiter G. Carrier capture into InAs/GaAs quantum dots via multiple optical phonon emission. Journal of Applied Physics. 2001;**89**(2):1180-1183. DOI: 10.1063/1.1333718

[12] Ohnesorge B, Albrecht M, Oshinowo J, Forchel A, Arakawa Y. Rapid carrier relaxation in self-assembled InxGa12xAs/GaAs quantum dots. Physical Review B. 1996;**54**: 11532-11538. DOI: 10.1103/physrevb.54.11532

[13] Narvaez GA, Bester G, Zunger A. Carrier relaxation mechanisms in self-assembled (In,Ga)As/GaAs quantum dots: Efficient P to S Auger relaxation of electrons. Physical Review B. 2006; **74**(7):075403/1-075403/7. DOI: 10.1103/PhysRevB.74.075403

[14] Gioannini M, Sevega A, Montrosset I. Simulations of differential gain and linewidth enhancement factor of quantum dot semiconductor lasers. Optical and Quantum Electronics. 2006; **38**:381-394. DOI: 10.1007/s11082-006-0038-1

[15] Gioannini M, Montrosset I. Numerical analysis of the frequency chirp in quantum-dot semiconductor lasers. IEEE Journal of Quantum Electronics. 2007;**43**:941-949. DOI: 10.1109/JQE.2007.904306

[16] Grillot F, Veselinov K, Gioannini M, Montrosset I, Even J, Piron R, et al. Spectral analysis of 1.55-μm InAs-InP(113)B quantum-dot lasers based on a multipopulation rate equations model. IEEE Journal of Quantum Electronics. 2009;**45**(7):872-878. DOI: 10.1109/JQE.2009.2013174

[17] Kashiri M, Asgari A. Modeling of carrier dynamics in InGaAs/GaAs self-assembled quantum dot lasers. Applied Optics. 2016;**55**(8):2042-2048. DOI: 10.1364/AO.55.002042

[18] Grundmann M. Nano-Optoelectronics, Concepts, Physics and Devices. New York: Springer; 2002. p. 442. DOI: 10.1007/978-3-642-56149-8

[19] Sugawara M, Hatori N, Ebe H, Ishida M. Modeling room temperature lasing spectra of 1.3-mself-assembled InAs/GaAs quantum-dot lasers: Homogeneous broadening of optical gain under current injection. Journal of Applied Physics. 2005;**97**(4): 43523-43527. DOI: 10.1063/1.1849426

[20] Sugawara M. Effect of carrier dynamics on quantum-dot laser performance and possibility of bi-exciton lasing. Proceedings of SPIE. 1998;**3283**:88-99. DOI: 10.1117/12.316640

[21] Berg TW, Bischoff S, Magnusdottir I, Mørk J. Ultrafast gain recovery and modulation limitations in self-assembled quantum-dot devices. IEEE Photonics Technology Letters. 2001;**13**(6):541-543. DOI: 10.1109/68.924013

[22] Razm-Pa M, Emami F. Effect of parameter variations on the static and dynamic behaviour of a self-assembled quantum dot laser using circuit-level modeling. Quantum Electronics. 2015; **45**(1):15-22. DOI: 10.1070/QE2015v045n01ABEH015264

[23] Ahmadi V, Yavari MH. Circuit-level implementation of semiconductor self-assembled quantum dot laser. IEEE Journal of Selected Topics in Quantum Electronics. 2009;**15**(3):774-779. DOI: 10.1109/JSTQE.2008.2011997

[24] Razm-Pa M, Emami F. Phonon bottleneck and auger recombination effect on the performance of a self-assembled quantum dot laser. SILICON. 2018;**10**:1513-1521. DOI: 10.1007/s12633-017-9634-x

Chapter 3

Improved Nanocomposite Materials and Their Applications

Tahira Mahmood, Abid Ullah and Rahmat Ali

Abstract

Nanotechnologies and nanocomposite materials have gained the attention of scientific community in recent years. Nanocomposite material consists of several phases where at least one, two, or three dimensions are in the nanometer range. Nanocomposites with advanced carbon nanostructures i.e., carbon nanotube (CNTs) and graphene, attachments have been regarded as promising prospects. CNTs and graphene-based improved nanocomposites are usually categorized into various classes based on different types of discontinues phases. The nanocomposites reinforced with carbon nanomaterials i.e., CNTs and graphene have been explored extensively for use as engineering materials in several demanding applications because of their excellent properties. The present book chapter has been prepared in three main sections. In the first portion, nanocomposites and carbon nanofillers i.e., CNTS and graphene have been presented. In the second part, different types of CNTs and graphene-based improved nanocomposites have been described with reported literature. In the third section, focus is on the applications of improved nanocomposites such as energy storage, antimicrobial activity, gene delivery, catalyzed organic reactions, radar adsorbing materials, actuators, wind turbine blades, pollutant removal, aerospace industry, and conductive plastics.

Keywords: nanocomposites, carbon nanoclusters, carbon nanotubes, graphene, improved nanocomposites, applications

1. Introduction

In modern technology, composites are one of the most essential materials, which are the aggregate of two or more materials having different physical and chemical properties discriminated by their interface. Therefore, unlike the individual materials, the composite materials exhibit a distinctive property [1]. Mostly composite materials consist of at least two components including a continuous matrix phase and discontinuous reinforcement material while the other consists of one or more discontinuous phases dispersed in one continuous phase. Generally, a discontinuous phase has more advanced mechanical properties than a continuous phase. Continuous phase is known as "matrix" while discontinuous phase is called "reinforcement" or reinforcing material. Based on the size of reinforcement in the structures, composites are commonly divided into three basic classes, named macrocomposites, microcomposites, and nanocomposites. Nanocomposites offer excellent features by the application of reinforcement in the composite below 100 nm in size phase [2].

1.1 Nanocomposites

Nanocomposites are multi-phasic materials, in which at least one phase show dimensions in the nano range (10–100 nm). In nanocomposites, interaction between matrix and reinforcement is very high due to high surface-to-volume ratio. The improved properties of nanocomposites depend on properties of each material, their relative amounts, and the overall geometry of the nanocomposites. Various materials possess different properties which when combined results in the formation of new material with additional advantages relevant to different areas of science and technology. They have high thermal and mechanical stability, multifunctional capabilities, chemical functionalization, and huge interphase zone. Generally, the nanocomposites show enhanced properties, such as high specific stiffness and strength, high toughness, low density, corrosion resistance, and thermal insulation [3]. Currently, nanocomposite materials have emerged as a suitable choice to overcome restrictions of different engineering materials. The amalgamation of nanoparticles into a matrix of materials like polymer, metal, or ceramics promote their properties such as excellent mechanical stability (in terms of strength, dimension stability, toughness, flexibility, Young's modulus, etc.), good optical activities, flame retardancy, low water/gas permeability, and high electro-thermal conductivity [4]. Nanocomposites are accepted at both the academic and industrial levels due to their extraordinary properties, distinctive design capacity, eco-friendly nature, easy fabrication, and cost-effectiveness. Nanocomposites have been commonly used in numerous applications due to their advanced properties. They are reported to be the materials of the 21st century in the vision of possessing design uniqueness and property permutation that is not found in conventional composites [5].

2. Carbon nanostructures

Carbon-based materials have a huge stimulus in encouraging the improvement of society due to their abundance on the earth and environmental kindliness and other merits. Over the past few decades, carbon materials such as CNTs and graphene class of materials have seen incredible growth due to the discovery of advanced nanostructures. The innovation and study of carbon nanofillers have played a major role in the development of nanocomposites. Based on their dimensions, researchers classified materials as zero-dimensional (0-D) nanoparticles or quantum dots, one-dimensional (1-D) nanobelts, nanowires or nanotubes, two-dimensional (2-D) nanoplates or nanodisks, and three-dimensional (3-D) nanocones or nanocoils as shown in **Figure 1**. The nano-structured materials (NSMs) have drawn extreme interest due to their structure, surface area, size effects, and considerably improve the performance of the composites [7]. Carbonaceous nanofillers such as carbon nanotubes (CNTs) and graphene play a potential role as compared to others due to their improved structural and functional properties such as high aspect ratio, high mechanical and electrical properties, etc. [8]. In the last few decades, CNTs and graphene have been considered the most substantial nanofiller to formulate advanced nanocomposites for both academic and industrial fields due to their several potential applications. The combination of polymer nanocomposites with graphene-related materials (GRMs)

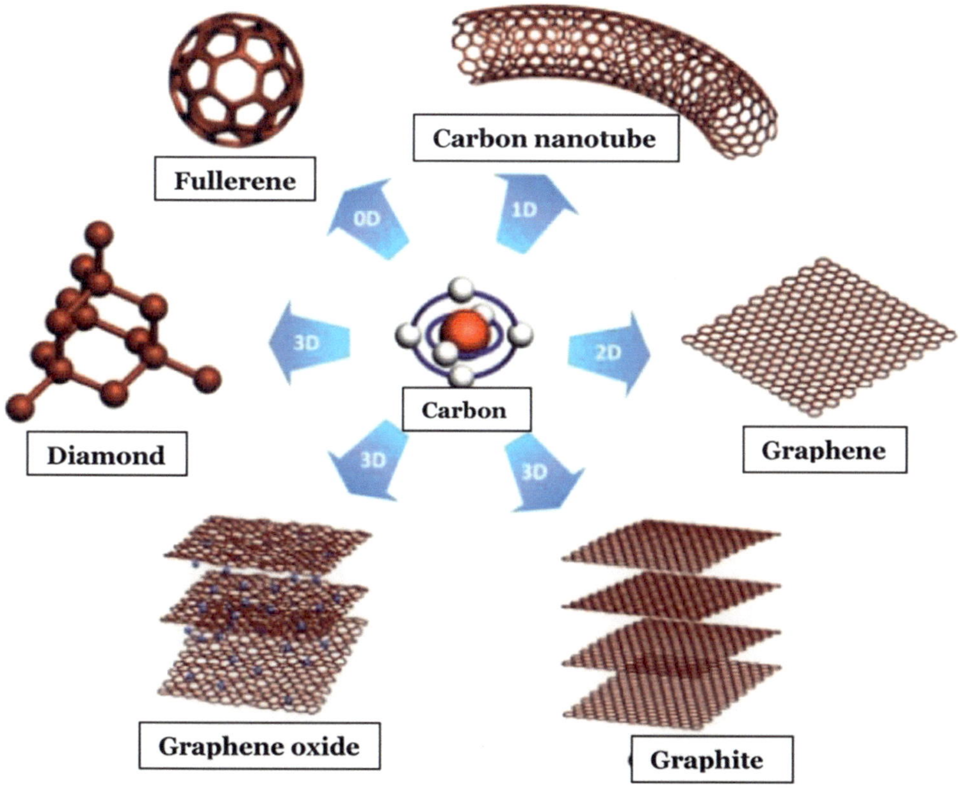

Figure 1.
Nano-carbon materials including 0D fullerene, 1D CNT, 2D graphene, 3D graphite, 3D graphene oxide, and 3D diamond are demonstrated [6].

or carbon nanotubes (CNTs) has been discovered as a result of their low mass density and excellent mechanical properties for use in engineering materials for various challenging purposes [9].

2.1 Carbon nanotubes

CNTs are one-dimensional carbon materials that are different from other carbon compounds, such as graphite, diamond, and fullerene (C_{60}, C_{70}, etc.), having an aspect ratio greater than 1000 [10]. In 1991, Iijima discovered carbon nanotubes [11], which brought innovatory changes in the field of polymer nanocomposites. Ajayan et al. [12], reported the first carbon nanotubes reinforced polymer nanocomposites. Basically, carbon nanotubes are graphene sheets having hexagonal structures which are rolled up into cylindrical form and rounded off with half shape of fullerene structure. The two types of carbon nanotubes are the single-walled nanotubes (SWNTs), which are single graphene sheets rolled into a cylinder and the other one is multi-walled nanotubes (MWNTs), in which numerous graphene layers are stacked into concentric layers in the form of cylinders with an interspacing of 0.34 nm (**Figure 2**). On the basis of atomic arrangement, the three types of structures are zigzag, armchair, and chiral (**Figure 3**) [13]. Properties of carbon nanotubes are greatly reliant on morphology, size, and diameter and maybe metallic or semiconducting depending on the atomic arrangement [14].

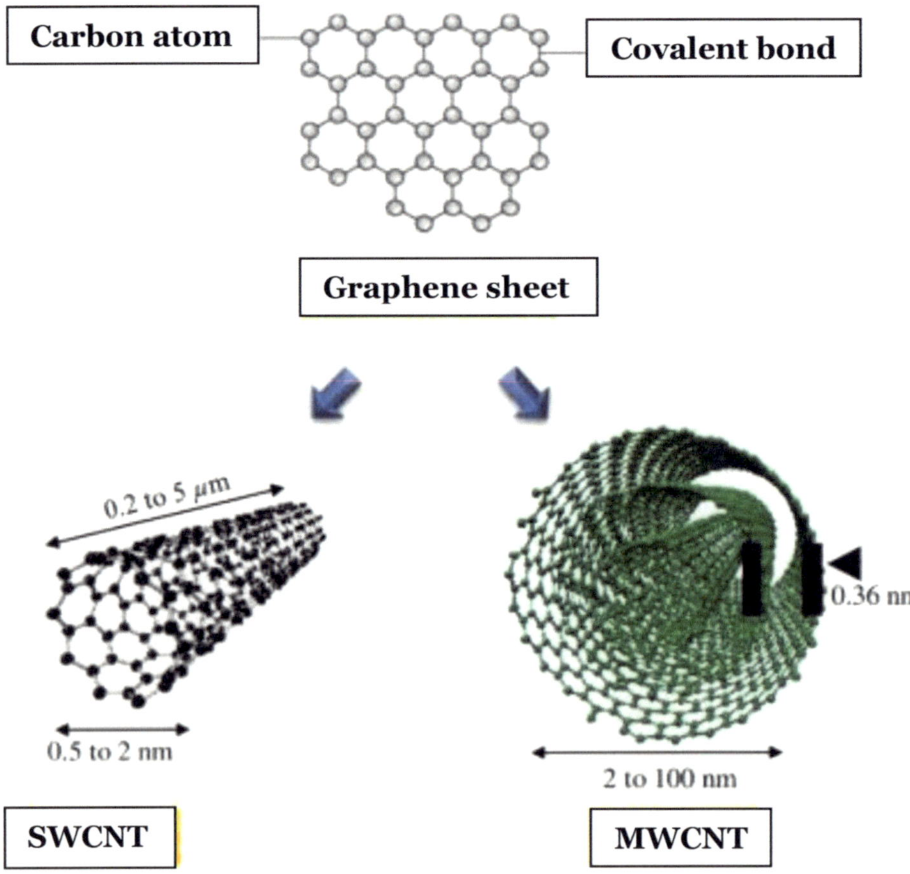

Figure 2.
The conceptual diagram showing the general dimensions of the length and width of single walled carbon nanotubes (SWCNTs) and multi-walled CNTs (MWCNTs [13].

2.2 Graphene

Graphene, a typical 2D material, has received incredible attention due to attractive features like very high specific surface area (2360 m^2g^{-1}), highest strength (≈130 GPa) and Young's modulus (≈1.0 TPa), best known thermal conductivity (TC, ≈5000 W m^{-1} K^{-1}), and electrical conductivity (10^8S m^{-1}). Therefore, graphene is the perfect nanofiller for improving the mechanical, electrical, thermal, and optical properties of polymers [15]. Through valuable interfacial stress transfer, graphene effectively improves the mechanical properties like tensile strength and Young's modulus of polymers [16, 17]. Graphene can competently strengthen brittle polymers by extending the crack propagation path in nanocomposite. The excellent electrical property of graphene can clearly improve the electrical conductivity of polymers for free electrons by building a conductive network [18]. The distinctive thermal conductance of graphene carries the excitement to formulate high-performance thermal conductive nanocomposites for application in high power density devices in thermal management [19, 20]. Due to distinct ultrahigh thermal conductivity, graphene is engaged to fabricate the thermal interface materials (TIMs) [21] and phase change energy storage composites (PCCs) [22].

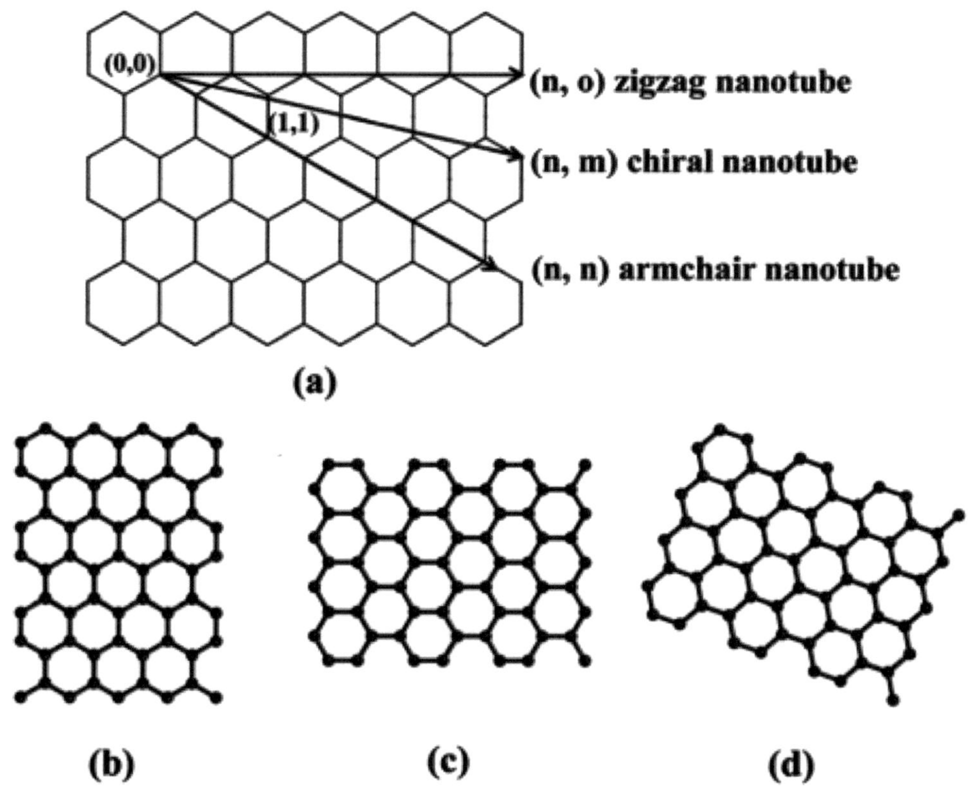

Figure 3.
Schematic representation of how a graphene sheet is rolled to form three chiralities of nanotubes: (b) zigzag, (c) armchair, and (d) chiral nanotubes [13].

2.3 Diamond

Diamond is a three-dimensional carbon material with a crystal structure called diamond cubic. Diamond is an outstanding carbon material because of its inert nature, high-thermal conductivity, stiffness, biocompatibility, and optical transparency. Nanodiamonds (NDs) are advanced carbon nanostructures with sp^3 hybridized carbon atoms bonded to form diamond-like cubic geometry. Their dimensions are in the range of 5 to 10^2 nm. Numerous properties of NDs are far better than bulk diamond and they present these properties on nanoscale [23]. The superior mechanical and thermal characteristics of ND make it a suitable nano-filler for carbon-based nanocomposites. The surface of common synthetic NDs has no functional groups. Nevertheless, an ND can be modified with a functional polymer or be functionalized with hydrogen/deuterium-terminated, halogenated, aminated, hydroxylated, and carboxylated by strong reagents and under severe conditions according to targeted applications and desired physicochemical properties [24].

2.4 Fullerenes

Fullerenes are a new class of carbon nanomaterials discovered in 1985 by Kroto et al. and get the Nobel prize award in chemistry for the year 1996 [25]. According to

Mukherjee et al. the diameter of fullerenes nanomaterials is ≤1 nm [26]. The fullerene family includes several atomic C_n clusters ($n > 20$), composed of carbon atoms on a spherical surface. They are closed-cage carbon molecules containing pentagonal and hexagonal rings with sp^2 hybridized carbon atoms bonded by covalent bonds. The carbon atoms are regularly arranged at the vertices of pentagons and hexagons surfaces. They have the formula C_{20+m}, with m being an integer number, and comprise a wide range of isomers and homologous series, from the most common and investigated C60 and C70 to the so-called higher fullerenes like C240, C540, and C720. The incorporation of fullerene nanofiller into numerous polymers has been achieved by physical and chemical methods [27]. In this way, the combination of distinctive features of fullerenes with physical properties of polymers may yield advanced polymeric materials with novel physicochemical characteristics. The attractive physiochemical properties of fullerene-based nanomaterials make them suitable materials to use in medicinal chemistry [28].

3. Improved carbon nanocomposites

In various fields of technology such as medical, sensors, computing etc. materials play important role for our comfort. The use of new materials with enhanced properties is a critical need due to demand for betterment. High-performance nanocomposites are recommended for industrial use like energy storage, damage sensing, aerospace and automobiles but have some limitations. The discovery of carbon-based nanofillers plays a vital role to overcome these limitations [29]. Combination of CNTs and Graphene extensively improves the properties of nanocomposites. CNTs and graphene are considered significant nanofillers to formulate advanced nanocomposites due to their several possible applications. At nanoscale, carbon nanotubes and graphene nanoplatelets (GnP) materials suggest an exclusive combination of elastic modulus (0.8–3 Tpa), thermal conductivity (3000–6000 W/m-K), and electrical resistivity (3–20 μΩ-cm) [30–33]. Some researchers have processed nanocomposites in the range of 0.2–5 weight % of CNTs into various polymer matrix materials [34]. These nanocomposites revealed 10–35% enhancement in properties such as modulus, strength, impact resistance, thermal conductivity, and electrical conductivity, compared to the matrix polymer.

3.1 Carbon nanotubes-based nanocomposites

Carbon nanotubes have been deemed as versatile building blocks to create a novel generation of nanocomposites desired for a variety of commercial applications.

3.1.1 Polymer/CNT nanocomposites

The nanoscale CNTs amalgamation into a polymer system highly modifies the properties of composites even at a particularly low content of filler. As explained earlier, CNTs are the strongest and hardest fibers ever known. The excellent mechanical and other physical properties of CNTs demonstrate huge potential applications of Polymer/CNT nanocomposites which are one of the most studied systems. Polymer matrix can be easily fabricated without disturbing CNTs by conventional manufacturing techniques resulting in cost reduction for mass production of nanocomposites in the future [35]. Paul et al. [36] synthesized polypyrrole (PPy) and MWCNTs nanocomposites with different compositions by chemical oxidative polymerization method. Polypyrrole (PPy)/MWCNT have been effectively used as supercapacitor devices. Béguin and his research group studied conducting polymer and CNTs based nano-electrodes with improved mechanical, thermal, and electrical properties [37]. PANI/ MWCNT nanocomposite with a specific capacity of 440 Fg^{-1} at 5 mVs^{-1} and capacitance retention of 93% after 1000 cycles was reported [38]. Lezak et al. [39] prepared polyaniline (PANI) as an intrinsically conducting polymer and poly (vinylidene fluoride) (PVDF) and MWCNTs.

3.1.2 Activated carbon/CNT nanocomposites

Activated carbon, also known as activated charcoal, is a carbon type that is treated with tiny volume holes to improve the surface area. One gram of activated carbon, due to its increased microporosity, has a surface area of more than 3000 m^2/g calculated by gas adsorption [40]. Activated carbon (AC)/CNTs nanocomposites are superior materials having AC as matrix material and CNTs as fillers. Numerous researchers have used activated carbon for production of CNT nanocomposites. Huq et al. [41] studied the preparation of AC and CNTs based supercapacitors by a superficial electrophoretic deposition (EPD) method. In this study, the as-prepared AC/CNT electrode had capacitance maintenance of 85% after 11,000 cycles. In EDLC electrode, activated carbon has been used for an extensive period due to its high capacitance, low cost, and long cycle life [42]. Qiu et al. [43] prepared activated carbon fibers (ACHFs) combined with carbon nanotubes and nickel nanoparticles (CNTs-Ni-ACHFs) by thermal reduction and chemical vapor deposition method. Usually activated carbon, due to its very high surface area, is used as an absorbent [44].

3.1.3 Metal oxide/CNT nanocomposites

Carbonaceous materials have high power and low energy density which cause restrictions in their general application. However, metal oxides, due to their high-energy density, are used as pseudo capacitor electrodes for supercapacitors [45]. Yuan et al. [46] developed a new method for CNTs coated magnesium oxide (MgO) nanoparticles to increase the interfacial bonding strength. Yuan et al. [47] examined a sandwich structured MoO_2 @TiO_2 @CNT nanocomposite by an easy two steps synthesis method under Ar/H_2 flow, controlled hydrolysis, and a subsequent heat treatment. Alam et al. [48] prepared BaMg0.5Co0.5TiFe10O19/MWCNT nanocomposites by varying the amount of MWCNTs (0, 4, 8, and 12 vol%). Nanocomposite with 8% vol. of MWCNTs performed best.

3.1.4 Carbon fibers/CNTs

A novel method was studied to graft carbon nanotubes over carbon fiber to form a CNT/CF [49]. Islam reported direct covalent bonded CNTs and CF without any catalyst or coupling agents through ester linkage. CNTs can be used to reinforce CFs to improve interfacial shear and impact strength [50]. Two methods are reported to attach CNTs with CF by physical adsorption (Van der waals interaction), which are weaker than chemical covalent bonding [51].

3.2 Graphene-based nanocomposites

Graphene-based nanostructured materials have distinctive 2D structures with high electronic mobility, exceptional electronic and thermal conductivities, excellent optical performance, good mechanical strength, and ultrahigh surface area as compared to other materials.

3.2.1 Polymer/graphene nanocomposite

Graphene nanofiller addition within the polymer has a promising application in biosensors, energy storage devices, photocatalysts, drug delivery. Recently, a wide range of processing methods has been studied for scattering both GNP and GO-derived fillers into polymer matrices. Controlled amount of nanomaterial by weight % and size is carefully taken into deliberation [52]. Salimikia et al. [53] synthesized a solid-phase microextraction fiber over polyaniline/graphene oxide nanomaterial using the electrospinning method and used it as sorbent for determination of nicotine. Farajvand et al. [54] prepared Graphene oxide/polyaniline nanocomposite and used it as an adsorbent to determine cadmium (II) ions in an aqueous solution.

3.2.2 Activated carbon/graphene nanocomposite

Activated carbon is considered as the center of research for commercial utilization. Adsorption properties of metal ions by AC/GR have been examined and a number of methods have been developed for synthesis of graphene/activated carbon nanosheet composite to make high-performance electrode material for supercapacitors. Many research groups have used activated carbon for preparation of graphene nanocomposites. Xin et al. [55] reported a new carbon nanocomposite material having graphene and activated carbon and used it for oxygen electrode (cathode) in Li-ion batteries. In the AC/GR, the graphene showed a three-dimensional (3D) arrangement having good electrical conductivity and exceptional mechanical strength and elasticity, while the AC coating on the graphene surface supplied several meso/micropores with diameters less than nanometers. Lu et al. [56] investigated an easy method to prepare a new catalyst by electrodepositing of Ag nanocrystals on the different polymer dyes, Poly (methylene blue) or Poly (4-(2-Pyridylazo)-Resorcinol) modified graphene carbon spheres (GS) hybrids which had advantages of both carbon spheres and graphene composite and were employed for detection of H_2O_2 as non-enzymatic electrochemical sensor. Hossain and Park [57] studied the hydrothermal method for the synthesis of glucose-treated reduced graphene oxide-activated carbon composites. Platinum nanoparticles were electrochemically deposited on a modified composite surface. Chitosan-glucose oxidase composites and Nafion were incorporated into modified surface of working electrode for the preparation of an extremely sensitive glucose sensor.

3.2.3 Metal oxide/graphene nanocomposite

In this type of nanocomposites, metal oxide particles are incorporated in graphene nanosheets. Metal oxide-based Graphene nanocomposite has attained the attention of scientific community as anode materials due to high kWh/cost and effective high-performance electrode material in an electrochemical supercapacitor. Beura et al. [58] prepared ZnO-based graphene nanocomposites by hydrothermal method and used it as a catalyst for the degradation of dyes. The band of the nanocomposites was 2.84 eV, while the photoluminescence lifetime increased from 15.05 to 21.60 ns. Photocatalytic activity of the composite material was investigated by both anionic and cationic dyes. Borah et al. [59] used an in-situ method to synthesize TiO_2/rGO. The prepared nanocomposites were used to catalyze the transesterification of waste cooking oil into biodiesel. Excellent catalytic activity was shown by the catalyst and 98% conversion of oil into biodiesel was seen at optimum reaction conditions. Wang et al. [60] presented a detailed summary of the research progress on the low-cost metal oxides/graphene nanocomposites (MOs/G) as anode materials for SIBs.

3.2.4 Metal/graphene nanocomposites

Various heavy metals such as Au, Fe, Cu, Ce, etc., have been introduced into graphene nanosheets to form different nanocomposites. Nanocomposite with ultra-low resistivity than conventional copper metal at room temperature is the next generation conductor. Several research groups have fabricated metals/ graphene nanosheets. Arukula and co-workers [61] prepared rGO/polyaniline (PANI)/Pt–Pd nanocomposite by wet reflux strategy. The prepared nanocomposites materials were used as potential anode catalysts with improved methanol oxidation tendency for direct methanol fuel cells (DMFCs). Xuan et al. [62] reported a 3D patterned porous laser-induced silver-based graphene nanocomposite. The prepared nanocomposites were used as an electrode, which showed high, uniform electrical conductivity even under mechanical deformations. Incorporation of platinum and gold nanoparticles on the 3D porous LIG importantly enhanced the electrochemical capacity for wearable glucose sensor applications. Zheng et al. [63] reported the quick and effective preparation and characterization of a novel nitrogen-doped graphene copper nanocomposite. The prepared nanocomposite showed superior electrical conductance of 538 W/m·K at room temperature, which is 138% greater than that of copper. The measured electrical resistance was 0.16 $\mu\Omega$ cm at 25°C which is much lower than that of copper. Gupta et al. [64] reported copper-based reduced graphene oxide nanocomposite for use as a catalyst. The copper-based reduced graphene oxide catalyst was easily recovered and used for seven consecutive cycles.

3.2.5 Fibers/graphene nanocomposites

In this type of nanocomposites, graphene is used as a filler while fibers are used as a matrix. Davoodi and co-workers [65] reported the preparation of polylactic acid and GO-based nanocomposite using the electrospinning method. The mechanical properties, surface chemical structure, and topology study of the nanofibers were performed. Jin et al. [66] used a facile method for the hybridization of polyaniline nanofibers (PANI NFs) on functionalized reduced graphene oxide (FrGO) films. The GO was first reduced and functionalized by sulfur to form FrGO. Hydrothermal method was used to hybridize FrGO and PANI NFs to form PANI NFs/FrGO composite films. The as-prepared nanocomposite films were uniform, flexible, and stable with a high specific capacitance of 692.0 F/g at 1 A g^{-1} and excellent capacitance retention of 53.5% at 40 A g^{-1}. Wan et al. [67] used a facile two-step method to prepare a ternary flexible nanocomposite material of bacterial cellulose/graphene/polyaniline (BC/GE/PANI). The prepared nanocomposite showed enhanced electrical conductivity of 1.7 ± 0.1 S/cm, which is greater than most of the polyaniline-based composites.

Matrix	Filler type	Filler fraction (wt %)	Young's modulus (GPa); polymers	Tensile strength (MPa); polymers	Young's modulus (GPa) and increment %; composites	Tensile strength (MPa) and increment %; composites
Epoxy	Pristine GNPs	3.0	1.48	46.46	1.64 (10.8%)	49.78 (7.1%)
Epoxy	Pristine MWCNTs	3.0	1.48	46.46	1.69 (14.2%)	54.48 (17.3%)
PVA	Modified MWCNTs	0.5	0.0166	19.11	0.0329 (98.2%)	34.60 (81.1%)
PVA	Graphene oxide	0.3	2.32	25.3	5.82 (150.9%)	63.0 (149.0%)
PES	Pristine MWCNTs	1.0	0.045	1.70	0.067 (48.9%)	2.38 (40.0%)
PES	Graphene oxide	1.0	1.16	30.79	1.95 (68.1%)	55.73 (81.0%)
HDPE	Modified MWCNTs	2.5	0.75	26.5	1.15 (53.3%)	34.5 (30.2%)
HDPE	Pristine GNPs	10	0.96	27.2	1.49 (55.2%)	33.4 (22.8%)

Table 1.
Comparison of mechanical properties of different polymers and their nanocomposites [68].

Table 1 shows the comparison of mechanical properties of some polymers and their nanocomposites.

4. Applications of improved nanocomposites

Applications of improved carbon nanocomposites are shown schematically in **Figure 4**.

4.1 Energy storage

Currently, Graphene materials find a number of applications in the field of renewable energy, particularly photoenergy field, including solar thermal conversion, solar electricity conversion, photocatalysis, etc. New technologies related to solar cells have been developed in which either the active medium or transparent/distributed electrode consists of Graphene materials. A newly developed 3D cross-linked graphene material working as an ideal solar thermal converter can obtain the efficiency of 80% and more than 80% under one sun intensity and the ambient sunlight respectively [69]. The structural design possessed by the material plays a significant role in improving the efficiency of energy conversion. The unique structure of graphene foam due to an array on its 3D skeleton, created by nanoplates of the graphene, provides a greater area for heat exchange. This enhances the efficiency of solar-thermal conversion up to 93.4%. Dye-sensitized solar cells (DSSC) consisting of a redox couple and a counter electrode are used extensively for solar-electrical energy conversion [70]. Dye-sensitized solar cells (DSSCs) composed of coloring molecules, natural liquid electrolytes, and nanocrystalline metal oxides

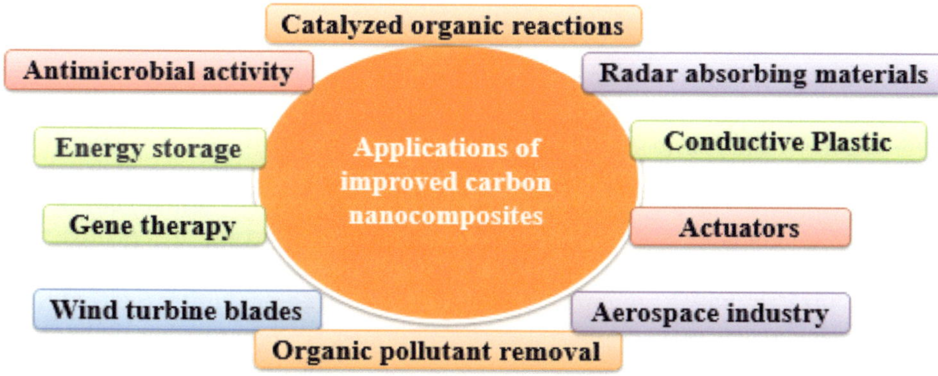

Figure 4.
Applications of improved carbon nanocomposites.

have shown greater performance in energy conversion and manufacturing costs and low energy. Nowadays, graphene-based electrodes exhibiting chemical stability and good conductivity, very large surface area, considerable high porosity, and electrocatalytic activity are used in DSSC. The application of this electrode led to the improvement in the performance and the reduction of the cast to a large extent. A super-capacitor designed by Stroller et al. [71] from the chemical modification of graphene material exhibited specific capacitance 135 F/g, 99 F/g, and 99 F/g in aqueous electrolytes, ionic electrolytes, and organic electrolytes respectively. High life cycle and high power have been shown by these types of storage devices. Zhang et al. [72] synthesized a useful stretchable electrode by the mechanical exfoliation of graphene prior to chemical treatment. These electrodes show high flexibility and compatibility in usage in various electrolytes. Moreover, graphene conducting polymer composites or graphene transition metal composites can be used to devise Super-capacitors.

4.2 Antimicrobial activity

Graphene nanomaterials possess intrinsic antimicrobial properties and also act as a platform to design antimicrobial nanocomposites having higher antimicrobial activity. Graphene is an ideal scaffold material owing to its huge surface area to anchor various sorts of macromolecules and nanoparticles. The attachment of diverse compounds such as quaternary phosphonium salts to graphene has greatly improved the antimicrobial properties. Researchers have shown great interest in Silver owing to its higher antimicrobial activity and studied it extensively to design graphene-based antimicrobial nanocomposites. In this section, the research carried out on the development of graphene–silver antimicrobial nanocomposites will be discussed. As the nanocomposites possess greater antimicrobial activity, graphene–silver nanocomposite is preferred over silver nanoparticles alone. Graphene–silver nanocomposites facilitate the leached silver ions from the nanoparticles to penetrate the cell owing to graphene which possesses the property to rupture a cell membrane. The purpose of this proposed mechanism was to explain the synergetic effect caused by silver and graphene existing together in the form of nanocomposite, and proteomic analysis of this effect of graphene–silver nanocomposites compared to silver nanoparticles alone [73] supported the mechanism as well.

4.3 Gene therapy

Gene therapy is a big breakthrough in medical science. Gene therapy is a new technique to treat various genetic diseases such as cystic fibrosis, Parkinson's, and various cancers. Efficient gene therapy includes a gene vector that protects the desired gene from nuclease degradation and allows cellular uptake of DNA with high transfection efficiency. The selection of an appropriate gene vector is the main obstacle in the development of gene therapy. The non-toxic nanocarrier in efficient gene therapy is Graphene and graphene-coated substrates. Graphene-based nanosheets have sp^2 hybridized orbitals, which are capable to interact with drugs and other molecules like nucleic acids such as DNA and RNA. Thus, they can be used for gene delivery or as carriers and protectors of probes involved in identifying miRNAs [74]. Hyunwoo et al. synthesized GO-based polyethyleneimine (PEI) composite and used it as an efficient gene delivery carrier in gene therapy [75]. Among different carriers, polyethyleneimine (PEI) has been typically recognized as the "golden standard" cationic polymer in gene transfection, because of its strong binding to DNA and RNA and effective uptake by cells. However, due to high cytotoxicity and poor biocompatibility, applications of PEI polymers in gene therapy is limited [76].

4.4 Catalyzed organic reactions

Carbon-based nanocomposite materials have been extensively used as heterogeneous catalysts in many organic chemical reactions. In chemical industries, less than 10% of the chemical reactions are still conducted without the addition of specific catalysts [77]. The catalytic products such as organic building blocks, pharmaceuticals, natural products, and agricultural derivatives are very valuable in chemical industries [78]. In many industrial chemical reactions, different types of supported and unsupported metal catalysts have been investigated. In the last few decades, the researchers gave more importance to using carbon nanostructure-based composites as heterogeneous catalysts in organic transformations. The highlighted advantages of carbon-based nanocomposites catalysts are high surface area, stability, fine dispersion, reusability, and easy recovery after completion of reactions. Furthermore, the introduction of metal nanoparticles onto the carbon support has shown more usefulness in carrying out the highly selective catalytic organic reactions [79]. In comparison with CNTs, graphene or GO has been preferred due to its low cost, large-scale preparation, and less health risk.

4.5 Radar absorbing materials

Electromagnetic wave absorbing materials are widely used in radar-absorbing technology. The absorbing of electromagnetic waves has many widespread applications, including minimizing the radar signature of a target, protection of human eyes, protective shielding of computers, consumer electronics, and optical sensors from intense laser pulses. Composites, in which polymer matrix having embedded multiwalled carbon nanotubes (MWCNTs), have been studied in multi-frequency detection mode instruments. Their applications in the microwave frequency range are anti-reflection, microwave absorbers, and electromagnetic interference shielding. The use of CNTs in radar-absorbing nanocomposite materials is more prominent due to their attractive properties i.e., electrical capacity, stiffness along large electromagnetic wave absorption tendency in the microwave range [80]. Zakharychev et al. [81]

have investigated the radar absorbing properties of epoxy binder and CNT nanocomposites in the frequency of 52–73 GHz.

4.6 Actuators

Shape memory alloys or liquid crystal elastomers have latent ability for activation under the right conditions. However, blending of two or more materials is required by other systems to report a new physical response leading to activation process. Recently, it has been proven that the polymer nanocomposite seems to be the best candidate for mechanical activation processes. The reported literature showed that most of the studies were concerned with the activation of the already described polymer matrix by introducing nanotubes. A new activator response was reported by Courty and their team by an electrical field due to the presence of CNTs in an elastomer polysiloxane [82]. Koerner et al. fabricated new polydimethylsiloxane (PDMS)/MWCNT nanocomposite material which showed a mechanical response to infrared radiation [83]. The mechanical response of (PDMS)/MWCNT was due to absorption of photons from irradiation and not due to slight heating of material. Moreover, the nature of activator mechanism is not known. Mylvaganam and Zhang [84] developed a method for the preparation of activating nanocomposite material i.e., polyimide and CNT nanocomposite.

4.7 Wind turbine blades

In the last few decades, it has been proved that the fastest-growing installed energy generation technology is wind energy. The energy department of US has set a goal for their researchers to produce at least 20% of its electrical energy up to 2030. To achieve the goal, the focus of the wind industry is to manufacture large blades. This is because the square of rotor radius increases the wind turbine energy output. This is a big challenge for the researchers to make such a large-sized blade with excellent mechanical properties i.e., excellent stiffness and strength, long fatigue life, and low weight. To enhance the strength of polymer matrices used for wind turbines, CNTs are the best candidate because of their excellent properties. Most of the previous work showed that CNTs used in polymer matrices have enhanced the strength and stiffness of composites materials. Recently this is also been reported that the use of CNT increased the fatigue and tensile properties of thermoset resins used in wind industry. To improve such type mechanical properties CNTs are the best candidates among the discovered fillers. It was reported that incorporating a small quantity of CNTs (0.2 wt%) improved the fatigue resistance of an epoxy system which is extensively used in the wind-energy industry. Tensile and dynamical mechanical analyses were performed on neat epoxy and on epoxy/CNT nanocomposite. The CNT composites had long fatigue life as compared to neat epoxies under the same conditions [85].

4.8 Organic pollutant removal

With rapid industrialization and untreated wastewater disposal, water pollution has become a serious threat to both flora and fauna. The contaminants which affect the water quality badly are synthetic organic dyes, petroleum, antibiotics, drugs, pesticides, polyaromatic hydrocarbons, heavy metal ions, etc., which have brought a lot of adverse effects on human health and social development. The most used method for decontamination of polluted water is the biological method. There are some limitations of the biological method i.e., some refractory organic and inorganic pollutants in wastewater are not effectively removed [86]. Therefore, some other methods have been developed to remove refractory organic and inorganic pollutants from wastewater such as adsorption and advanced oxidation processes (APS). In the last few decades, CNTs and graphene-based materials have been widely used by many researchers in wastewater treatment. Carbon nanomaterials-based nanocomposites can be used as an adsorbent in adsorption studies while in AOPS they are used as effective catalysts due to their unique physical and chemical properties. In a word, the CNTs-based materials may have potential application prospects in water treatment [87].

4.9 Aerospace industry

In recent years, the dependence of humans on speedy transportation has boosted many folds. The high demands make the research of this area very fast and more fascinating. In aerospace industry, keen interest is shown in the mechanical, thermal, chemical along with electrical and biodegradable properties. In chemical properties, the focus is to stop the corrosion of the system. The lightweight of the aerospace structure is vital, but the role of mechanical properties cannot be ignored. The role of mechanical properties is crucial for design like impact and scratch tolerance, toughness, strength, etc. The high thermal emissivity decreased solar absorption, electric conductivity, and resistance for radiations are also the key parameters, which are kept in focus while designing the aircrafts [88]. In the last few decades, the use of various epoxy resins in the aerospace industry had enhanced enormously. Almost in all these applications, carbon nanostructures are considered as advanced materials, which may be used incorporation with other nanomaterials or simply as the replacement of them [89]. In certain nanocomposites, carbon fibers were replaced by CNTs, the reason was to reduce the weight of composite materials. The use of epoxy nanocomposites is valued in aircraft/aerospace industry due to its high strength-to-weight ratio and enhanced temperature resistance. The best example is the wingtip fairings of Lockheed's F-35 which is made of CNT/epoxy nanocomposite. In the present situation, the application of CNT-based epoxy nanocomposites in the aeronautical and aerospace industry is obvious from their use in the space shuttle and advanced commercial aircraft (such as Boeing 787 and Airbus A380). In 2010, NASA published a report which showed the road maps for the future applications of nanomaterials and utilization of CNTs in aerospace industries. This report reveals that CNTs have tremendous potential to be used in aerospace applications which includes reduced vehicle mass, superior functionality, improved self-healing characteristics, improved control and damage tolerance, durability, and greater thermal protection. The use of CNT in the aerospace industry represents a very bright picture [90].

4.10 Conductive plastic

Plastic is one of the important materials used in the modern world as a metal replacement. The mechanical properties of plastics have been improved up to a maximum extent, but their limitations are arising when electrical conductivity is needed, as plastics are mostly nonelectrical conductors. To remove this deficiency different filling agents like carbon black and bigger graphite fibers have been incorporated in plastic polymers [91]. The loading of polymers with traditional filler to enhance the electrical conductivity results in weighty parts with high degraded construction properties

[92]. For this reason, carbon nanostructures i.e., CNT is exceptional with the greatest carbon fiber aspect ratio. In addition, their intrinsic inclination to shape cords provides naturally very lengthy leading routes even at extremely low loads [93]. This property of CNTs makes it the best choice to use in applications, such as gasket, enclosure, composite EMI (electromagnetic interference), electrostatic dissipation (ESD), shielding, and other usage coatings, low-observance radar absorption materials, and conductive (even transparent) antistatic and material coatings [94].

5. Conclusion

In recent years, the CNTs and graphene-based improved nanocomposites have been found to play a significant role in a wide range of potential applications. The incorporation of CNTs and graphene nanofillers into a matrix of particular materials either polymer activated carbon, metal, metal oxide, and fibers, upgrades their novel properties, such as excellent mechanical stability (in terms of strength, toughness, flexibility, Young's modulus, dimension stability, etc.), good optical features, flame retardancy, low water/gas permeability, and high electro-thermal conductivity. The improved nanocomposites material is known as the material of the 21st century because of their widespread applications in different fields such as energy storage, medical, wastewater treatment, aerospace, etc.

Acknowledgements

The authors would like to convey their gratefulness to National Centre of Excellence in Physical Chemistry, University of Peshawar for providing us necessary support and facilities to carry out this study.

Author details

Tahira Mahmood*, Abid Ullah and Rahmat Ali
National Centre of Excellence in Physical Chemistry, University of Peshawar, Peshawar, Pakistan

*Address all correspondence to: tahiramahmood@uop.edu.pk

References

[1] Saleh TA, Parthasarathy P, Irfan M. Advanced functional polymer nanocomposites and their use in water ultra-purification. Trends in Environmental Analytical Chemistry. 2019;**24**:e00067. DOI: 10.1016/j.teac.2019.e00067

[2] Ates B, Koytepe S, Ulu A, Gurses C, Thakur VK. Chemistry, structures, and advanced applications of nanocomposites from biorenewable resources. Chemical Reviews. 2020;**120**(17):9304-9362. DOI: 10.1021/acs.chemrev.9b00553

[3] Omanović-Mikličanin E, Badnjević A, Kazlagić A, Hajlovac M. Nanocomposites: A brief review. Health and Technology. 2020;**10**(1):51-59. DOI: 10.1007/s12553-019-00380-x

[4] Kumar S, Nehra M, Dilbaghi N, Tankeshwar K, Kim KH. Recent advances and remaining challenges for polymeric nanocomposites in healthcare applications. Progress in Polymer Science. 2018;**80**:1-38. DOI: 10.1016/j.progpolymsci.2018.03.001

[5] Camargo PH, Satyanarayana KG, Wypych F. Nanocomposites: Synthesis, structure, properties and new application opportunities. Materials Research. 2009;**12**(1):1-39. DOI: 10.1590/S1516-14392009000100002

[6] Chae SH, Lee YH. Carbon nanotubes and graphene towards soft electronics. Nano Convergence. 2014;**1**(1):1-26. DOI: 10.1186/s40580-014-0015-5

[7] Kumar A, Sharma K. Amit rai Dixit. A review on the mechanical properties of polymer composites reinforced by carbon nanotubes and graphene. Carbon Letters. 2021;**31**:149-165. DOI: 10.1007/s42823-020-00161-x

[8] Mittal G, Dhand V, Rhee KY, Park SJ, Lee WR. A review on carbon nanotubes and graphene as fillers in reinforced polymer nanocomposites. Journal of Industrial and Engineering Chemistry. 2015;**21**:11-25. DOI: 10.1016/j.jiec.2014.03.022

[9] Papageorgiou DG, Li Z, Liu M, Kinloch IA, Young RJ. Mechanisms of mechanical reinforcement by graphene and carbon nanotubes in polymer nanocomposites. Nanoscale. 2020;**12**(4):2228-2267. DOI: 10.1039/C9NR06952F

[10] Ma PC, Siddiqui NA, Marom G, Kim JK. Dispersion and functionalization of carbon nanotubes for polymer-based nanocomposites: A review. Composites Part A: Applied Science and Manufacturing. 2010;**41**(10):1345-1367. DOI: 10.1016/j.compositesa.2010.07.003

[11] Iijima S. Helical microtubules of graphitic carbon. Nature. 1991;**354**(6348):56-58. DOI: 10.1038/354056a0

[12] Ajayan PM, Stephan O, Colliex C, Trauth D. Aligned carbon nanotube arrays formed by cutting a polymer resin—nanotube composite. Science. 1994;**265**(5176):1212-1214. DOI: 10.1126/science.265.5176.1212

[13] Patel DK, Kim H-B, Dutta SD, Ganguly K, Lim K-T. Materials. 2020;**13**:1679. DOI: 10.3390/ma13071679

[14] Mittal G, Dhand V, Rhee KY, Park SJ, Lee WR. A review on carbon nanotubes and graphene as fillers in reinforced polymer nanocomposites. Journal of Industrial and Engineering Chemistry. 2015;**21**:11-25. DOI: 10.1016/j.jiec.2014.03.022

[15] Sun X, Huang C, Wang L, Liang L, Cheng Y, Fei W, et al. Recent progress in graphene/polymer nanocomposites. Advanced Materials. 2021;**33**(6):2001105. DOI: 10.1002/adma.202001105

[16] Legge EJ, Paton KR, Wywijas M, McMahon G, Pemberton R, Kumar N, et al. Determining the level and location of functional groups on few-layer graphene and their effect on the mechanical properties of nanocomposites. ACS Applied Materials & Interfaces. 2020;**12**(11):13481-13493. DOI: 10.1021/acsami.9b22144

[17] Gong L, Kinloch IA, Young RJ, Riaz I, Jalil R, Novoselov KS. Interfacial stress transfer in a graphene monolayer nanocomposite. Advanced Materials. 2010;**22**(24):2694-2697. DOI: 10.1002/adma.200904264

[18] Kumar P. Ultrathin 2D nanomaterials for electromagnetic interference shielding. Advanced Materials Interfaces. 2019;**6**(24):1901454. DOI: 10.1002/admi.201901454

[19] Zeng H, Wu J, Ma Y, Ye Y, Liu J, Li X, et al. Scalable approach to construct self-assembled graphene-based films with an ordered structure for thermal management. ACS applied Materials & Interfaces. 2018;**10**(48):41690-41698. DOI: 10.1021/acsami.8b13808

[20] Feng W, Qin M, Feng Y. Toward highly thermally conductive all-carbon composites: Structure control. Carbon. 2016;**109**:575-597. DOI: 10.1016/j.carbon.2016.08.059

[21] Shi G, Araby S, Gibson CT, Meng Q, Zhu S, Ma J. Graphene platelets and their polymer composites: Fabrication, structure, properties, and applications. Advanced Functional Materials. 2018;**28**(19):1706705. DOI: 10.1002/adfm.201706705

[22] Allahbakhsh A, Arjmand M. Graphene-based phase change composites for energy harvesting and storage: State of the art and future prospects. Carbon. 2019;**148**:441-480. DOI: 10.1016/j.carbon.2019.04.009

[23] Budama-Kilinc Y, Ozdemir B, Zorlu T, Gok B, Egil AC. Nanobiomaterials for neural regenerative medicine. In: Neural Regenerative Nanomedicine. Amsterdam, The Netherlands: Elsevier; 2020. pp. 25-45. DOI: 10.1016/B978-0-12-820223-4.00002-4

[24] Zhang BT, Zheng X, Li HF, Lin JM. Application of carbon-based nanomaterials in sample preparation: A review. Analytica Chimica Acta. 2013;**784**:1-7. DOI: 10.1016/j.aca.2013.03.054

[25] Kroto HW, Heath JR, O'Brien SC, Curl RF, Smalley RE. C 60: buckminsterfullerene. Nature. 1985;**318**(6042):162-163. DOI: 10.1038/318162a0

[26] Mukherjee A, Majumdar S, Servin AD, Pagano L, Dhankher OP, White JC. Carbon nanomaterials in agriculture: A critical review. Frontiers in Plant Science. 2016;**7**:172. DOI: 10.3389/fpls.2016.00172

[27] Kausar A. Advances in polymer/fullerene nanocomposite: A review on essential features and applications. Polymer-Plastics Technology and Engineering. 2017;**56**(6):594-605. DOI: 10.1080/03602559.2016.1233278

[28] Reza EK, Malekia A, de la Guardia M, Bani MS, Chenab KK, Pashazadeh-Panahid P, et al. Carbon based nanomaterials for tissue engineering of bone: Building new bone on small black scaffolds. Journal of Advanced Research. 2019;**18**:185-201. DOI: 10.1016/j.jare.2019.03.011

[29] Kumar A, Sharma K, Dixit AR. A review on the mechanical properties of polymer composites reinforced by carbon nanotubes and graphene. Carbon Letters. 2021;**31**(2):149-165. DOI: 10.1007/s42823-020-00161-x

[30] Dresselhaus MS, Dresselhaus G, Charlier JC, Hernandez E. Electronic, thermal and mechanical properties of carbon nanotubes. Philosophical transactions of the Royal Society of London. Series A: Mathematical, Physical and Engineering Sciences. 2004;**362**(1823):2065-2098. DOI: 10.1098/rsta.2004.1430

[31] De Volder MF, Tawfick SH, Baughman RH, Hart AJ. Carbon nanotubes: Present and future commercial applications. Science. 2013;**339**(6119):535-539. DOI: 10.1126/science.1222453

[32] Murali R, Brenner K, Yang Y, Beck T, Meindl JD. Resistivity of graphene nanoribbon interconnects. IEEE Electron Device Letters. 2009;**30**(6):611-613. DOI: 10.1109/LED.2009.2020182

[33] Balandin AA, Ghosh S, Bao W, Calizo I, Teweldebrhan D, Miao F, et al. Superior thermal conductivity of single-layer graphene. Nano Letters. 2008;**8**(3):902-907. DOI: 10.1021/nl0731872

[34] Rawal S, Brantley J, Karabudak N. Development of carbon nanotube-based composite for spacecraft components. In: Proceedings of 6th International Conference on Recent Advances in Space Technology (RAST), Turkiye, 12-14 June, 2013. Piscataway, New Jersey, United States: Institute of Electrical and Electronics Engineers (IEEE); 2013. pp. 13-19. DOI: 10.1109/RAST.2013.6581186

[35] Ma PC, Siddiqui NA, Marom G, Kim JK. Dispersion and functionalization of carbon nanotubes for polymer-based nanocomposites: A review. Composites Part A: Applied Science and Manufacturing. 2010;**41**(10):1345-1367. DOI: 10.1016/j.compositesa.2010.07.003

[36] Paul S, Lee YS, Choi J, Kang YC, Kim DW. Synthesis and electrochemical characterization of polypyrrole/multi-walled carbon nanotube composite electrodes for supercapacitor applications. Bulletin of the Korean Chemical Society. 2010;**31**(5):1228-1232. DOI: 10.5012/bkcs.2010.31.5.1228

[37] Béguin F, Szostak K, Lillo-Rodenas M, Frackowiak E. Carbon nanotubes as backbones for composite electrodes of supercapacitors. In: AIP Conference Proceedings 2004 Sep 29. Vol. 723, No. 1. Melville, New York, United States: American Institute of Physics (AIP); 2004. pp. 460-464. DOI: 10.1063/1.1812129

[38] Otrokhov G, Pankratov D, Shumakovich G, Khlupova M, Zeifman Y, Vasil'eva I, et al. Enzymatic synthesis of polyaniline/multi-walled carbon nanotube composite with core shell structure and its electrochemical characterization for supercapacitor application. Electrochimica Acta. 2014;**123**:151-157. DOI: 10.1016/j.electacta.2013.12.089

[39] Blaszczyk-Lezak I, Desmaret V, Mijangos C. Electrically conducting polymer nanostructures confined in anodized aluminum oxide templates (AAO). eXPRESS. Polymer Letters. 2016;**10**(3):259. DOI: 10.3144/expresspolymlett.2016.24

[40] Georgakilas V, Tiwari JN, Kemp KC, Perman JA, Bourlinos AB, Kim KS, et al. Noncovalent functionalization of graphene and graphene oxide for energy materials,

biosensing, catalytic, and biomedical applications. Chemical Reviews. 2016;**116**(9):5464-5519. DOI: 10.1021/acs.chemrev.5b00620

[41] Huq MM, Hsieh CT, Ho CY. Preparation of carbon nanotube-activated carbon hybrid electrodes by electrophoretic deposition for supercapacitor applications. Diamond and Related Materials. 2016;**62**:58-64. DOI: 10.1016/j.diamond.2015.12.014

[42] Khomenko V, Raymundo-Piñero E, Béguin F. High-energy density graphite/AC capacitor in organic electrolyte. Journal of Power Sources. 2008;**177**(2):643-651. DOI: 10.1016/j.jpowsour.2007.11.101

[43] Qiu J, Wu X, Qiu T. High electromagnetic wave absorbing performance of activated hollow carbon fibers decorated with CNTs and Ni nanoparticles. Ceramics International. 2016;**42**(4):5278-5285. DOI: 10.1016/j.ceramint.2015.12.056

[44] Gangupomu RH, Sattler ML, Ramirez D. Comparative study of carbon nanotubes and granular activated carbon: Physicochemical properties and adsorption capacities. Journal of Hazardous Materials. 2016;**302**:362-374. DOI: 10.1016/j.jhazmat.2015.09.002

[45] Lv Q, Wang S, Sun H, Luo J, Xiao J, Xiao J, et al. Solid-state thin-film supercapacitors with ultrafast charge/discharge based on N-doped-carbon-tubes/Au-nanoparticles-doped-MnO2 nanocomposites. Nano Letters. 2016;**16**(1):40-47. DOI: 10.1021/acs.nanolett.5b02489

[46] Yuan QH, Zeng XS, Liu Y, Luo L, Wu JB, Wang YC, et al. Microstructure and mechanical properties of AZ91 alloy reinforced by carbon nanotubes coated with MgO. Carbon. 2016;**96**:843-855. DOI: 10.1016/j.carbon.2015.10.018

[47] Yuan D, Yang W, Ni J, Gao L. Sandwich structured MoO2@ TiO2@ CNT nanocomposites with high-rate performance for lithium ion batteries. Electrochimica Acta. 2015;**163**:57-63. DOI: 10.1016/j.electacta.2015.02.149

[48] Alam RS, Moradi M, Nikmanesh H. Influence of multi-walled carbon nanotubes (MWCNTs) volume percentage on the magnetic and microwave absorbing properties of BaMg0. 5Co0. 5TiFe10O19/MWCNTs nanocomposites. Materials Research Bulletin. 2016;**73**:261-267. DOI: 10.1016/j.materresbull.2015.09.016

[49] Islam MS, Deng Y, Tong L, Faisal SN, Roy AK, Minett AI, et al. Grafting carbon nanotubes directly onto carbon fibers for superior mechanical stability: Towards next generation aerospace composites and energy storage applications. Carbon. 2016;**96**:701-710. DOI: 10.1016/j.carbon.2015.10.002

[50] Wang C, He X, Tong L, Peng Q, Wang R, Li Y, et al. Theoretical prediction and experimental verification of pulling carbon nanotubes from carbon fiber prepared by chemical grafting method. Composites Part A: Applied Science and Manufacturing. 2013;**50**:1-10. DOI: 10.1016/j.compositesa.2013.03.008

[51] Bekyarova E, Thostenson ET, Yu A, Kim H, Gao J, Tang J, et al. Multiscale carbon nanotube– carbon fiber reinforcement for advanced epoxy composites. Langmuir. 2007;**23**(7):3970-3974. DOI: 10.1021/la062743p

[52] Sharma B, Malik P, Jain P. Biopolymer reinforced nanocomposites: A comprehensive review. Materials Today Communications. 2018;**16**:353-363. DOI: 10.1016/j.mtcomm.2018.07.004

[53] Salimikia I, Heydari R, Yazdankhah F. Polyaniline/graphene

oxide nanocomposite as a sorbent for extraction and determination of nicotine using headspace solid-phase microextraction and gas chromatography–flame ionization detector. Journal of the Iranian Chemical Society. 2018;**15**(7):1593-1601. DOI: 10.1007/s13738-018-1357-4

[54] Farajvand M, Farajzadeh K, Faghani G. Synthesis of graphene oxide/polyaniline nanocomposite for measuring cadmium (II) by solid phase extraction combined with dispersive liquid-liquid microextraction. Materials Research Express. 2018;**5**(7):075017. DOI: 10.1088/2053-1591/aace3a

[55] Xing LL, Wu X, Huang KJ. High-performance supercapacitor based on three-dimensional flower-shaped Li4Ti5O12-graphene hybrid and pine needles derived honeycomb carbon. Journal of Colloid and Interface Science. 2018;**529**:171-179. DOI: 10.1016/j.jcis.2018.06.007

[56] Lu B, Yuan X, Ren Y, Shi Q, Wang S, Dong J, et al. Cost-effective three dimensional Ag/polymer dyes/graphene-carbon spheres hybrids for high performance nonenzymatic sensor and its application in living cell H2O2 detection. Bioelectrochemistry. 2018;**123**:103-111. DOI: 10.1016/j.bioelechem.2018.05.001

[57] Hossain MF, Park JY. Plain to point network reduced graphene oxide-activated carbon composites decorated with platinum nanoparticles for urine glucose detection. Scientific Reports 2016;6(1):1-0. DOI: 10.1038/srep 21009.

[58] Beura R, Thangadurai P. Effect of Sn doping in ZnO on the photocatalytic activity of ZnO-graphene nanocomposite with improved activity. Journal of Environmental Chemical Engineering. 2018;**6**(4):5087-5100. DOI: 10.1016/j.jece.2018.07.049

[59] Borah MJ, Devi A, Saikia RA, Deka D. Biodiesel production from waste cooking oil catalyzed by in-situ decorated TiO2 on reduced graphene oxide nanocomposite. Energy. 2018;**158**:881-889. DOI: 10.1016/j.energy.2018.06.079

[60] Wang AY, Chaudhary M, Lin TW. Enhancing the stability and capacitance of vanadium oxide nanoribbons/3D-graphene binder-free electrode by using VOSO4 as redox-active electrolyte. Chemical Engineering Journal. 2019;**355**:830-839. DOI: 10.1016/j.cej.2018.08.214

[61] Arukula R, Vinothkannan M, Kim AR, Yoo DJ. Cumulative effect of bimetallic alloy, conductive polymer and graphene toward electrooxidation of methanol: An efficient anode catalyst for direct methanol fuel cells. Journal of Alloys and Compounds. 2019;**771**:477-488. DOI: 10.1016/j.jallcom.2018.08.303

[62] Xuan X, Kim JY, Hui X, Das PS, Yoon HS, Park JY. A highly stretchable and conductive 3D porous graphene metal nanocomposite based electrochemical-physiological hybrid biosensor. Biosensors and Bioelectronics. 2018;**120**:160-167. DOI: 10.1016/j.bios.2018.07.071

[63] Zheng L, Zheng H, Huo D, Wu F, Shao L, Zheng P, et al. N-doped graphene-based copper nanocomposite with ultralow electrical resistivity and high thermal conductivity. Scientific Reports. 2018;**8**(1):1-7. DOI: 10.1038/s41598-018-27667-9

[64] Gupta A, Jamatia R, Patil RA, Ma YR, Pal AK. Copper oxide/reduced graphene oxide nanocomposite-catalyzed

synthesis of flavanones and flavanones with triazole hybrid molecules in one pot: A green and sustainable approach. ACS Omega. 2018;**3**(7):7288-7299. DOI: 10.1021/acsomega.8b00334

[65] Davoodi AH, Mazinani S, Sharif F, Ranaei-Siadat SO. GO nanosheets localization by morphological study on PLA-GO electrospun nanocomposite nanofibers. Journal of Polymer Research. 2018;**25**(9):1-11

[66] Jin K, Zhang W, Wang Y, Guo X, Chen Z, Li L, et al. In–situ hybridization of polyaniline nanofibers on functionalized reduced graphene oxide films for high-performance supercapacitor. Electrochimica Acta. 2018;**285**:221-229. DOI: 10.1016/j.electacta.2018.07.220

[67] Wan Y, Li J, Yang Z, Ao H, Xiong L, Luo H. Simultaneously depositing polyaniline onto bacterial cellulose nanofibers and graphene nanosheets toward electrically conductive nanocomposites. Current Applied Physics. 2018;**18**(8):933-940. DOI: 10.1016/j.cap.2018.05.008

[68] Su X, Wang R, Li X, Araby S, Kuan HC, Naeem M, et al. A comparative study of polymer nanocomposites containing multi-walled carbon nanotubes and graphene nanoplatelets. Nano Materials Science. 2021. (Article in Press) DOI: 10.1016/j.nanoms.2021.08.003

[69] Yang Y, Zhao R, Zhang T, Zhao K, Xiao P, Ma Y, et al. Graphene-based standalone solar energy converter for water desalination and purification. ACS Nano. 2018;**12**(1):829-835. DOI: 10.1021/acsnano.7b08196

[70] Mao S, Lu G, Chen J. Three-dimensional graphene-based composites for energy applications. Nanoscale. 2015;**7**(16):6924-6943. DOI: 10.1039/C4NR06609J

[71] Stoller MD, Park S, Zhu Y, An J, Ruoff RS. Graphene-based ultracapacitors. Nano Letters. 2008;**8**(10):3498-3502. DOI: 10.1021/nl802558y

[72] Zhang LL, Zhou R, Zhao XS. Graphene-based materials as supercapacitor electrodes. Journal of Materials Chemistry. 2010;**20**(29):5983-5992

[73] Perreault F, De Faria AF, Elimelech M. Environmental applications of graphene-based nanomaterials. Chemical Society Reviews. 2015;**44**(16):5861-5896. DOI: 10.1039/c5cs00021a

[74] Madni A, Noreen S, Maqbool I, Rehman F, Batool A, Kashif PM, et al. Graphene-based nanocomposites: Synthesis and their theranostic applications. Journal of Drug Targeting. 2018;**26**(10):858-883. DOI: 10.1080/1061186X.2018.1437920

[75] Kim KS, Zhao Y, Jang H, Lee SY, Kim JM, Kim KS, et al. Large-scale pattern growth of graphene films for stretchable transparent electrodes. Nature. 2009;**457**(7230):706-710

[76] Burt JL, Elechiguerra JL, Reyes-Gasga J, Montejano-Carrizales JM, Jose-Yacaman M. Beyond Archimedean solids: Star polyhedral gold nanocrystals. Journal of Crystal Growth. 2005;**285**(4):681-691. DOI: 10.1016/j.jcrysgro.2005.09.060

[77] Clark JH. Solid acids for green chemistry. Accounts of Chemical Research. 2002;**35**:791-797. DOI: 10.1021/ar010072a

[78] Baur JA, Sinclair DA. Therapeutic potential of resveratrol: The in vivo evidence. Nature Reviews Drug

Discovery. 2006;**5**:493-506. DOI: 10.1038/nrd2060

[79] Machadoa BF, Serp P. Graphene-based materials for catalysis. Catalysis Science & Technology. 2012;**2**:54-75. DOI: 10.1039/C1CY00361E

[80] Ates M, Eker AA, Eker B. Carbon nanotube-based nanocomposites and their applications. Journal of Adhesion Science and Technology. 2017;**31**(18):1977-1997. DOI: 10.1080/01694243.2017.1295625

[81] Zakharychev EA, Razov EN, Semchikov Yu D, Zakharycheva NS, Kabina MA, Bakina LI, et al. Radar absorbing properties of carbon nanotubes/polymer composites in the V-band. Bulletin of Materials Science. 2016;**39**(2):451-456. DOI: 10.1007/s12034-016-1168-0

[82] Courty S, Mine J, Tajbakhsh AR, Terentjev EM. Nematic elastomers with aligned carbon nanotubes: New electromechanical actuators. EPL (Europhysics Letters). 2003;**64**(5):654. DOI: 10.1209/epl/i2003-00277-9

[83] Koerner H, Price G, Pearce NA, Alexander M, Vaia RA. Remotely actuated polymer nanocomposites—Stress-recovery of carbon-nanotube-filled thermoplastic elastomers. Nature Materials. 2004;**3**(2):115-120. DOI: 10.1038/nmat1059

[84] Mylvaganam K, Zhang LC. Fabrication and application of polymer composites comprising carbon nanotubes. Recent Patents on Nanotechnology. 2007;**1**(1):59-65. DOI: 10.2174/187221007779814826

[85] Kausar A, Rafique I, Muhammad B. Review of applications of polymer/carbon nanotubes and epoxy/CNT composites. Polymer-Plastics Technology and Engineering. 2016;**55**(11):1167-1191. DOI: 10.1080/03602559.2016.1163588

[86] Zhang H, Ji Q, Lai L, Yao G, Lai B. Degradation of p-nitrophenol (PNP) in aqueous solution by mFe/Cu-air-PS system. Chinese Chemical Letters. 2019;**30**(5):1129-1132. DOI: 10.1016/j.cclet.2019.01.025

[87] Peng J, He Y, Zhou C, Su S, Lai B. The carbon nanotubes-based materials and their applications for organic pollutant removal: A critical review. Chinese Chemical Letters. 2021;**32**(5):1626-1636. DOI: 10.1016/j.cclet.2020.10.026

[88] Blackford R. Performance demands on aerospace paints relative to environmental legislation. Aircraft Engineering and Aerospace Technology. 1998;**70**(66):451-455. DOI: 10.1108/00022669810243421

[89] Heimann M, Wirts-Ruetters M, Boehme B, Wolter KJ. Investigations of carbon nanotubes epoxy composites for electronics packaging. In: 58th Electronic Components and Technology Conference, May 27, 2008. Piscataway, New Jersey, United States: Institute of Electrical and Electronics Engineers (IEEE); 2008. pp. 1731-1736. DOI: 10.1109/ECTC.2008.4550214

[90] Iqbal A, Saeed A, Ul-Hamid A. A review featuring the fundamentals and advancements of polymer/CNT nanocomposite application in aerospace industry. Polymer Bulletin. 2021;**78**(1):539-557. DOI: 10.1007/s00289-019-03096-0

[91] Kim D, Hirayama Y, Liu Z, Kwon H, Kobashi M, Takagi K. Highly conductive Al/Al interfaces in ultrafine grained Al compact prepared by low oxygen powder metallurgy technique. Nanomaterials. 2021;**11**(5):1182. DOI: 10.3390/nano11051182

[92] Zou Q, Xiong SW, Jiang MY, Chen LY, Zheng K, Fu PG, et al. Highly thermally conductive and eco-friendly OH-h-BN/chitosan nanocomposites by constructing a honeycomb thermal network. Carbohydrate Polymers. 2021;**266**:118127. DOI: 10.1016/j.carbpol.2021.118127

[93] Zavanelli N, Yeo WH. Advances in screen printing of conductive nanomaterials for stretchable electronics. ACS Omega. 2021;**6**(14):9344-9351. DOI: 10.1021/acsomega.1c00638

[94] Veeman D, Shree MV, Sureshkumar P, Jagadeesha T, Natrayan L, Ravichandran M, et al. Sustainable development of carbon nanocomposites: Synthesis and classification for environmental remediation. Journal of Nanomaterials. 2021;**2021**:21. Article ID: 5840645. DOI: 10.1155/2021/5840645

Chapter 4

Nanocomposites Thin Films: Manufacturing and Applications

Weslley Rick Viana Sampaio, Petteson Linniker Carvalho Serra, Noelio Oliveira Dantas, Rômulo Ríbeiro Magalhães de Sousa and Anielle Christine Almeida Silva

Abstract

Thin films of nanocomposite materials arouse a lot of interest due to their excellent mechanical, electrical, optical, tribological properties and also by the vast field of application. This chapter covers some techniques of thin films growth, such as the processes of physical vapor deposition, such as magnetron sputtering; the processes of chemical vapor deposition; layer-by-layer; among other techniques. Additionally, relevant features and applications of some nanocomposites thin films are presented. The wide variety of thin films growth techniques have allowed the development of several devices including those that act as: transistors, actuators, sensors, solar cells, devices with shape memory effect, organic light-emitting diodes (OLEDs), thermoelectric devices.

Keywords: thin films, coatings, nanocomposites, PVD films, CVD films

1. Introduction

The possibility of improving the surface properties of a material preserving the original properties of the bulk is a major interest in the application of thin films. In general, the purpose of coatings is to obtain films of high hardness and resistant to wear and corrosion deposited on the surface of metallic materials. A class of thin films that has been gaining prominence is that of nanocomposites thin films (NTFs) that in addition to the advantage allow the union of different materials whose combination of properties allows obtaining specific characteristics for use in each application [1–4].

The NTFs are widely applied in the automotive, chemical, aerospace, electronics, and biomedical industries due to their attractive properties such as high hardness, good tribological properties, excellent resistance to corrosion and oxidation at high temperatures [5–7].

An important fact to be observed is the synergism of the properties of nanocomposite materials. Often a monolithic thin film is not able to gather specific properties in isolation, and the complement of these properties can be obtained by inserting another type of material that presents the characteristics that were missing.

As an example, we can mention the Cr-C nanocomposite thin film that coats devices exposed to the action of seawater. C improves adhesion, toughness and decreases surface roughness while Cr exhibits excellent anticorrosion properties [8]. Thus, we can note that the combination of elements Cr and C provides the obtaining of a thin film that brings together the best characteristics of the two elements.

There are several techniques for NTF growth. In general, these techniques are inserted in the following groups: (a) processes of physical vapor deposition (PVD) and (b) chemical vapor deposition (CVD) processes. In PVD processes, the material is converted to its vapor phase in a vacuum chamber and condensed on the substrate surface in the form of a thin film. In CVD processes, there is the spraying of the substance to be deposited, where the vapor is thermally decomposed into atoms and molecules, which can react with other gases, vapors, or liquids in order to produce a solid film on the substrate surface [9–12].

Some control parameters differentiate well the techniques belonging to these groups, which makes in many cases a technique more interesting or more appropriate than another. For example, in conventional CVD processes, the treatment temperature is high, while PVD processes occur at lower temperatures [13–15].

This chapter presents some techniques for NTF growth based on the processes of physical vapor deposition, chemical vapor deposition, and layer-by-layer technique. For each case the advantages, disadvantages and applications are mentioned. We hope that this brief chapter will encourage researchers to study the NTF in terms of improving their properties and in the knowledge or development of other thin films growth techniques.

2. Nanocomposites (thin films)

The nanocomposites are multiphase solid materials obtained by insertion of nanoscale crystals dispersive in a second phase. The most common nanocrystals are based on hard phases such as transition metal nitrides and carbides. The second-phase materials usually present in two types: (a) the hard phase, where the nanocomposite exhibits ultrahardness, but demonstrates brittle failure at high loads; and (b) the soft phase, where the resulting nanocomposite has high hardness and a super-tough structure [8].

The ratio control of these phases allows the obtaining of nanocomposites whose properties of one of the phases complement those of the other phase resulting in materials that can be applied in several areas. In the field of coatings, NTFs arouse a lot of interest due to the various combinations of properties. In the next paragraphs, we will highlight some studies in NTF in terms of the properties obtained in each case.

Lakra et al. [16] carried out the fabrication and electrochemical characterization of cobalt oxide (Co_3O_4)/graphene nanosheets nanocomposite deposited on stainless steel substrates as efficient electrode for supercapacitor application. The supercapacitor can achieve a power density in the range of 0.3–0.8 kW/kg with energy density of 2.4–9.8 Wh/kg, and it shows good electrochemical stability with a loss of 5% in capacitance after 1000 cycles. For the authors, the performance of the device can be improved by optimizing appropriate concentration of CO_3O_4 and graphene in the nanocomposite. The better performance was attributed to the synergistic effect of graphene and CO_3O_4 in the composite.

Scharf et al. [17] analyzed the synthesis, structure, and friction behavior of titanium doped tungsten disulfide (Ti-WS2) nanocomposite solid lubricant thin films deposited on Si (100) substrates. The pure WS2 thin films show columnar-plate morphologies with porosity between the plates, which leads to early fracture at room and high temperatures. With the addition of Ti, the room temperature friction coefficient measurements show some improvement when low amounts of Ti are added to WS2. This behavior was also identified for high-temperature (500°C) friction tests.

Dang et al. [18] studied the influence of Ag contents on structure and tribological properties of TiSiN-Ag nanocomposite coatings deposited on Ti-6Al-4V substrates. In this study, the authors observed that the variation of the Ag content directly influences the hardness, friction coefficient, and wear resistance. For low Ag content (1.4 at.%), the coating exhibited high hardness (36 GPa), but poor wear resistance. As the silver content increased from 5.3 to 8.7 at.%, the films exhibited small variation in hardness and greater layer homogeneity. A better performance in terms of wear resistance in artificial sea water was observed for a silver content of 5.3 at.%, while in ambient air, the wear resistance was higher for the coating with 7.9 at.% Ag. Further increasing the Ag content (21.0 at.%), there is a very large loss in hardness although possessing low friction coefficient both in ambient air and in artificial seawater.

3. Thin film growth techniques

The thin films growth can be carried out through several techniques where they can be deposited as vapor, liquid, solid phases, or a combination of these phases through plasma enhancement, ion bombardment, self-assembly, chemical treatment, among others [19, 20]. In this chapter, four thin film deposition techniques will be presented: magnetron sputtering, cathodic cage plasma deposition (CCPD), PECVD, and layer-by-layer.

3.1 Magnetron sputtering

Within the PVD processes, the magnetron sputtering technique is one of the most used in research with nanocomposites [21, 22]. The conventional sputtering technique has very low deposition rates, and an alternative found to increase this rate is the magnetron sputtering technique [23, 24].

The mentioned fact occurs because magnets are positioned in the vicinity of the target to act as electron traps. Due to electron trapping in the region close to the target, the plasma will also be restricted to this region resulting in a higher sputtering rate [24].

Figure 1 shows the principle of sputtering with a planar magnetron where it is possible to observe the confinement of electrons in the region close to the magnet. During the process, the target is intentionally positioned close to this region to increase the deposition rate.

The advantages of this technique include low plasma impedance and thus high discharge currents from 1 A to 100 A (depending on cathode length) at typical voltages around 500 V; deposition rates in the range from 1 to 10 nm/s; low thermal load to the substrate; dense and well-adherent coatings; large variety of film materials available (nearly all metals and compounds). Some disadvantages of this technique include improvement requirement of target utilization; stabilization of the reactive process in the transition regime [25]. The magnetron sputtering technique can be applied for wear-resistant coatings, low friction coatings, corrosion-resistant coatings, decorative coatings, and coatings with specific optical or electrical properties [21–25]. The **Figure 2** shows schematic diagram of magnetron sputtering process.

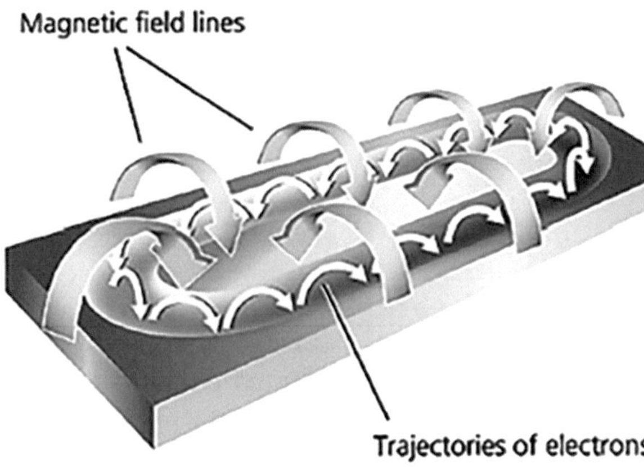

Figure 1.
The principle of magnetron sputtering [25].

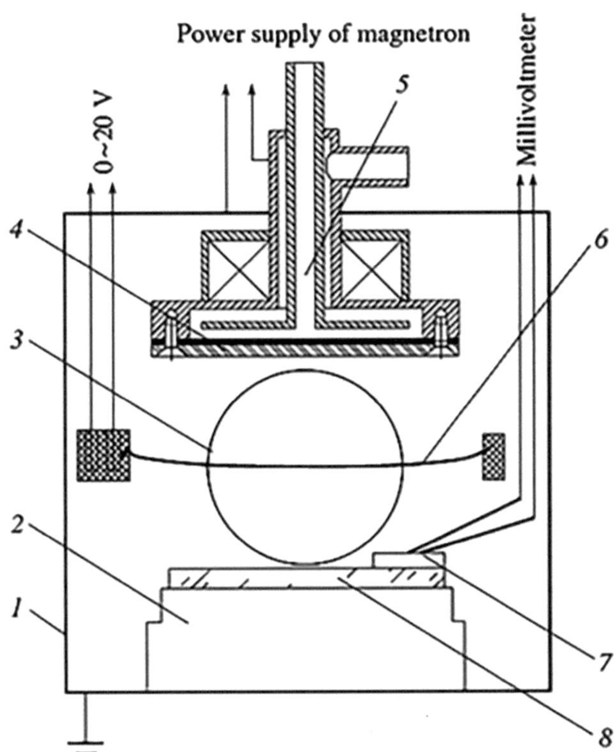

Figure 2.
Schematic diagram of magnetron sputtering process: (1) working chamber, (2) height adjustable substrate holder, (3) quartz window, (4) target to be sputtered, (5) magnetron, (6) molybdenum wire, (7) thermocouple, and (8) substrate [24].

3.2 Cathodic cage plasma deposition (CCPD)

This technique is a variation of the reactive sputtering process; it consists of the use of a cage with well-defined geometry where the effect of cylindrical multicathodes is used to obtain coatings and three-dimensional surface treatments. In this technique, the cage functions as a cathode in which a potential difference that acts only in the cage is applied (for the case of deposition), because the sample remains isolated on an alumina disc [26–28]. **Figure 3** shows a schematic representation of the spatial arrangement of the sample and cage in the sample holder.

During treatment, it is possible to observe the formation of plasma on the surface of the cage as well as the luminous intensification of the cage in each hole of the cage. **Figure 4a** shows the visual aspect of the plasma formed on the cathodic cage.

The process is carried out in a reactor and conducted to very low working pressures (~ 2.5 mbar) with controlled treatment atmosphere (Ar, H_2, N_2, O, for example). During the sputtering process, the atoms ejected from the surface of the cage can combine with the reactive gas of the plasma atmosphere, forming compounds that deposit on the surface of the sample. In this sense, the cage must be made of the material to be deposited. The technique of CCPD allows both nitriding and deposition of a thin film on a substrate and thus a better adhesion between the film and the substrate [30].

For the case of NTF, an example of arrangement for deposition of thin films would be the use of a graphite cylinder with titanium cover. **Figure 4b** shows the example of the cited arrangement. In this example, the film will present the synergism between the properties of the C from the graphite cylinder representing the soft phase and the Ti due to the cover, representing the hard phase.

The advantages of this technique include treatment of parts with complex geometries; minimizing the edge effect and opening of arcs in relation to the conventional sputtering technique; obtaining uniform layers; allowing to carry out both the nitriding process and the deposition of thin films. Some disadvantages of this technique include requiring a vacuum procedure; the production process of the cages can be complex depending on the type of material. The CCPD technique can be used for oxide and nitride deposition, deposition, desorption, and diffusion. It allows the deposition of thin films of high hardness, high wear, and corrosion resistance [26–30].

3.3 Plasma-enhanced chemical vapor deposition (PECVD)

The PECVD technique is a form of CVD, which uses plasma to enhance/enable the reactivity of organic/inorganic chemical monomers for the deposition of thin films. In this technique, it is possible to use precursors in solid, liquid, or gaseous form for

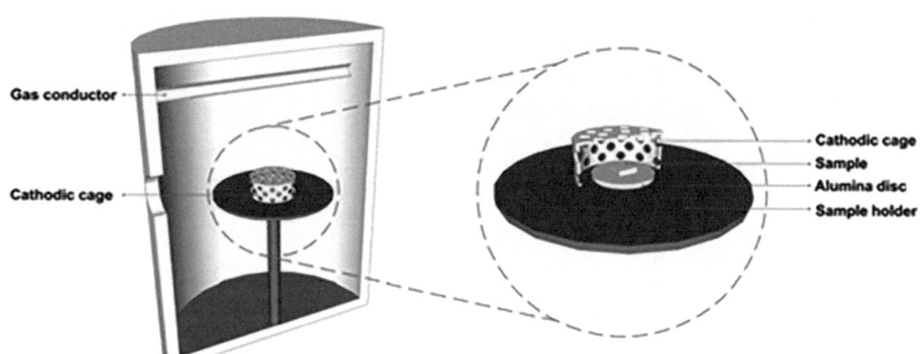

Figure 3.
Sample and cathodic cage arrangement over the sample holder [29].

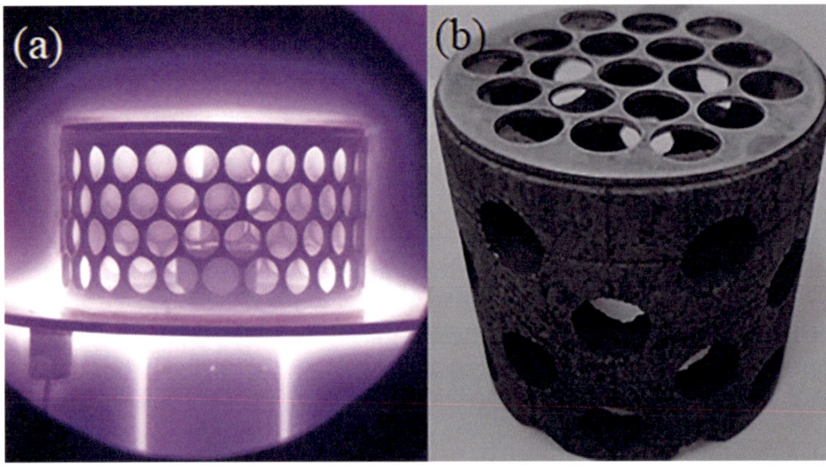

Figure 4.
(a) Visual aspect of plasma formation on the surface of the cage. and (b) Example of arrangement for treatment by CCPD (graphite cylinder and Ti cover).

the deposition of thin films [31]. The precursors are typically decomposed by temperature or a combination of temperature and plasma chemistry [32].

In conventional CVD processes, high treatment temperatures are used and the use of plasma by PECVD technique activates the chemical vapor precursor, and it allows the deposition to progress at a lower temperature than is normally used for CVD [32].

A schematic representation of the standard PECVD thin film deposition process is shown in **Figure 5**.

The advantages of this technique include highly cross-linked, uniform, robust films with good surface adhesion; compatible with solid, liquid, and gas monomers; leading to higher deposition rates and high density of cross-linking; working with low and atmospheric pressures. As a disadvantage, this technique may suffer from high internal stresses, which can cause film delamination. This technique can be used for highly cross-linked, uniform, robust films with good surface adhesion; corrosion and wear resistance; deposition of various types of materials with different microstructures [31–34].

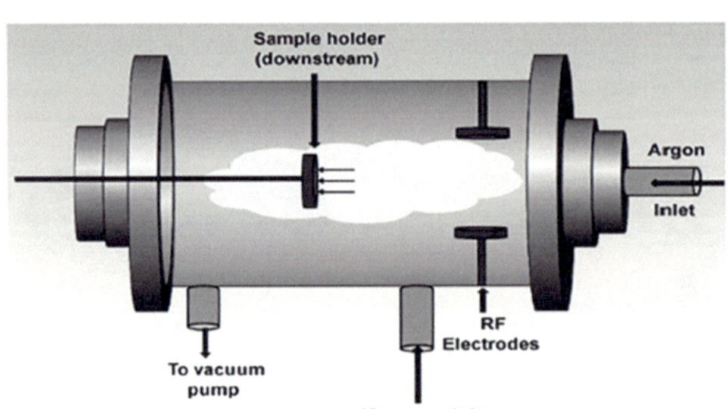

Figure 5.
Schematic representation of a standard chamber used in the PECVD process [31].

3.4 Layer-by-layer (LbL)

Layer-by-layer (LbL) assembly is a prevalent method for coating substrates with functional thin films [35]. It can be applied for coating substrates with polymers, colloids, biomolecules, and even cells, which offers superior control and versatility when compared with other thin film deposition techniques [36]. The versatility is due to the possibility of integrating different materials, creating defined layer sequences, orienting anisotropic materials within layers [37].

The process is based on a sequence of steps that leads to the exposure of the substrate to the material that will be deposited on the thin film. In general, more than one coating layer is formed, that is, multilayer thin films are obtained. Some technologies for films deposition by LbL include immersive; spin; spray; electromagnetic; and fluidic assembly [35, 36]. Immersive LbL assembly is the most widely used method, and it is typically performed by manually immersing a planar substrate into a solution of the desired material followed by three washing steps to remove unbound material [35]. **Figure 6** shows a schematic representation of the immersive LbL assembly method.

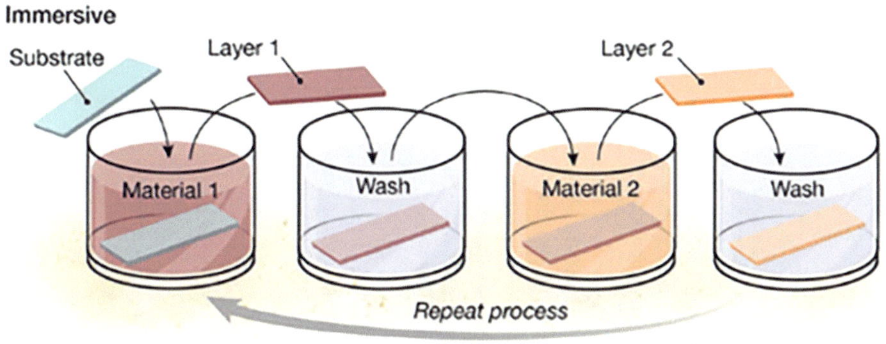

Figure 6.
Schematic representation of immersive LbL assembly. Adapted from Richardson et al. [35].

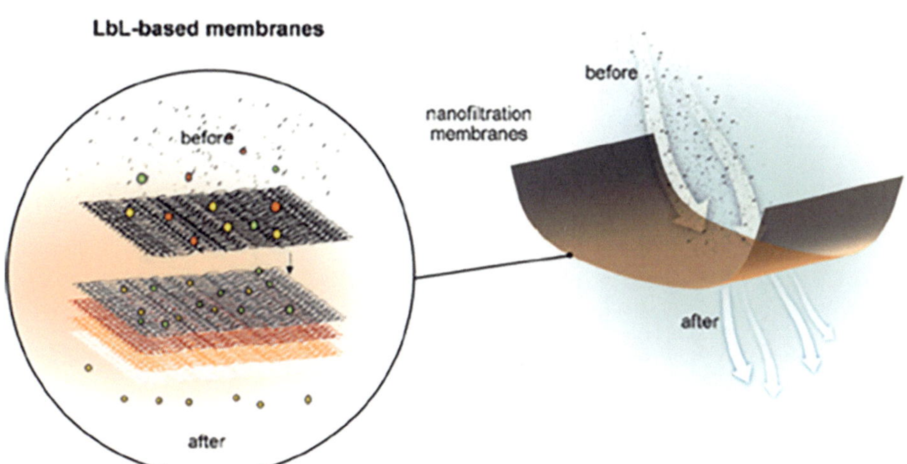

Figure 7.
Layer-by-layer assembly used for membranes for gases and liquids separation [37].

Technique	Advantages	Disadvantages	Applications	References
Magnetron Sputtering	Low plasma impedance/deposition rates in the range from 1 to 10 nm/s/low thermal load to the substrate/dense and well-adherent coatings/large variety of film materials available (nearly all metals and compounds)	Improvement requirement of target utilization/stabilization of the reactive process in the transition regime	Hard, wear-resistant coatings/low friction coatings/corrosion-resistant coatings/decorative coatings/coatings with specific optical or electrical properties	[21–25]
CCPD	Treatment of parts with complex geometries/minimizes the edge effect/uniform layers/both nitriding and deposition process/low deposition temperature	Requires a vacuum procedure/production process of cages can be complex	Oxide and nitride deposition/deposition/desorption and diffusion/thin films of high hardness/high wear and corrosion resistance	[26–30]
PECVD	Highly cross-linked, uniform, robust films with good surface adhesion/compatible with solid, liquid, and gas monomers/leads to higher deposition rates and high density of cross-linking/works with low and atmospheric pressures	May suffer from high internal stresses, which can cause film delamination	Highly cross-linked, uniform, robust films with good surface adhesion/corrosion and wear resistance/deposition of various types of materials with different microstructures	[31–34]
LbL	Simple and versatile process/applied straightforwardly on almost any surface/nanoscale functional films/applied to planar surfaces, spherical particles, inside complex geometries	Requires time for atomistic relaxation at the interfaces	LEDs and OLEDs device manufacturing/water purification/solar cells; optics/batteries/drug delivery/tissue engineering; functional fabrics/antibacterial coatings/biosensors/photonics applications/liquid or gas separation	[35–39]

Table 1.
Summary of some thin film deposition techniques used for NTF growth showing the main advantages, disadvantages, and applications.

The advantages of this technique include simple and versatile process; applied straightforwardly on almost any surface; nanoscale functional films; applied to planar surfaces, spherical particles, inside pores, and onto other more complex geometries; among others. The LbL process requires time for atomistic relaxation at the interfaces. may suffer from high internal stresses, which can cause film delamination. This technique can be used for light-emitting diodes (LEDs) and organic light-emitting diodes (OLEDs) device manufacturing; water purification; solar cells; optics; batteries; drug delivery; tissue engineering; functional fabrics; antibacterial coatings; biosensors; photonics applications; liquid or gas separation; among others [35–39].

Schematic representation of an application of coating deposited by the LbL technique with the function of membrane for liquid or gas separation is shown in **Figure 7**.

3.5 Summary table

The thin film deposition techniques presented in this chapter present advantages, disadvantages, and several applications that are commented in **Table 1**.

4. Conclusion

Nanocomposite coatings can be applied in several areas to improve properties such as hardness, wear, and corrosion resistance, tribological properties, among others. The synergistic effect exhibited in nanocomposite thin films shows the importance of mixing materials where specific properties are obtained but which would be difficult or even impossible for individual materials to exhibit. To obtain the synergism of the properties, it is necessary to establish the appropriate treatment parameters and select the most appropriate film deposition technique for a given application. Thin film growth techniques can be performed by several routes, which leads to the development of many variations in thin film deposition methods. Some of these methods/techniques were seen in this chapter and have application field in several areas. We hope that this chapter stimulates research on nanocomposites thin films in terms of studying their properties, developing, or improving thin films and the techniques for growing these films.

Acknowledgements

The authors thank the research funding agencies CNPq, CAPES, and FAPEAL for the financial support.

Conflict of interest

The authors declare no conflict of interest.

Author details

Weslley Rick Viana Sampaio[1,2], Petteson Linniker Carvalho Serra[3],
Noelio Oliveira Dantas[1], Rômulo Ríbeiro Magalhães de Sousa[4]
and Anielle Christine Almeida Silva[1,2,5]*

1 Laboratório de Novos Materiais Nanoestruturados e Funcionais, Physics Institute, Federal University of Alagoas (UFAL), Maceió, Brazil

2 Programa de Pós-Graduação em Materiais, Federal University of Alagoas (UFAL), Maceió, Brazil

3 Federal Institute of Education, Science and Technology of Piauí (IFPI), Teresina, Brazil

4 Federal University of Piauí, (UFPI), Teresina, Brazil

5 Programa de Pós-Graduação da Rede Nordeste de Biotecnologia (RENORBIO), Federal University of Alagoas, Maceió, Brazil

*Address all correspondence to: acalmeida@fis.ufal.br

References

[1] Kim YS et al. Investigation of structure and mechanical properties of TiZrHfNiCuCo high entropy alloy thin films synthesized by magnetron sputtering. Journal of Alloys and Compounds. 2019;**797**:834-841

[2] Braeckman BR et al. The nanostructure and mechanical properties of nanocomposite Nbx-CoCrCuFeNi thin films. Scripta Materialia. 2017;**139**:155-158

[3] von Fieandt K, Riekehr L, Osinger B, Fritze S, Lewin E. Influence of N content on structure and mechanical properties of multicomponent Al-Cr-Nb-Y-Zr based thin films by reactive magnetron sputtering. Surface & Coatings Technology. 2020;**389**:125614

[4] Vanapalli S, Naidu KL, Krishna MG, Padmanabhan KA. Growth and mechanical properties of TiN-AuN2 nanocomposite thin films on IN718 alloy substrates. Materials Letters. 2018;**210**:101-104

[5] Bhattacharya AB, Raju AT, Chatterjee T, Naskar K. Development and characterizations of ultra-high molecular weight EPDM/PP based TPV nanocomposites for automotive applications. Polymer Composites. 2020;**41**:4950-4962

[6] Garcés JM, Moll DJ, Bicerano J, Fibiger R, McLeod DG. Polymeric nanocomposites for automotive applications. Advanced Materials. 2000;**12**:1835-1839

[7] Qi W, Zhang X, Wang H. Self-assembled polymer nanocomposites for biomedical application. Current Opinion in Colloid & Interface Science. 2018;**35**:36-41

[8] Zhao D, Jiang X, Wang Y, Duan W, Wang L. Microstructure evolution, wear and corrosion resistance of Cr-C nanocomposite coatings in seawater. Applied Surface Science. 2018;**457**:914-924

[9] Jansson U, Lewin E. Sputter deposition of transition-metal carbide films—A critical review from a chemical perspective. Thin Solid Films. 2013;**536**:1-24

[10] Lai GS, Lau WJ, Goh PS, Karaman M, Gürsoy M, Ismail AF. Development of thin film nanocomposite membrane incorporated with plasma enhanced chemical vapor deposition-modified hydrous manganese oxide for nanofiltration process. Composites Part B. 2019;**176**:107328

[11] Poltorak C, Stüber M, Leiste H, Bergmaier A, Ulrich S. Study of (Ti,Zr) C:H/a-C:H nanocomposite thin film formation by low temperature reactive high power impulse magnetron sputtering. Surface & Coatings Technology. 2020;**398**:125958

[12] Alfonso E, Olaya J, Cubillos G. Thin film growth through sputtering technique and its applications. In: Crystallization—Science and Technology. London, United Kingdom: IntechOpen; 2012

[13] Sproul WD. Multilayer, multicomponent, and multiphase physical vapor deposition coatings for enhanced performance. Journal of Vacuum Science & Technology A. 1994;**12**:1595

[14] Wu L et al. Structure and mechanical properties of PVD and CVD TiAlSiN coatings deposited on cemented carbide. Crystals. 2021;**11**:598

[15] Palaniyappan S et al. Yttria-coated tungsten fibers for use in tungsten fiber-reinforced composites: A comparative study on PVD vs. CVD Routes, Coatings. 2021;**11**:1128

[16] Lakra R, Kumar R, Sahoo P, Sharma D, Thatoi D, Soam A. Facile synthesis of cobalt oxide and graphene nanosheets nanocomposite for aqueous supercapacitor application. Carbon Trends. 2022;**7**:100144

[17] Scharf T, Rajendran A, Banerjee R, Sequeda F. Growth, structure and friction behavior of titanium doped tungsten disulphide (Ti-WS2) nanocomposite thin films. Thin Solid Films. 2009;**517**:5666-5675

[18] Dang C, Li J, Wang Y, Yang Y, Wang Y, Chen J. Influence of Ag contents on structure and tribological properties of TiSiN-Ag nanocomposite coatings on Ti–6Al–4V. Applied Surface Science. 2017;**394**:613-624

[19] Devasahayam S, Hussain CM. Thin-film nanocomposite devices for renewable energy current status and challenges. Sustainable Materials and Technologies. 2020;**26**:e00233

[20] Depla D, Mahieu S, Greene JE. Chapter 5—Sputter deposition processes. In: Handbook of Deposition Technologies for Films and Coatings. 3rd ed. Boston: William Andrew Publishing; 2010. pp. 253-296

[21] Wei R. Plasma enhanced magnetron sputter deposition of Ti–Si–C–N based nanocomposite coatings. Surface & Coatings Technology. 2008;**203**:538-544

[22] Musil J, Vlcek J. Magnetron sputtering of hard nanocomposite coatings and their properties. Surface and Coatings Technology. 2001;**142-144**:557-566

[23] Kelly PJ, Arnell RD. Magnetron sputtering: A review of recent developments and applications. Vacuum. 2000;**56**:159-172

[24] Kostanovskii AV, Zhilyakov LA, Pronkin AA, Kirillin AV. Preparation of diamond-like films in the process of magnetron sputtering of graphite target. Teplofizika Vysokikh Temperature. 2009;**47**:141-143

[25] Brauer G, Szyszka B, Vergohl M, Bandorf R. Magnetron sputtering—Milestones of 30 years. Vacuum. 2010;**84**:1354-1359

[26] Nogueira Junior W et al. Surface modification of AISI-304 steel by ZnO synthesis using cathodic cage plasma deposition. Materials Research Express. 2021;**8**:096403

[27] Naeem M et al. Synthesis of molybdenum oxide on AISI-316 steel using cathodic cage plasma deposition at cathodic and floating potential. Surface and Coatings Technology. 2021;**406**:126650

[28] Libório MS et al. Surface modification of M2 steel by combination of cathodic cage plasma deposition and magnetron sputtered MoS2-TiN multilayer coatings. Surface & Coatings Technology. 2020;**384**:125327

[29] Costa PMO et al. Influence of Hastelloy's cathodic cage plasma deposition on corrosion resistance of AISI 304 stainless steel and of AISI D6 tool steel. Materials Research. 2021;**24**:e20200267

[30] Sousa RRM et al. Deposition of TiO2 film on duplex stainless steel substrate using the cathodic cage plasma technique. Materials Research. 2016;**19**:1207-1212

[31] Vasudev MC et al. Exploration of plasma-enhanced chemical vapor

deposition as a method for thin-film fabrication with biological applications. ACS Applied Materials Interfaces. 2013;**5**:3983-3994

[32] Mattox DM. Chapter 3—Plasmas and plasma enhanced CVD. In: The Foundations of Vacuum Coating Technology. Second ed. Norwich, NY: William Andrew; 2018. pp. 61-86

[33] Martinu L, Zabeida O, Klemberg-Sapieha JE. Chapter 9—Plasma enhanced chemical vapor deposition of functional coatings. In: Handbook of Deposition Technologies for Films and Coatings. Third ed. Norwich, NY: William Andrew; 2010. pp. 392-465

[34] Ghodselahi T et al. Co-deposition process of RF-Sputtering and RF-PECVD of copper/carbon nanocomposite films. Surface and Coatings Technology. 2008;**202**:2731-2736

[35] Richardson JJ, Björnmalm M, Caruso F. Technology-driven layer-by-layer assembly of nanofilms. Science. 2015;**348**:2491

[36] Richardson JJ et al. Innovation in Layer-by-Layer Assembly. Chemical Reviews. 2016;**116**:14828-14867

[37] Zhao S et al. The future of layer-by-layer assembly: A tribute to ACS nano associate editor Helmuth Möhwald. ACS Nano. 2019;**13**:6151-6169

[38] Sunny S et al. Lubricant-infused nanoparticulate coatings assembled by layer-by-layer deposition. Advanced Functional Materials. 2014;**24**: 6658-6667

[39] Villiers MM et al. Introduction to nanocoatings produced by layer-by-layer (LbL) self-assembly. Advanced Drug Delivery Reviews. 2011;**63**:701-715

Section 2

Nanocomposite Materials for Biomedical Applications

Chapter 5

Nanocomposite Biomaterials for Tissue Engineering and Regenerative Medicine Applications

Shuai Liu, Rurong Lin, Chunyi Pu, Jianxing Huang, Jie Zhang and Honghao Hou

Abstract

Nanocomposites are materials that are usually created by introducing appropriate nanoparticles into a macroscopic matrix, enabling the resulting bulk nanocomposites remarkable characteristics in electrical, thermal conductivity, mechanical, optical, magnetic properties, and so on. Such nanocomposite materials are of great importance for biomedical applications, particularly promising for tissue engineering scaffolds. Recent trends in the nanocomposites field show bio-based/environmentally friendly materials to be among the components in these nanocomposite materials. Particular attention has been paid to the use of bio-based/biodegradable polymers as a matrix component in nanocomposite applications, because of their great widespread potential and advantages over other traditional synthetic materials. In this chapter, we focus on the current research trends of the tissue engineering scaffolds based on nanocomposite materials and mainly introduce the properties, types, manufacturing techniques, and tissue engineering applications of various nanocomposite biomaterials. Besides, challenges and prospects associated with nanocomposite biomaterials for the tissue engineering field were discussed. We believe that this chapter provides a new envision for building functional nanocomposite materials for broad biomedical applications.

Keywords: nanocomposite, nanomaterial, biomedical applications, tissue engineering, biodegradable, biocompatible

1. Introduction

Nanomaterials are becoming a new strategy to control and manipulate the nano and macro properties of hydrogels without hindering the exchange of nutrients with the surrounding environment [1]. Nanomaterials refer to synthetic or natural materials whose size does not exceed 100 nm, but for better medical applications, a diameter in the range of 10–100 nm is more required. Materials with a diameter of less than 10 nm are reactive and toxic due to their large specific surface area and activity and materials larger than 100 nm can even cause embolism. In addition, it has been shown that nanotopology rather than microtopology is the main influence factor on cell structure and arrangement [2]. Usually, natural tissues also contain nano-scale substances, such as proteins, which can directly interact with cells. The nanocomposite hydrogel designed with a biomimetic structure is more suitable for host cell colonization. Therefore, nanomaterials have high biological activity (for example, cell-binding motifs), which are conducive to protein adsorption and cell adhesion and proliferation [3]. The properties of the obtained nanocomposite hydrogel can be the sum of the functions of the nanomaterial and the hydrogel, or the result of better properties, such as improved stiffness, stretchability, and even higher cell compatibility, which may be derived from synergistic interactions [4]. Liu et al. introduced hydroxyapatite (HAp) into the hydrogel network, which improved the modulus and toughness of the hydrogel, and had a qualitative impact on the differentiation behavior of mesenchymal stem cells (MSCs) [5]. Nanoparticles (NPs) can be cross-linked agents with polymers both chemically and physically to achieve dynamic system behavior and form shear-thinning hydrogels. The resulting hydrogels exhibit a temporarily reduced viscosity under shear stress and can be injected through a plug flow system, and then return to their original viscosity under low shear owing to the electrostatic interactions [3]. Because of these characteristics, they can be further applied to 3D bioprinting. Moreover, due to their structural characteristics, nanocomposite hydrogels can easily present a gradient biomolecules delivery by (porous) NPs. A nanocomposite hydrogel composed of a poly(amide amine) network and mesoporous silica NPs capable of releasing chemokines has been reported. They have been observed to play an important role in tissue regeneration *in vivo* through their influence on (MSCs) (**Figure 1**) [7].

Nanocomposites are defined as a combination of materials or phases, where one or more components is more concentrated and provide support, and other components enhance the performance of the prepared composite by adding extra characteristics [8]. In recent years, there have been more and more research toward nanocomposites by closely mimicking the biological environment and natural matrix. Especially, the properties of nanocomposites can be modified to meet the functional requirements of each tissue, which makes it an excellent choice for tissue engineering applications [9, 10]. Nanoscale scaffolds are needed to provide a suitable niche for interactions between cells and the extracellular matrix (ECM) and guide cell behaviors [11]. The emergence of nanocomposite materials provides more meaningful characteristics, combinations, and unique design possibilities for scaffolds, and finally provides a revolutionary platform for tissue engineering.

Nanocomposites involving biodegradable and bioactive properties have been regarded as strategies for tissue engineering and regeneration medicine [12]. Nanoscale fillers can greatly change the physical properties of the polymer matrix, thereby achieving improved engineering design of biomaterials. Compared with traditional ones, nanoparticles with a larger surface area, can form a tighter interface with the polymer matrix, provide better mechanical properties, while outstanding mechanical conductivity and biocompatibility of the filler, thereby affecting protein adsorption, cell adhesion, proliferation, and differentiation [13].

The components used to prepare the nanocomposite can be of natural or synthetic origin. The components obtained from natural sources include polysaccharides from microorganisms, such as chitin/chitosan, starch, alginate, and cellulose, while biopolymers from animal proteins include wool, silk, gelatin, and collagen. These polymers have been widely used in biomedical applications for their physical and chemical similarities to natural ECM [12]. These polymers provide favorable support

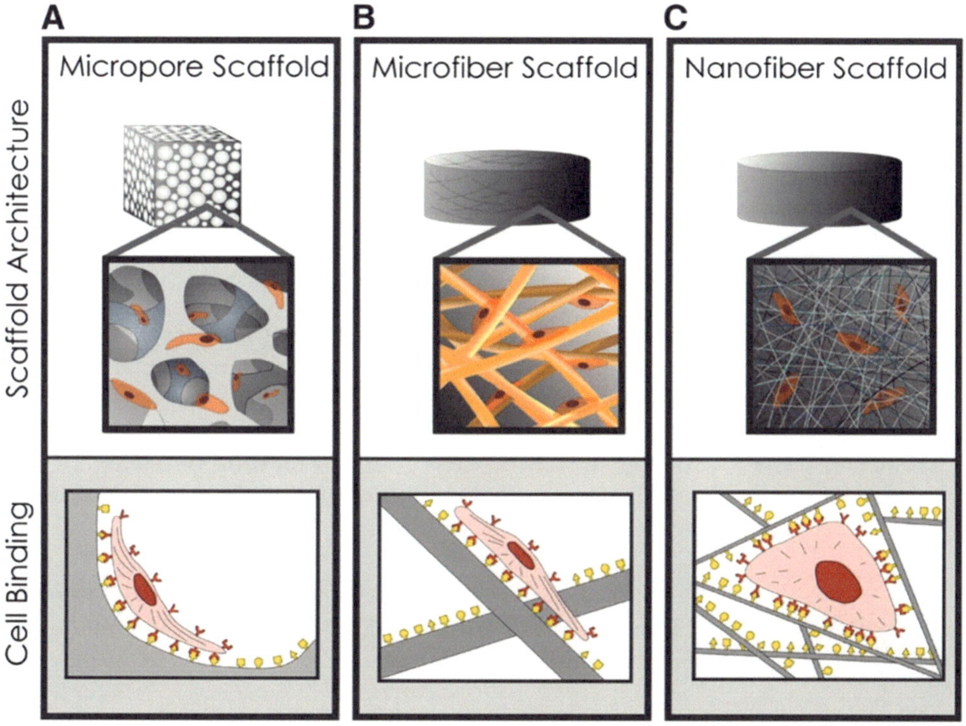

Figure 1.
Various topographical architecture of nanocomposite scaffolds affects the binding and spreading of seed cells [6].

for living systems, including outstanding biological properties, adjustable degradation rates, and faster tissue regeneration [14, 15]. However, some disadvantages, including the risk of pathogen transmission, allergic reactions, poor mechanical strength, and high cost have also restricted their applications [16]. In recent years, researchers have begun to overcome these problems related to natural nanocomposite scaffolds. In tissue engineering, the developed scaffold can be used as a substitute for ECM. Scaffolds formulated with biodegradable polymers, cells, growth factors, and appropriate biochemical signals can repair or replace damaged tissues by providing the environment and conditions that enable cells to secrete their own new natural ECM [17, 18].

Tissue engineering and regenerative medicine (TERM) is a multidisciplinary research field that uses the principles of chemistry, biology, and engineering science to study the regeneration process of damaged tissues or organs [19]. Tissue engineering requires a combination of molecular biology and materials design for the urgent need of providing temporary artificial substrates for cell seeding. In general, the scaffold should exhibit high porosity, appropriate pore size, biocompatibility, and appropriate degradation rate [20]. Besides, the scaffold also needs to provide sufficient mechanical support to maintain the stress and load generated during the regeneration process *in vitro* or *in vivo*. For the purpose of improving mechanical properties and cell adhesion and proliferation, the incorporation of nanoparticles (such as apatite components, carbon nanostructures, and metal nanoparticles) has been extensively studied. Polymer/layered nanocomposites such as HAp, carbon nanotubes, and layered silicates have become the focus of attention in academia and industry [21, 22]. The introduction of a small amount of high-aspect-ratio silicate nanoparticles can significantly improve the mechanical and physical properties of the polymer matrix.

This chapter mainly introduces the current research trends of nanocomposite materials used in tissue engineering, the properties, types, manufacturing techniques, and applications of ideal nanocomposite scaffolds in various tissue engineering fields.

2. Properties and types of nanocomposite materials for tissue engineering

2.1 Properties and selection criteria for ideal nanocomposite scaffolds for tissue engineering

In the past research, various nanocomposite scaffolds have been prepared using different manufacturing techniques. In a broad sense, the ideal nanocomposite scaffold should meet the following demands:

1. Biocompatible: It provides an ECM-like environment for cells' physiological behavior, such as proliferation, migration, differentiation. Its degradation must be nontoxic and easily adsorbed or eliminated from the body.

2. Biodegradability: It must possess a controllable rate of degradation to maintain the balance between newly formed tissues or organs and the degradation (resorption) of the nanocomposite scaffold.

3. Nutrient delivery: Scaffolds should contain a large number of interconnected pores to facilitate the transport of nutrients and regulatory factors.

4. Mechanical factor: It must have sufficient mechanical integrity to adapt to the surgical process and meet the corresponding tissue tolerance. There should be a balance between the porosity, stiffness, and hardness. Stiffness determines the direction of cell differentiation to a certain extent and avoids insufficient vascularization caused by excessive porosity.

5. Pore geometry: The pore distribution of the scaffold should be conducive to the morphology of the cells. Pore size must be in the critical range larger than the lower limit so that cells can migrate easily and smaller than the upper limit to have enough surface area or binding sites.

6. Surface topography: It must have adjustable surface topography, hydrophilicity, surface energy, etc. This allows the control of cell adhesion, migration, intracellular signaling, as well as *in vivo* cell recruitment.

Of course, the final criteria for choosing a nanocomposite scaffold depends on the type of tissue to be treated. These nanocomposite scaffolds must be resistant to damage during the implantation process, as damage can lead to necrosis and inflammation. For example, nanofibers scaffolds for tendon tissue engineering should be arranged in parallel to imitate the natural tissues arrangement structure. In cardiac tissue engineering, nanocomposite scaffold materials should have a certain electrical conductivity, and gradually promote the repair and regeneration of myocardial infarction damaged tissue under the conditions of satisfying electrophysiology [23]. For neural tissue engineering, scaffolds with higher flexibility are needed to allow cells to adhere, migrate and differentiate, and they need to mimic the geometry of natural nerves [24]. The scaffold required for bone tissue engineering should have the characteristics of providing multiple growth factors for different stages of bone tissue regeneration, corrosion resistance, osteoconductivity, and shape controllability [25, 26]. It also needs to create multiscale layered structures at the nanometer level to add growth factors that promote vascularization and provide a surface for stem cells to initiate bone repair and regeneration [15].

According to the type of tissue to be repaired, degradation rate and surface morphology are also important criteria for stent selection. There should be coordination between the degradation time of the scaffold and the time required for natural tissue replacement, it can be implanted for a long time without multiple operations and avoid subsequent surgical removal. The degradation rate can be controlled by introducing nanoparticles into the scaffold. On the other hand, the scaffolds used for skin and heart repair usually require planar scaffolds, the scaffolds used for bone tissue repair are usually cube-shaped or disc-shaped, while the tissue regeneration of nerves, blood vessels, and trachea require tubular scaffolds [27].

2.2 Types of nanocomposite scaffolds

2.2.1 Hydrogel

Hydrogel, a highly hydrated 3D network, has a similar structure and composition with ECM and become preferred tissue engineering scaffolds due to their adjustable network structure, outstanding biocompatibility, effective mass transfer, and the ability to encapsulate cells and biological factors [28]. These properties are affected by the degree of crosslinking of polymer chains, molecular arrangement, and the amount of water they absorb [29]. Hydrogels typically exhibit a hydrophilic network porous structure with interconnected pores (>10 μm) to allow gas penetration, nutrient delivery and provide sufficient space for cell attachment and interaction. In addition, the porosity of the hydrogel can be controlled by preparation methods (phase separation, gas foaming, solvent evaporation, etc.), types of raw materials, and the polymers concentrations [30]. The hydrogel nanocomposite scaffold can be designed in different sizes and shapes, such as patches, microspheres, sheets, hollow tubes, to satisfy their unique practical application [31]. From the perspective of regeneration medicine, hydrogels can be used as the cell niche and provide stimuli to accelerate tissue regeneration, and able to act as the delivery vehicle for biologically active molecules.

Nanocomposite hydrogels were first reported in 2002 when Haraguchi et al. introduced exfoliated clay flakes into poly(N-isopropylacrylamide) (PNIPAM) to form a unique organic/inorganic network [32]. Compared to conventional ones, nanocomposite hydrogels have enhanced chemical, physical, electrical, and biological properties, which are mainly attributed to the improved interaction between the polymer network and nanoparticles. However, nanoparticles themselves lack some essential properties, such as biodegradation and stimulated reactivity, which can be ameliorated by incorporating multiphase in nanocomposite hydrogels (**Figure 2**).

2.2.1.1 Inorganic nanomaterial based nanocomposite hydrogels

Over the last decades, significant progress has been made in the study of the impact of introducing inorganic nanomaterials into tissue engineering applications in natural and synthetic polymer networks. Inorganic nanomaterials, such as HAp, metal oxide nanoparticles, bioglass, and calcium phosphate NPs, are based

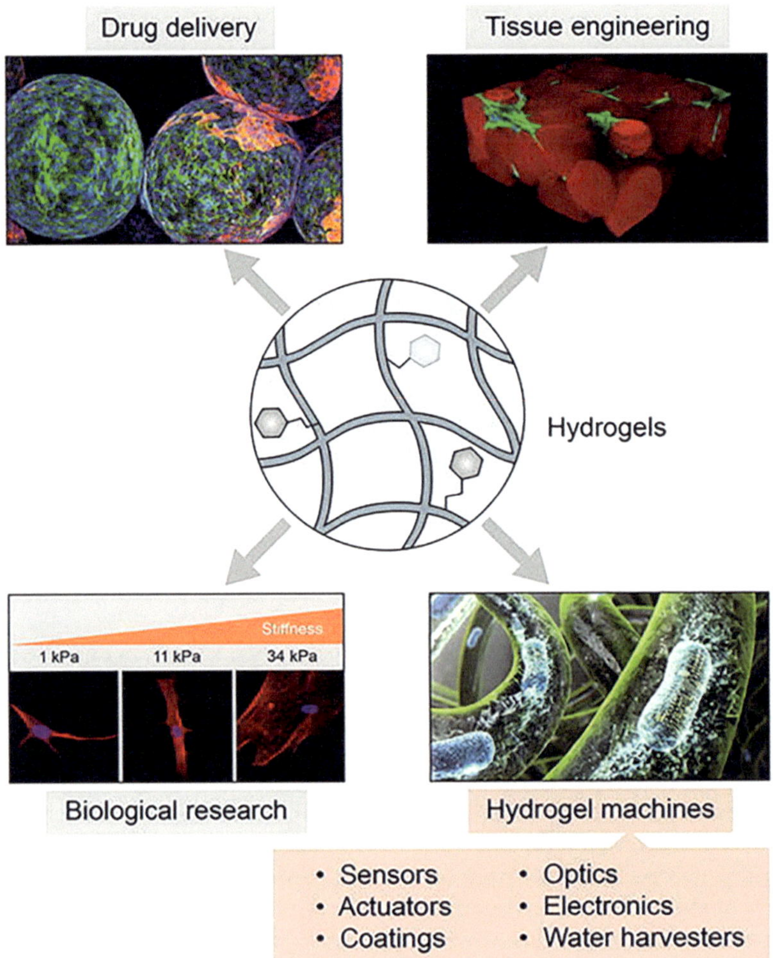

Figure 2.
Hydrogels for different applications such as tissue engineering and other biomedical researches [33].

on materials found in biological tissues and show significant biological activity. NPs confer and enhance signal (mechanics, bioelectricity, etc.) conductivity, while polymers provide flexibility and stability to nanocomposite structures. Surface chemistry favorable for protein adsorption and enhanced matrix stiffness makes these nanocomposites ideal for tissue engineering scaffolds.

2.2.1.2 Organic nanomaterial based nanocomposite hydrogels

Organic nanomaterials, such as dendrimers, liposomes, hyperbranched polymers, and micelles, are some typical polymer NPs that have a wide range of applications in tissue engineering, drug delivery, and cancer immunotherapy. For example, the properties of polymer NPs (such as conductivity) can be adjusted to achieve the repair of myocardial infarction. Particle size and shape are the most widely studied and reported properties, as they affect blood circulation time, cellular uptake, antigen presentation, and T-cell immunity. *In vivo* studies of nanocomposite hydrogels have shown that changing the shape of nanoparticles can lead to changes in the immune

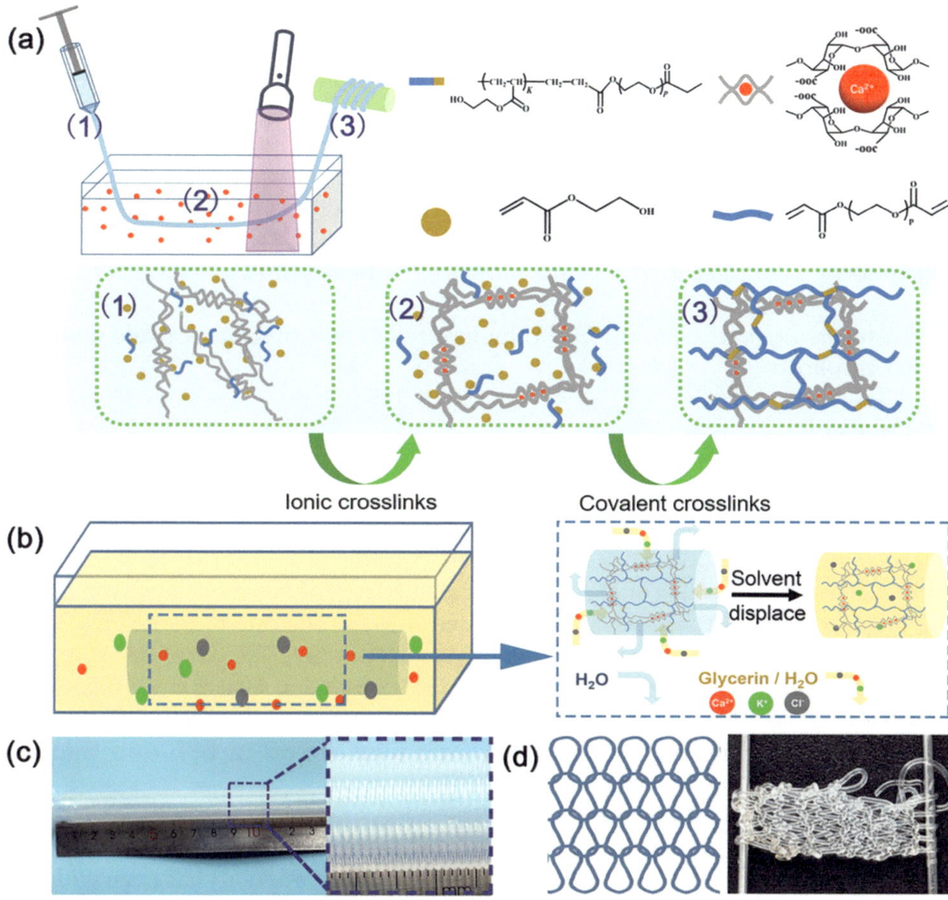

Figure 3.
Design and fabrication of organohydrogel fibers [34]. (a) Molecular design of hybrid crosslinked polymeric network in organohydrogel fibers and schematic of the wet-spinning process and molecular evolution of hydrogel fibers. (b) Schematic of the preparation of organohydrogel fibers from hydrogel fibers by displacement solvent. (c) Photograph of a long single fiber collected on a continuously winding drum spool. (d) Schematic and photograph of an organohydrogel-fiber knitted textile.

response. Compared with larger sizes, smaller size nanoparticles show better immune responses due to their larger specific surface area and higher reactivity (**Figure 3**).

2.2.2 Fibers

Fibrous scaffolds, a considerable option of mimicking the nanostructure of natural ECM, possess a more favorable morphology compared to porous scaffolds and hydrogels [35]. The nanofibers show similarity to the collagen fiber network, whose diameter distributes in the range of 50–500 nm. Besides, nanoscale fiber scaffolds with well-controlled pattern structures have received special attention in enhancing cell functions, such as cell adhesion, migration, proliferation, and differentiation, for their isotropic structure, uniform fiber size, and pore distribution, which also decide their mechanical properties [36]. Nanofiber scaffolds have been used in tissue engineering for heart, bone, cartilage, ligament, skeletal muscle, skin, and nerve tissue engineering, and as a carrier for the controlled delivery of drugs, proteins, and even DNA [37].

Nanocomposite fiber scaffolds can be obtained from polymers and nano-sized phases by molecular self-assembly, phase separation, and electrospinning manufacturing technology. Compared to the other two methods, electrospinning methods can produce fibers with controllable pore size, fiber size, and stiffness, making it the most widely studied technique [38]. Moreover, incorporating nanoparticles into electrospun fibers can improve biomimetic scaffolds because cell-matrix interactions are strongly affected by the presence of chemical cues that support cell attachment, proliferation, and differentiation.

2.3 Preparation of tissue engineering nanocomposite scaffold materials

The processing method of the above-mentioned nanocomposite scaffold is summarized as follows:

2.3.1 Blending

This mixing method is the simplest preparation method and has many advantages. We can obtain NPs with specialized size by varying the stirring speed, material concentration, and preparation period. Additionally, the preparation process is easy for the separate preparation and cross-linking of NPs. Filippi et al. mixed PEG-functionalized iron oxide (II, III) nanoparticles with an average particle size of 15 nm and cells in sterile Tris-buffered saline to form a nanocomposite hydrogel network [39]. The obtained hydrogel has a smooth texture, possesses excellent mechanical properties, such as stress relaxation and high elastic modulus. However, the asymmetric distribution of nanoparticles in nanocomposite hydrogels and their diffusion behavior in solution needs to be further studied.

2.3.2 Solvent casting/particulate leaching

A combination of solvent casting and particle leaching methods has been widely used to successfully manufacture 3D porous scaffolds [40]. It is a process of dispersing salts in polymers that are dissolved in organic solvents. The solvent is removed and then the salt crystals are leached out with water to form the porous scaffold. Utilizing this process, people can make a scaffold with a porosity value of up to 93% and an average pore diameter of about 500 μm by changing the size of the salt crystals and salt/polymer ratio [41]. However, this method cannot control certain key variables, such as pore shape and pore interconnectivity (**Figure 4**).

2.3.3 Electrospinning

Electrospinning is a special form of electrostatic atomization of polymer fluids. The biggest difference between the two is that the electrostatic spray uses a low-viscosity fluid, and the product is in the form of nanoparticles, which are mainly used for surface spraying. Electrospinning is a fluid with a higher viscosity, and the product is a nanofiber membrane, which is mainly used to prepare materials with three-dimensional shapes. Electrospinning is essentially a stretching process to generate long nanofibers of uniform diameter from a polymer solution in the electric field [42]. As the electrostatic interaction between the positively charged polymer and the collector increases, the droplets of the polymer solution become finer and stretch further, resulting in an ejection of fibers out of the solution.

This technique allows the production of bio-sized fibers with higher porosity and high surface area-to-volume ratio, making them promising candidates for tissue

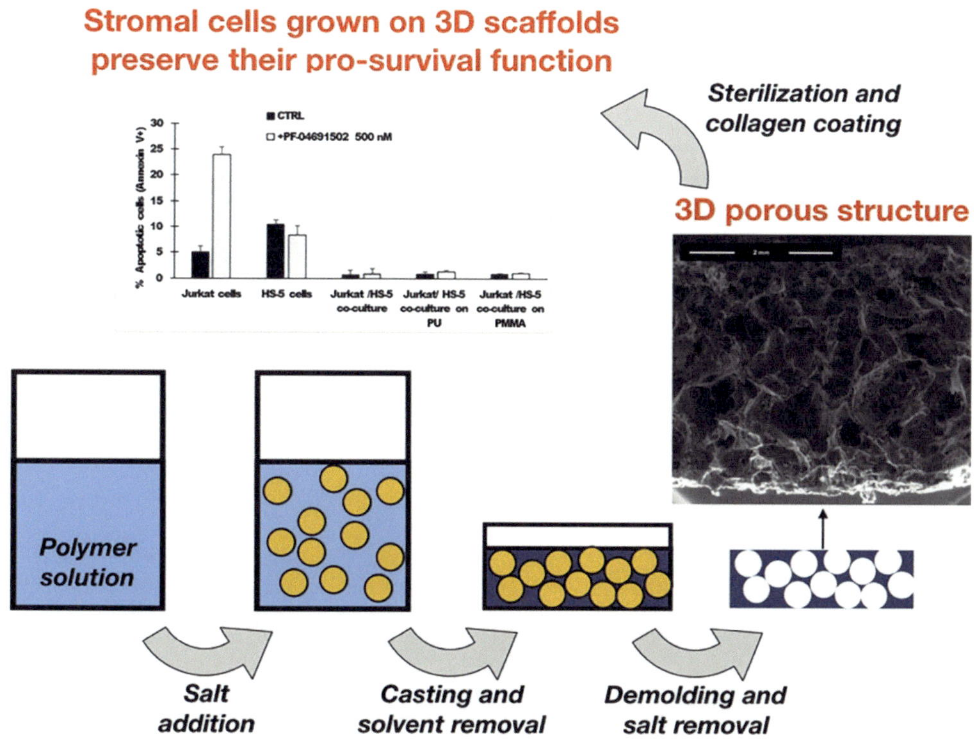

Figure 4.
Solvent casting/particulate leaching process [40].

engineering and drug delivery applications. The increased cell surface area provides more cell attachment sites and allows for effective cell adhesion. Compared with synthetic polymers, natural polymers are generally less suitable for spinning. Therefore, for natural polymers, the polymer and solvent concentration optimization must be carefully optimized (**Figure 5**) [43, 44].

Electrospinning technology has been used to develop tissue engineering scaffolds. For example, hepatocyte-like cells from human mesenchymal stem cells (hMSCs) were observed to aggregate on PLLA co-PCL collagen (PLACL/collagen) nanofiber scaffolds to form functional liver spheres. The results indicate that the bioengineered PLACL/collagen nanofiber scaffold may be a promising candidate for the treatment of damaged hepatocytes in advanced liver failure [45]. Researchers have also used electrospinning to manufacture different types of nanocomposite scaffolds for tissue engineering applications, such as polyurethane/cellulose fiber scaffolds, and polyethylene terephthalate (PET) scaffolds [46–48]. The main advantage of nanofibers prepared by electrospinning technology is that it allows easy transport of nutrients and waste across the scaffold. However, there are still some limitations, the use of cytotoxic solvents and a wide range of optimization parameters, such as applied voltage, flow rate, and travel distance to obtain ordered nanofibers, urgently need to be resolved.

2.3.4 Freeze drying/emulsification

The principle of freeze-drying is actually the sublimation process and a promising technique for preparing scaffolds, in which the frozen water is directly

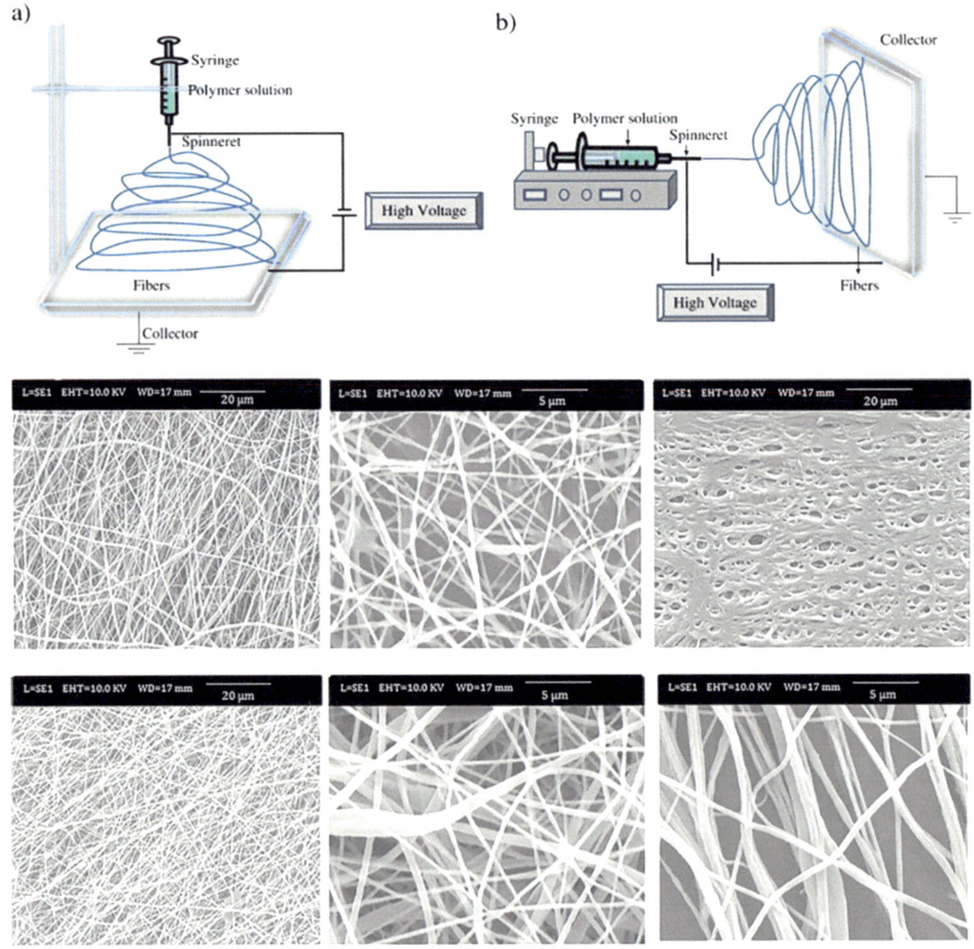

Figure 5.
Schematic diagram of set up of electrospinning apparatus (a) typical vertical set up and (b) horizontal set up of electrospinning apparatus and (bottom column) SEM micrographs of representative electrospinning fibrous materials [42, 43].

transformed from solid to gas without liquefaction. Thus, obtained highly porous interconnected polymer structure is used as a scaffold for tissue engineering [49]. The obtained scaffold usually has a high porosity, but it can be further adjusted as needed by varying the freezing method, the amount of water, the polymer concentration, the size of the ice (solvent) crystals, and the pH value of the solution as needed. Freeze drying is widely used to prepare nanocomposite scaffolds, but some of the limitations associated with this process include long preparation time, high-energy consumption, and the possible formation of closed cells by gas foaming. The obtained scaffold usually has a high porosity, but it can be further adjusted as needed by varying changing the freezing method, the amount of water, the polymer concentration, the size of the ice (solvent) crystals, and the pH value of the solution as needed. Freeze drying is widely used to prepare nanocomposite scaffolds, but some of the limitations associated with this process include long time, high-energy consumption, the use of cytotoxic solvents, and the formation of closed cells by gas foaming (**Figure 6**) [51].

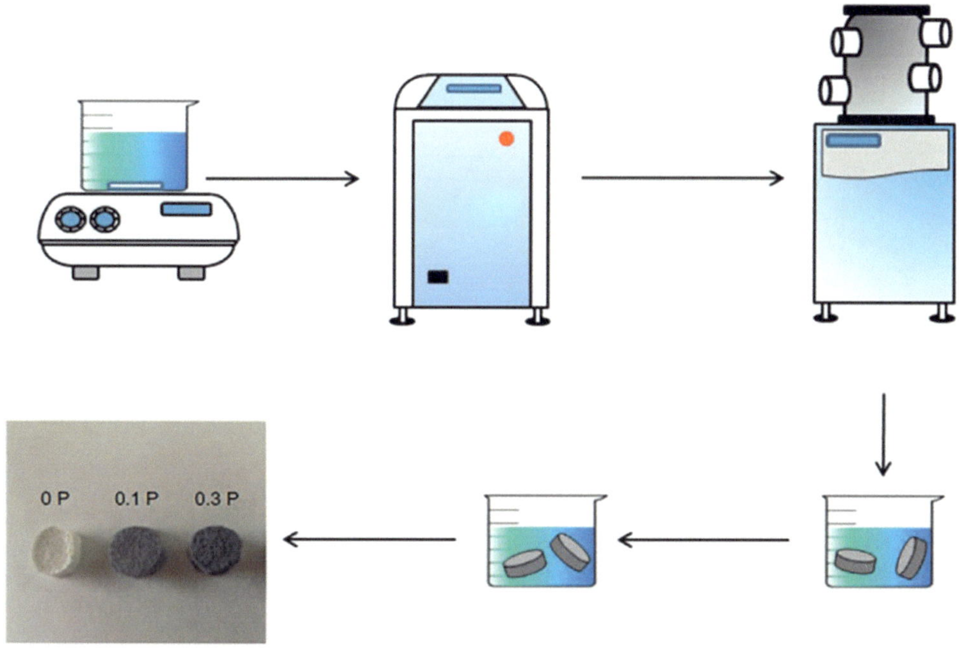

Figure 6.
Schematic representation of freeze-drying process [50].

The freeze-drying technique has been successfully used in many studies related to tissue engineering, such as chitin-chitosan/zirconium oxide (ZrO_2) and chitosan/gelatin/nanosilicon dioxide ($nSiO_2$)) composite material [52]. In one such study, freeze-drying methods were used to produce a chitosan 3D scaffold. The results obtained by dispersing nHA and fucoidan in a chitosan matrix were found to be suitable for cell growth and nutritional supplementation. *In vitro* results show that mesenchymal stem cells (PMSCs) derived from periosteum grow well in nanocomposites, which implies the potential for tissue engineering.

2.3.5 3D printing

Traditional manufacturing techniques, such as electrospinning, freeze-drying, and solvent casting, have certain defects, including limited control over pore size, fiber arrangement, and pore interconnection, which can lead to poor nutrient transport or reduced cell survival and migration rates. 3D printing is one of the new methods to obtain highly ordered scaffolds. It provides a highly controllable and precise design for the internal structure and surface of the bracket. It can deposit cells and biological materials in a way that mimics the structure of biological tissues [53]. The principle of printing is based on a polymer solution containing cells or cell aggregates "biological ink," which is deposited layer by layer on a substrate to generate 3D structures, such as tissues or organs (**Figure** 7).

3D printing has been used to prepare a nanocomposite scaffold containing nano-HA and chondrogenic transforming growth factor-β1 (TGF-β1) in the highly porous subchondral bone layer. In the cartilage layer, the prepared scaffolds showed osteogenic differentiation, high biocompatibility with hMSC, and mechanical properties required for osteochondral tissue regeneration [55]. 3D printing is promising, but the

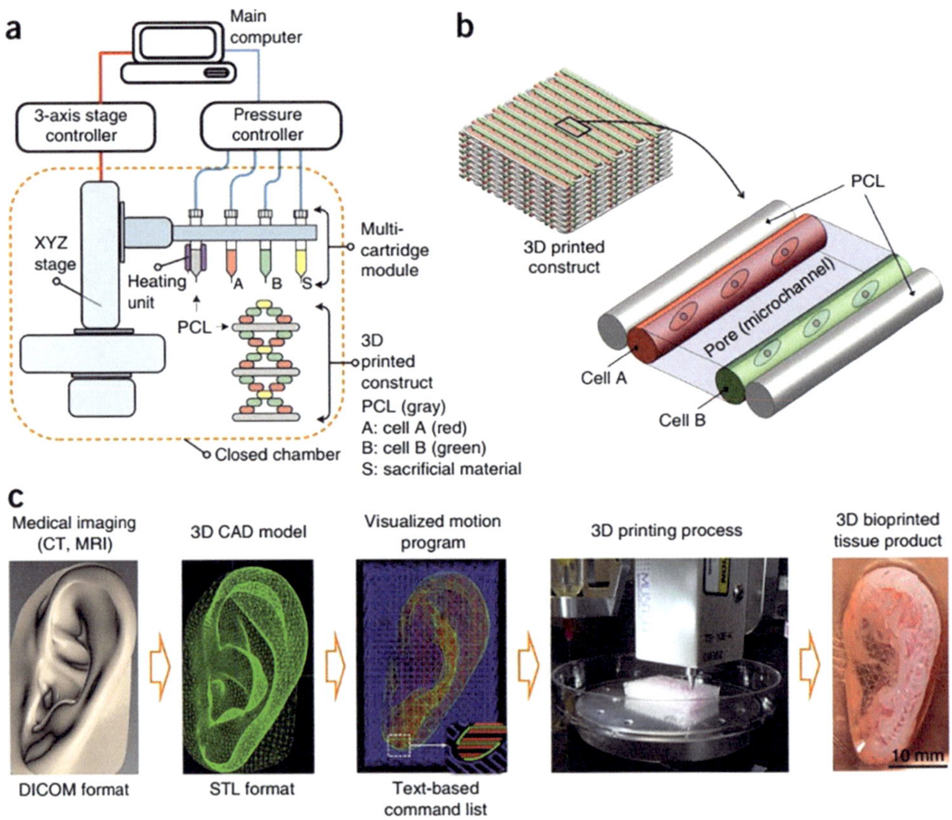

Figure 7.
Integrated tissue-organ printer (ITOP) system [54]. (a) The ITOP system consists of three major units: (i) 3-axis stage/controller, (ii) dispensing module including multi-cartridge and pneumatic pressure controller and (iii) a closed acrylic chamber with temperature controller and humidifier. (b) Illustration of basic patterning of 3D architecture including multiple cell-laden hydrogels and supporting PCL polymer. (c) CAD/CAM process for automated printing of 3D shape imitating target tissue or organ. A 3D CAD model developed from medical image data generates a visualized motion program, which includes instructions for XYZ stage movements and actuating pneumatic pressure to achieve 3D printing.

application of this technique in tissue engineering still needs to be explored due to high device costs, limited available materials, and low mechanical strength.

2.3.6 Self-assembly/self-organizing

It is an effective method to generate well-organized and stable supramolecular structures through the spontaneously rearranging of molecules under thermodynamic equilibrium conditions. Self-assembly can occur spontaneously in nature through hydrogen and ionic bonds, hydrophobicity, etc., like the self-assembly of phospholipid bilayer membranes in cells. Molecular self-assembly is a useful method for generating supramolecular structures that rely relies on the potential of molecules to spontaneously rearrange themselves into well-organized and stable structures under thermodynamic equilibrium conditions. Such molecular interaction is non-covalent and can occur through hydrogen and ionic bonds, hydrophobicity, van der Waals forces, metal coordination, and electromagnetic interactions [56]. Self-assembly can occur spontaneously in nature, just like the self-assembly of lipid

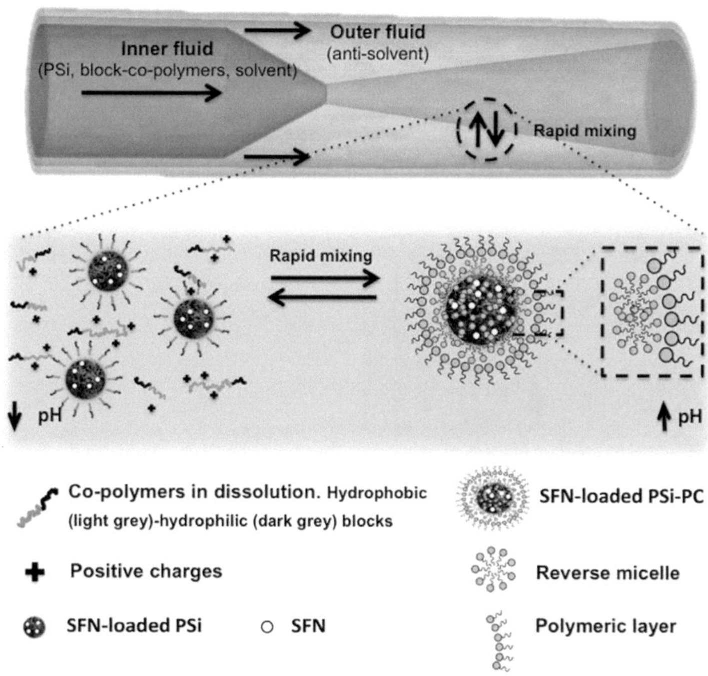

Figure 8.
Schematic representation of self-assembly/self-organizing [58].

Materials	Preparation methods	Applications	Advantages	Disadvantages	Refs
Hydrogel, fibrous materials	Blending or doping	Controlled drug delivery, tissue engineering, cell induction	Stress relaxation and high elastic modulus	Asymmetric distribution	[39, 59]
Nanofiber membrane	Electrospinning	Controlled drug delivery, enzyme immobilization, wound dressings, tissue engineering, biosensors, filtration	Tunable porosity, high surface area-to-volume ratio	Cytotoxicity, diversified parameters	[43, 60]
3D scaffold, hydrogel	3D printing	Tissue engineering, organ transplantation, biomaterial, growth factor delivery	High precise internal and surface structure	High device costs, limited available materials , low mechanical strength	[61, 62]
Hydrogel, nano/micro-particle	Microfluidics	Precise drug delivery, microfluidic biochip, growth factor delivery	High throughput production rate, ultrahigh drug loading degree	Clogging and limited architectural features, low-volume production	[63–66]

Materials	Preparation methods	Applications	Advantages	Disadvantages	Refs
Nanofiber	Self-assembly	Tissue engineering , biomaterial	Enhanced mechanical properties and biocompatibility, controlled intermolecular forces	Structure restricted, few biocompatible and biodegradable characteristics	[67, 68]
Hydrogel	Cross-linking/binding	Tissue engineering, shape memory, controlled drug delivery	Excellent biocompatibility, higher intensity and willfulness	Lower modulus	[69]
Nano/ micro-particle	Ink-jet or electrospray	Tissue regeneration , controlled drug delivery, biosensors	Higher biological activity, convenient delivery	Limited thickness	[70, 71]
3D scaffold	Freeze-drying/ emulsification	Tissue engineering, regenerative medicine	High and tunable porosity	Long time, high-energy consumption, cytotoxic solvents	[52]

Table 1.
Nanocomposites biomedical materials for tissue engineering applications.

bilayer membranes in cells. Molecular self-assembly is a strategy to develop nanofiber materials with tissue engineering potential [57]. At the molecular scale, the precise and controlled application of intermolecular forces can lead to new and previously unachievable nanostructures. Thus, with proper design, the mechanical properties and release characteristics of the assembled material can be tailored specifically for its intended use through appropriate design (**Figure 8** and **Table 1**).

3. Application for tissue engineering

See **Figures 9** and **10**.

3.1 Cardiovascular tissue engineering

Nowadays, cardiovascular disease has become a major issue affecting the health of countless people. Myocardial infarction (MI) is the leading cause of morbidity and mortality worldwide [74]. MI is caused by local obstruction of myocardial blood flow, which leads to permanent damage to the heart pump function and gradually develops into chronic heart failure. Due to the limited regeneration capacity of myocardial tissue, heart transplantation is the only solution for patients with advanced heart failure to restore heart function. However, the shortage of heart donors and the high cost of surgery procedures make it out of reach for many patients (**Figure 11**).

Cardiac tissue engineering is a promising approach to effectively replace or repair damaged myocardium [78]. The approach of cardiac tissue engineering is to create potential 3D scaffolds that mimic ECM to reconstruct injured myocardium.

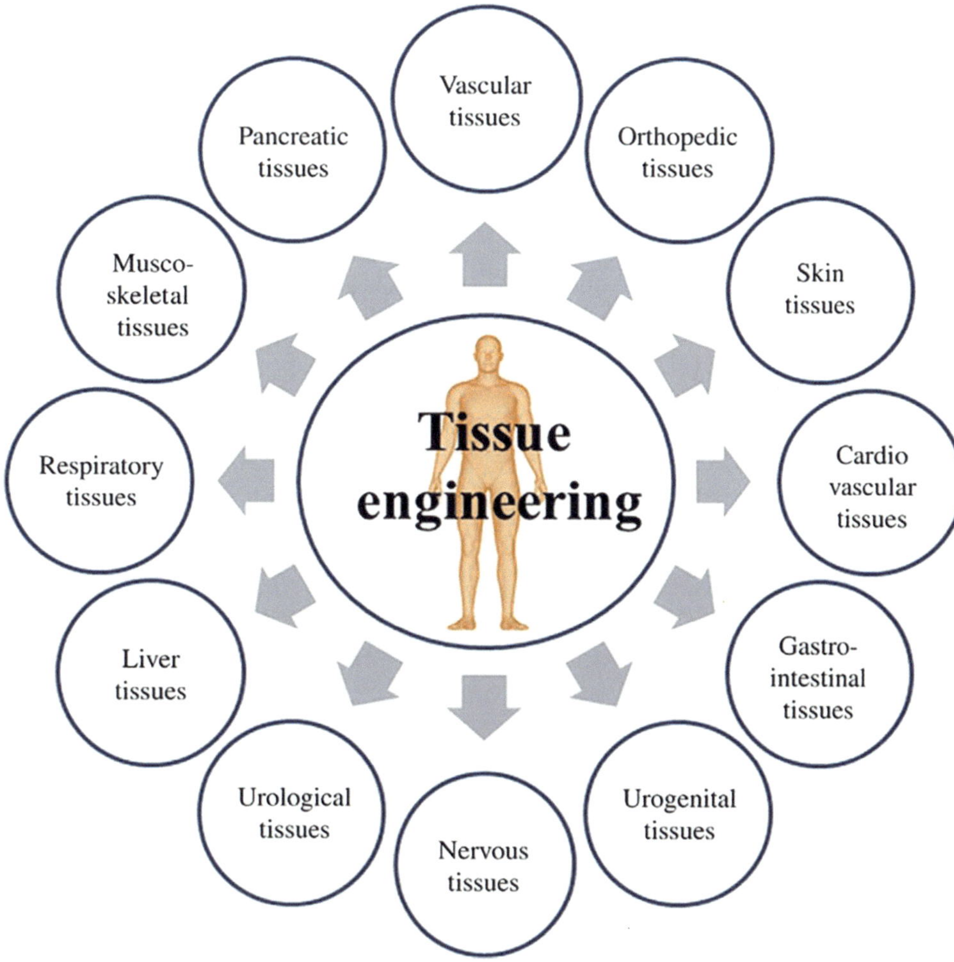

Figure 9.
Different types of damaged body tissues repaired by applications of nanocomposites biomaterial-based tissue engineering [72].

Cardiac tissue engineering should meet some essential requirements, such as biodegradability, porosity, and mechanical/conductive properties, to match healthy cardiac tissue. Based on these characteristics, nanocomposite hydrogels containing conductive nanomaterials are considered attractive strategies for cardiacregeneration [79]. There are many reports on the use of electroactive nanocomposite hydrogels for cardiac tissue engineering.

Polypyrrole (Ppy) is widely used in cardiac tissue engineering due to its outstanding electrical conductivity and biocompatibility. Qiu et al. prepared a variety of conductive hydrogel myocardial patches containing Ppy NPs and used them to repair myocardial defects. Some animal experiments results showed that the fractional shortening and ejection fraction are elevated by about 50% and that the infarct size is reduced by 42.6% [77]. Metal-based nanomaterials are also widely used in cardiac tissue engineering, Chen et al. prepared a hyaluronic acid-based injectable hydrogel containing Au NPs to repair myocardial defects by loading human induced pluripotent stem cells [80]. Additionally, Dong et al. introduced Au NPs into silk-based hydrogel (SF/ECM) for cell proliferation and expansion of cardiomyocytes [81]. The uniform distribution of Au NPs in the matrix

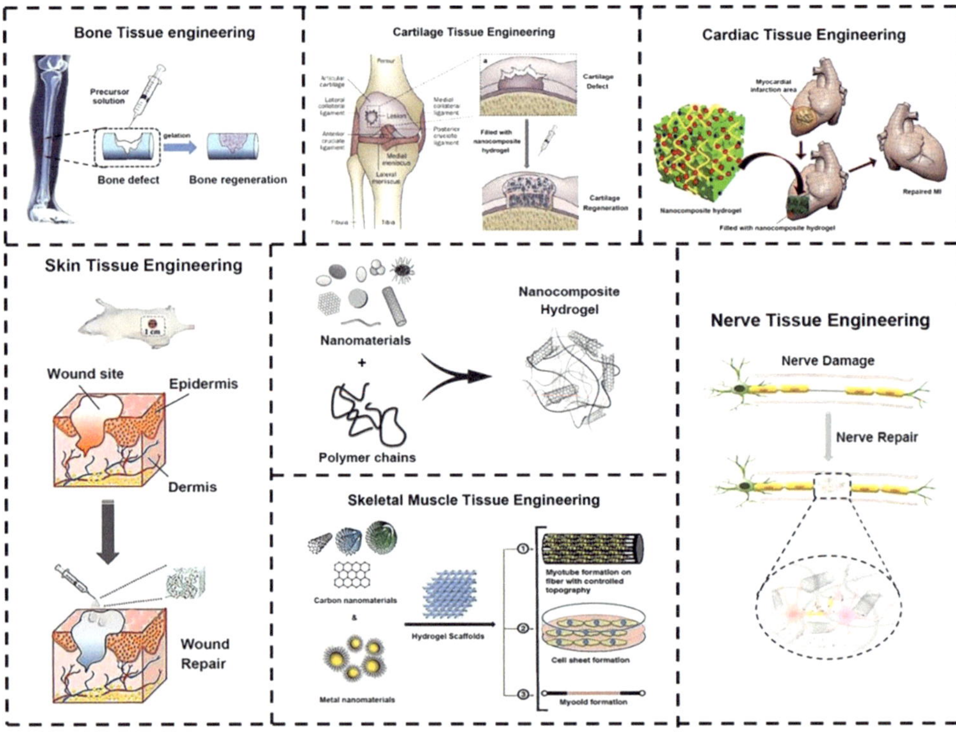

Figure 10.
Various nanocomposite hydrogels were designed and prepared for different types of tissue engineering [73].

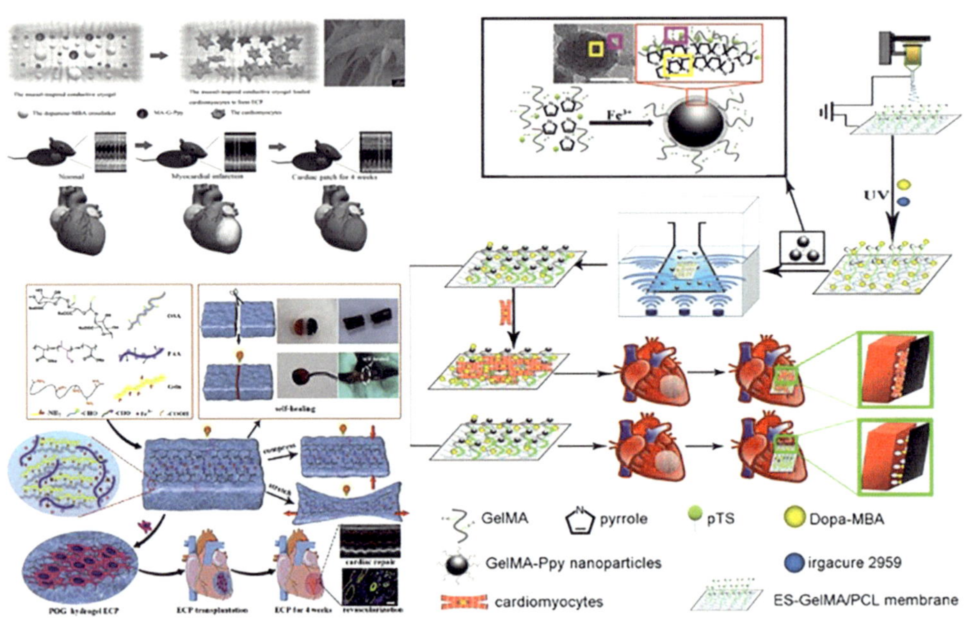

Figure 11.
Schematic illustrating of various nanocomposite conductive hydrogel and their applications in myocardial infarction repair [75–77].

can provide favorable electrical conductivity and biological effects for cardiac repair. In addition to Au NPs, Liu et al. prepared an injectable PEGylated chitosan hydrogel scaffold for cardiac repair by introducing titanium dioxide (TiO_2) nanoparticles [82]. The TiO_2 nanocomposite hydrogel significantly enhances the functionalization of the cardiomyocytes, resulting in excellent synchronous contraction by increasing the expression of α-actinin and connexin CX-43. Results of cardiac markers confirmed the formation of interconnected cardiac layers within these nanocomposite hydrogels and the formation of cell-hydrogel matrix interactions. These nanocomposites are well suited for cardiac regeneration and provide a new platform for cardiac tissue engineering.

Mineral-based bioactive nanomaterials, such as SiO_2, P_2O_5, and CaO, can also be used to effectively improve composite hydrogels for cardiac tissue engineering. It has been proven that these materials can stimulate cells to secrete a large number of angiogenic factors, thereby promoting the formation of blood vessels in engineering scaffold [83–85]. To address the lack of functional blood vessels in engineered tissue and low survival rates of injected cells in cardiac tissue engineering, Barabadi et al. [86] introduced bioactive glass nanoparticles into the gelatin-collagen hydrogel to improve the biological properties. In addition, *in vivo* evaluation experiments confirmed that this nanocomposite hydrogel scaffold can effectively promote the formation of capillaries, reduce scar area, and improve cardiac function.

Although electroactive nanocomposite hydrogels can achieve satisfying biocompatibility and durability for the generation of cardiac microtissues *in vitro*, there is still space for improvement in terms of mechanical properties and electrical conductivity [87]. In addition, the goals of high-level cell attachment, viability, endothelialization, and recruitment of cardiac progenitor cells have not yet been fully achieved. Therefore, future research should aim to address these issues.

3.2 Bone tissue engineering

Among all tissues in the human body, cartilage and bone are some of the most extensively researched tissues in tissue engineering because of their high regeneration potential. Bone graft materials, due to their osteoinductive and osteoconductive properties, have been used to repair fractures and other defects [88]. However, there may be risks of disease metastasis, infection, chronic pain, immunogenicity, and inadequate supply (**Figure 12**).

Bone tissues, connective tissues, that are composed primarily of cells and extracellular matrix. These tissues are hard tissues that can withstand repeated mechanical stimulation without harm or loss of human functions. Bone tissue is closely related to the movement of the human body. Once damaged, it will cause great inconvenience to the patient's life. Based on this, the key goal of bone tissue engineering is to develop a tough and natural 3D microenvironment with the physiological environment required to form the target tissue [6, 90]. In recent years, more and more nanocomposite hydrogels have been developed for bone tissue engineering, especially bone and cartilage tissue engineering.

Inspired by the nature of bones, there is a need for three-dimensional hierarchical structures and nanocomposites that can contain multiple levels of tissues, that is, from the macroscopic tissue arrangement to the molecular arrangement of proteins [91]. These nanostructured materials can provide enhanced mechanical properties and allow the proper transduction of mechanical stimuli to the cellular level. Bioabsorbable β-TCP can improve the clinical application of pure HAp to achieve

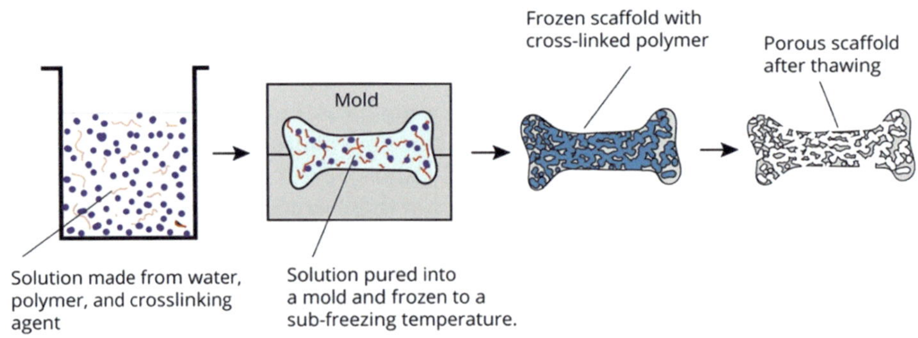

Figure 12.
Cryotropic gelation scaffold for bone tissue engineering [89].

better bone regeneration. The main attraction of these materials is that they can bind well with host tissues to form a robust interface. However, these materials are limited to non-load-bearing applications due to their poor mechanical properties.

The mineral composition of bone is similar to HAp, but it contains other ions in composition, which can better prepare biological materials. Yazaki et al. developed the incorporation of carbonate or fluoride into the DNA-fibronectin-apatite composite layer for tissue engineering to adjust the solubility of the layer [92]. The incorporation of carbonate increases the effect of gene transfer on the efficiency layer, while fluoride reduces the efficiency and delays the time of gene transfer in a dose-dependent manner. In addition, manganese (M_n), detected as a minor component of teeth and bones, modulates bone remodeling.

The scaffold is designed to mimic the 3D support structure of the ECM in the surrounding bone tissue and offers the following advantages—(i) porous interconnected structure, that afford the transport of quality, nutrition, and regulatory factors to allow cell survival, proliferation and differentiation, (ii) sufficient mechanical properties for the support of cells, (iii) controllable degradation and minimal inflammation or toxicity to the body [93]. Besides, the scaffold has the desirable properties of transferring cells to the defect site, limiting cell loss, and even recruiting the body's own cells, rather than simply injecting cells into the defect site [94].

The challenge is to ensure good compatibility of the scaffold while maintaining the porous structure and mechanical properties. It is important to achieve a uniform distribution of NPs within the scaffold. Methods for maximizing the distribution of NPs include precipitating nanoparticles *in situ* in a polymer matrix or using dispersants. For example, the *in situ* synthesis method is used to prepare nanocomposite scaffolds of SF and CaPs. Phosphate ions are added to the calcium chloride solution in which SF is dissolved, and then the diammonium hydrogen phosphate solution is added by salt immersion/freeze drying technology [95]. The scaffold exhibits highly interconnected macropores with a size of approximately 500 μm and a micropore size range of 10–100 μm under SEM.

3.3 Skin tissue engineering

The skin is the first line of defense against infection and its injuries (such as burns) may cause serious health problems. Skin tissue engineering is a promising method of skin regeneration. In addition to the proper mechanical properties, the scaffold material for skin tissue engineering should also possess antibacterial and anti-infective properties, and can play a role in drug delivery to promote wound repair. To achieve

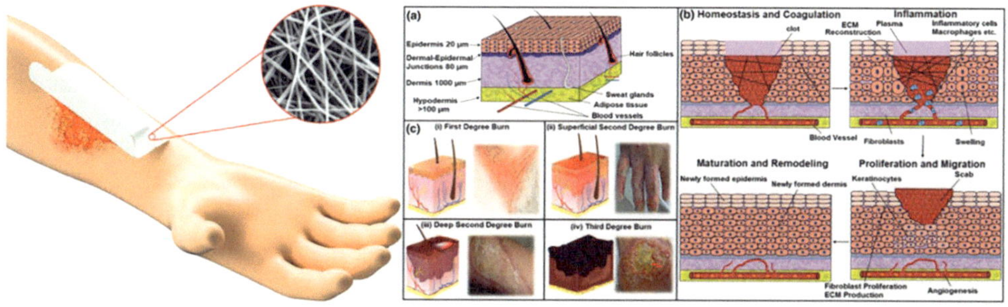

Figure 13.
Skin healing process and mechanism using the nanocomposite wound dressing [97].

this goal, nanocomposite hydrogels are widely used in skin tissue engineering. Studies have shown that zinc oxide (ZnO) NPs exhibit strong antibacterial activity and have no side effects on normal tissue. Rakhshaei et al. introduced ZnO nanoparticles into chitosan/gelatin hydrogel (CS-gel/ZnO) to endow the hydrogel with additional antibacterial and drug delivery properties [96]. The ZnO NPs on the surface of the scaffold extremely enhance the flexibility of the scaffold. These nanocomposite hydrogels can effectively destroy the cell walls of gram-negative bacteria to achieve antibacterial properties, which is due to ROS and Zn^{2+} in the nanocomposite hydrogel can attack the negatively charged bacterial cell wall, causing bacterial death (**Figure 13**).

In addition to ZnO nanoparticles, graphene-based nanomaterials are also used as fillers to prepare nanocomposite hydrogels for skin tissue engineering. Narayanan et al. mixed rGO nanosheets with PAAm to form nanocomposite hydrogels which have antibacterial activity and can promote the formation of biofilms [98]. Besides, it is confirmed that the ginsenoside molecules in the nanocomposite scaffold can be released slowly to achieve antioxidant effects. Xu et al. introduced ginsenoside Rg3 (GS-Rg3) into an electrospun fiber of polyglutamic acid, this nanocomposite hydrogel exhibits fast tissue repair and inhibits the excessive scar formation [99]. The results show that this nanocomposite hydrogel can be used as a wound dressing in skin tissue engineering.

3.4 Other tissue engineering applications

In addition to the aforementioned fields, nanocomposite scaffolds also have a wide range of applications in other areas, such as muscle tissue and nerve tissue.

3.4.1 Skeletal muscle tissue engineering

Skeletal muscle is one of the most abundant tissues in the body, possessing a complex structure. In addition to supporting, connecting, nourishing, and protecting muscle tissue, skeletal muscle also has the function of regulating the activity of muscle fiber groups. Extensive skeletal muscle defects caused by trauma or tumor ablation can cause movement disorders and organ dysfunction, which can lead to pain in patients [100]. Restoration of the original function of skeletal muscle is limited and fibrosis and scarring may occur for serious injuries of a mass loss of more than 20% [101].

The purpose of skeletal muscle tissue engineering is to replicate the natural structure and function of muscle *in vitro* and to transplant this tissue to the damaged area, since the behavior of myogenic cells is regulated by the flexibility and strength of the scaffold, such scaffold systems in skeletal muscle tissue engineering should have mechanical elasticity. Nanocomposite hydrogels with conductive components have proven to be good candidates as 3D biomaterials for skeletal muscle tissue engineering. In the field of skeletal muscle regeneration, graphene and carbon nanotubes are the two most widely used carbon nanomaterials in skeletal muscle regeneration. To further optimize the mechanical properties of the scaffold, Patel et al. combined graphene with chitosan and gellan gum to develop a graphene-polysaccharide nanocomposite hydrogel scaffold [102]. This nanocomposite hydrogel can positively affect myoblasts, and as an ideal multifunctional biomaterial, it has the mechanical properties of natural skeletal muscle tissue and the ideal conductivity for transmitting electrical signals to cells.

Metal oxides can also be used in skeletal muscle tissue engineering as nanoreinforcing materials. For example, Tognato et al. introduced iron oxide (Fe_2O_3) nanoparticles into gelatin methacrylic acid (GelMA) [103]. In an external magnetic field, the magnetic Fe_2O_3 nanoparticles can be aligned, which then caused the seed cells to line up in the same direction. In particular, in the absence of differentiation media, the skeletal myoblasts in the nanocomposite hydrogel can differentiate into myotubes. One of the challenges of current research *in vitro* skeletal muscle regeneration is the lack of functional vascular structure, which also exists in most areas of tissue engineering today. Therefore, the direction of skeletal muscle tissue engineering in the future is to overcome the challenge above to promote the formation of muscle-like intravascular structures.

3.4.2 Nervous tissue engineering

The function of the nervous system is to receive information from cells in different parts of the body, process the received information, and send signals to other cells and organs to elicit appropriate responses [104]. Nervous system damage caused by ischemia, chemical, mechanical or thermal factors is very common in our current society. Surgical strategies, such as nerve repair and autologous nerve transplantation, are widely used in nervous system injuries to achieve optimal recovery. However, the prognosis is not ideal. Recent studies have shown that only 40–50% of patients recover their motor function after receiving autologous nerve transplantation [105]. Therefore, tissue engineering technology is considered an ideal method to repair nerve damage.

The goal of nervous tissue engineering is to manufacture nerve graft substitutes for the treatment of nerve damage and achieve long-term functional recovery. Among them, graphene-based nanocomposite hydrogels can be used as a viable option to promote nerve regeneration. Huang et al. prepared a nanocomposite hydrogel based on graphene and polyurethane [106]. This polyurethane hydrogel can improve the growth of neural stem cells and the differentiation of neurons. Qiao and his colleagues introduced GO into poly(acrylic acid) (PAA) and formed an electrically responsive nanocomposite hydrogel through *in situ* polymerization [107]. In particular, scientists also proved that carbon nanotube (CNT)-based polyethylene glycol (PEG) nanocomposite hydrogels can increase total neurite growth and average neurite length through electrical stimulation.

However, the influence of these nanocomposite hydrogels on the activity of neural stem cells needs to be further studied.

4. Conclusion

Currently, these reported nanocomposite hydrogels can mimic certain characteristics of natural tissues to a certain extent. However, there are still many challenges in applying them as tissue engineering scaffold biomaterials in clinical practice. Although some new functional nanomaterials have been introduced into nanocomposite scaffolds, the relatively poor interfacial interaction between polymer chains and nanomaterials and the uneven dispersion of nanomaterials in the matrix significantly limit their application. At present, processing technologies, such as 3D printing, electrospinning, and microfluidic reactors, have been rapidly developed, providing a great number of opportunities to quickly obtain nanocomposite scaffolds with anisotropic structures.

5. Future perspectives

We expect that the discussion in this chapter on the effects of different nanomaterials on the physical and chemical properties of nanocomposite hydrogels and their eventual application in different types of tissue engineering will be useful for the design, fabrication, and production of new generation nanocomposite hydrogel materials. Nanocomposite hydrogels are a new generation of 3D biomaterials for many different applications, including kinds of tissue engineering. The introduction of different nanomaterials in a polymer hydrogel network leads to the formation of nanocomposite systems with different functions. For instance, the addition of conductive nanomaterials into the polymer hydrogel matrix, such as Ppy, carbon nanotubes, graphene, and metal-based nanoparticles, not only provides mechanically strong hydrogels but also imparts conductivity to the hydrogel. These nanocomposite hydrogels can be used for the regeneration and repair of electroactive tissues, such as nerve, cardiac, and skeletal muscle tissues. Additionally, the introduction of rigid nanomaterials, such as nHAp, tricalcium phosphate, bioglass into the hydrogel matrix, can enhance its mechanical properties and promote the occurrence of mineralization inside the scaffolds. These nanocomposite hydrogels can be used for bone or cartilage tissues repair. In addition to these, some nanocomposite hydrogels can even be used in skin and vascular tissue engineering.

We believe that high-performance nanocomposite hydrogels with excellent mechanical properties, ordered structures, and novel functions will soon be used in tissue engineering applications.

Author details

Shuai Liu, Rurong Lin, Chunyi Pu, Jianxing Huang, Jie Zhang and Honghao Hou*
Guangdong Provincial Key Laboratory of Construction and Detection in Tissue Engineering, Southern Medical University, Guangzhou, Guangdong, China

*Address all correspondence to: ss.hhh89@hotmail.com

References

[1] Han X, Xu H, Che L, Sha D, Huang C, Meng T, et al. Application of inorganic nanocomposite hydrogels in bone tissue engineering. iScience. 2020;**23**(12)

[2] Kim YH, Yang X, Shi L, Lanham SA, Dawson JI. Bisphosphonate nanoclay edge-site interactions facilitate hydrogel self-assembly and sustained growth factor localization. Nature Communications. 2020;**11**(1):1365

[3] Xavier JR, Thakur T, Desai P, Jaiswal MK, Gaharwar AK. Bioactive Nanoengineered hydrogels for bone tissue engineering: A growth-factor-free approach. ACS Nano. 2015;**9**(3):3109-3118

[4] Wang Z, Zhao J, Tang W, Hu L, Chen X, Su Y, et al. Multifunctional nanoengineered hydrogels consisting of black phosphorus nanosheets upregulate bone formation. Small. 2019;**15**(41)1901560

[5] Liu S, Li P, Liu X, Wang P, Ye Z. Bioinspired mineral-polymeric hybrid hyaluronic acid/poly (γ-glutamic acid) hydrogels as tunable scaffolds for stem cells differentiation. Carbohydrate Polymers. 2021;**1-3**:118048

[6] Stevens MM, George JH. Exploring and engineering the cell surface interface. Science. 2005;**310**(5751):1135-1138

[7] Piantanida E, Alonci G, Bertucci A, Cola LD. Design of nanocomposite injectable hydrogels for minimally invasive surgery. Accounts of Chemical Research. 2019;**52**(8)2101-2112

[8] Fernandes EM, Pires RA, Mano JF, Rui LR. Bionanocomposites from lignocellulosic resources: Properties, applications and future trends for their use in the biomedical field. Progress in Polymer Science. 2013;**38**(10-11):1415-1441

[9] Hule RA, Pochan DJ. Polymer nanocomposites for biomedical applications. MRS Bulletin. 2007;**32**(4): 354-358

[10] Murugan R, Ramakrishna S. Development of nanocomposites for bone grafting. Composites Science and Technology. 2005;**65**(15/16):2385-2406

[11] Wheeldon I, Farhadi A, Bick AG, Jabbari E, Khademhosseini A. Nanoscale tissue engineering: Spatial control over cell-materials interactions. Nanotechnology. 2011;**22**(21):212001

[12] Pina S, Oliveira JM, Reis RL. Natural-based nanocomposites for bone tissue engineering and regenerative medicine: A review. Advanced Materials. 2015;**27**(7):1143-1169

[13] Bonfield W, Grynpas M, Tully A, Bowman J, Abram J. Hydroxyapatite reinforced polyethylene--a mechanically compatible implant material for bone replacement. Biomaterials. 1981;**2**(3):185-186

[14] Okamoto M, John B. Synthetic biopolymer nanocomposites for tissue engineering scaffolds. Progress in Polymer Science. 2013;**38**(10-11): 1487-1503

[15] Mcmahon RE, Wang L, Skoracki R, Mathur AB. Development of nanomaterials for bone repair and regeneration. Journal of Biomedical Materials Research Part B Applied Biomaterials. 2013;**101B**(2):387-397

[16] Basha RY, Sampath K, Doble M. Design of biocomposite materials for bone tissue regeneration. Materials Science & Engineering C. 2015;**57**(DEC): 452-463

[17] Chan BP, Leong KW. Scaffolding in tissue engineering: General approaches and tissue-specific considerations. European Spine Journal. 2008;**17**(s4):467-479

[18] Atala A. Technology insight: Applications of tissue engineering and biological substitutes in urology. Nature Clinical Practice Urology. 2005;**2**(3):143-149

[19] Zaffagnini S, Giordano G, Vascellari A, Bruni D, Neri MP, Iacono F, et al. Arthroscopic collagen meniscus implant results at 6 to 8 years follow up. Knee Surgery, Sports Traumatology, Arthroscopy. 2007;**15**(2):175-183

[20] Quirk RA, France RM, Shakesheff KM, Howdle SM. Supercritical fluid technologies and tissue engineering scaffolds. Current Opinion in Solid State & Materials Science. 2004;**8**(3-4):313-321

[21] Tsivintzelis I, Marras SI, Zuburtikudis I, Panayiotou C. Porous poly(−lactic acid) nanocomposite scaffolds prepared by phase inversion using supercritical CO as antisolvent. Polymer. 2007;**48**(21):6311-6318

[22] Ray SS, Okamoto M. Polymer/layered silicate nanocomposites: A review from preparation to processing. Progress in Polymer Science. 2003;**28**(11):1539-1641

[23] Song X, Mei J, Ye G, Wang L, Ananth A, Yu L, et al. In situ pPy-modification of chitosan porous membrane from mussel shell as a cardiac patch to repair myocardial infarction. Applied Materials Today. 2019;**15**:87-99

[24] Subramanian A, Krishnan UM, Sethuraman S. Development of biomaterial scaffold for nerve tissue engineering: Biomaterial mediated neural regeneration. Journal of Biomedical Science. 2009;**16**(1):108-108

[25] Sahoo NG, Pan YZ, Li L, He CB. Nanocomposites for bone tissue regeneration. Nanomedicine. 2013;**8**(4):639-653

[26] Khan WS, Rayan F, Dhinsa BS, Marsh D. An osteoconductive, osteoinductive, and osteogenic tissue-engineered product for trauma and orthopaedic surgery: How far are we? Stem Cells International. 2012;**2012**(1687-966X):236231

[27] Patel M, Fisher JP. Biomaterial scaffolds in pediatric tissue engineering. Pediatric Research. 2008;**63**(5):497-501

[28] Slaughter BV, Khurshid SS, Fisher OZ, Khademhosseini A, Peppas NA. Hydrogels in regenerative medicine. Advanced Materials. 2010;**21**(32-33):3307-3329

[29] Hoffman A. Hydrogels for biomedical applications. Advanced Drug Delivery Reviews. 2002;**54**(1):3-12

[30] Annabi N, Nichol JW, Zhong X, Ji C, Dehghani F. Controlling the porosity and microarchitecture of hydrogels for tissue engineering. Tissue Engineering Part B Reviews. 2010;**16**(4):371-383

[31] Haraguchi K. Nanocomposite gels: New advanced functional soft materials. Macromolecular Symposia. 2007;**256**(1):120-130

[32] Haraguchi K, Takehisa T. Nanocomposite hydrogels: A unique organic–inorganic network structure with extraordinary mechanical, optical, and swelling/deswelling properties. Advanced Materials. 2002;**14**(16):1120-1124

[33] Liu X, Liu J, Lin S, Zhao X. Hydrogel machines. Materials Today. 2020;**36**:102-124

[34] Song J, Chen S, Sun L, Guo Y, Zhang L, Wang S, et al. Mechanically

and electronically robust transparent organohydrogel fibers. Advanced Materials. 2020;**32**(8):1906994

[35] Alvarez-Barreto JF, Linehan SM, Shambaugh RL, Sikavitsas VI. Flow perfusion improves seeding of tissue engineering scaffolds with different architectures. Annals of Biomedical Engineering. 2007;**35**(3):429

[36] Nair LS, Bhattacharyya S, Laurencin CT. Development of novel tissue engineering scaffolds via electrospinning. Expert Opinion on Biological Therapy. 2004;**4**(5):659-668

[37] Vasita R, Katti DS. Nanofibers and their applications in tissue engineering. International Journal of Nanomedicine. 2006;**1**(1):15-30

[38] Hong JK, Madihally SV. Next generation of electrosprayed fibers for tissue regeneration. Tissue Engineering Part B Reviews. 2011;**17**(2):125-142

[39] Filippi M, Dasen B, Guerrero J, Garello F, Isu G, Born G, et al. Magnetic nanocomposite hydrogels and static magnetic field stimulate the osteoblastic and vasculogenic profile of adipose-derived cells. Biomaterials. 2019;**223**:119468

[40] Antonella S, Jessika B, Daniele D'A, Laura A, Tullia M, Sandra M, et al. Development of solvent-casting particulate leaching (SCPL) polymer scaffolds as improved three-dimensional supports to mimic the bone marrow niche. Materials Science and Engineering C. 2019;**96**:153-165

[41] Zhu N, Chen X. Biofabrication of Tissue Scaffolds. Advances in Biomaterials Science and Biomedical Applications. InTech; 2013

[42] Rui LR, Neves NM, Mano JF, Gomes ME, Azevedo HS. Natural-Based Polymers for Biomedical Applications. Woodhead Publishing, CRC Press; 2008:781-802

[43] Bhardwaj N, Kundu SC. Electrospinning: A fascinating fiber fabrication technique. Biotechnology Advances. 2010;**28**(3):325-347

[44] Rahmati M, Mills DK, Urbanska AM, Saeb MR, Venugopal JR, Ramakrishna S, et al. Electrospinning for tissue engineering applications. Progress in Materials Science. 2021;**117**:100721

[45] Bishi DK, Mathapati S, Venugopal JR, Guhathakurta S, Cherian KM, Ramakrishna S, et al. Trans-differentiation of human mesenchymal stem cells generates functional hepatospheres on poly(l-lactic acid)-co-poly(ε-caprolactone)/collagen nanofibrous scaffolds. Journal of Materials Chemistry B. 2013;**1**(32):3972-3984

[46] Chen PH, Liao HC, Hsu SH, Chen RS, Wu MC, Yang YF, et al. A novel polyurethane/cellulose fibrous scaffold for cardiac tissue engineering. RSC Advances. 2014;**5**(9):6932-6939

[47] Gustafsson Y, Haag J, Jungebluth P, Lundin V, Macchiarini P. Viability and proliferation of rat MSCs on adhesion protein-modified PET and PU scaffolds. Biomaterials. 2012;**33**(32):8094-8103

[48] Tan A, Farhatnia Y, Seifalian AM. Polyhedral oligomeric silsesquioxane poly(carbonate-urea) urethane (POSS-PCU): Applications in nanotechnology and regenerative medicine. Critical Reviews in Biomedical Engineering. 2013;**41**(6):495-513

[49] Madihally SV, Matthew HW. Porous chitosan scaffolds for tissue engineering. Biomaterials. 1999;**20**(12):1133

[50] Shahini A, Yazdimamaghani M, WalkerKJ,EastmanMA,Hatami-MarbiniH, Smith BJ, et al. 3D conductive nanocomposite scaffold for bone tissue engineering. International Journal of Nanomedicine. 2014;**9**:167

[51] Ho MH, Kuo PY, Hsieh HJ, Hsien TY, Hou LT, Lai JY, et al. Preparation of porous scaffolds by using freeze-extraction and freeze-gelation methods. Biomaterials. 2004;**25**(1):129-138

[52] Kavya KC, Jayakumar R, Nair S, Chennazhi KP. Fabrication and characterization of chitosan/gelatin/ $nSiO_2$ composite scaffold for bone tissue engineering. International Journal of Biological Macromolecules. 2013;**59**:255-263

[53] Derakhshanfar S, Mbeleck R, Xu K, et al. 3D bioprinting for biomedical devices and tissue engineering: A review of recent trends and advances. Bioactive Materials. 2018;**3**(2):144-156

[54] Kang H-W, Lee SJ, Ko IK, Kengla C, Yoo JJ, Atala A. A 3D bioprinting system to produce human-scale tissue constructs with structural integrity. Nature Biotechnology. 2016;**34**(3):312-319

[55] Castro NJ, Patel R, Grace LZ. Design of a novel 3D printed bioactive nanocomposite scaffold for improved osteochondral regeneration. Cellular and Molecular Bioengineering. 2015;**8**(3):416-432

[56] Ball P. Materials science. Polymers made to measure. Nature. 1994;**367**(6461):323-324

[57] Wang X, Ding B, Li B. Biomimetic electrospun nanofibrous structures for tissue engineering. Materials Today. 2013;**16**(6):229-241

[58] Herranz-Blanco B, Liu D, Mäkilä E, Shahbazi MA, Ginestar E, Zhang H, et al. On-chip self-assembly of a smart hybrid nanocomposite for antitumoral applications. Advanced Functional Materials. 2015;**25**(10):1488-1497

[59] Cha GD, Lee WH, Lim C, Choi MK, Kim DH. Materials engineering, processing, and device application of hydrogel nanocomposites. Nanoscale. 2020;**12**(19):10456-10473

[60] Levengood SL, Zhang M. Chitosan-based scaffolds for bone tissue engineering. Journal of Materials Chemistry B. 2014;**2**(21):3161-3184

[61] Castro NJ, Patel R, Zhang LG. Design of a novel 3D printed bioactive nanocomposite scaffold for improved osteochondral regeneration. Cellular and Molecular Bioengineering. 2015;**8**(3):416-432

[62] Leu Alexa R, Ianchis R, Savu D, Temelie M, Trica B, Serafim A, et al. 3D printing of alginate-natural clay hydrogel-based nanocomposites. Gels (Basel, Switzerland). 2021;**7**(4):211

[63] Liu D, Zhang H, Cito S, Fan J, Mäkilä E, Salonen J, et al. Core/shell nanocomposites produced by superfast sequential microfluidic nanoprecipitation. Nano Letters. 2017;**17**(2):606-614

[64] Udoh CE, Cabral JT, Garbin V. Nanocomposite capsules with directional, pulsed nanoparticle release. Science Advances. 2017;**3**(12): eaao3353

[65] Mahmoudi Z, Mohammadnejad J, Razavi Bazaz S, Abouei Mehrizi A, Saidijam M, Dinarvand R, et al. Promoted chondrogenesis of hMCSs with controlled release of TGF-β3 via microfluidics synthesized alginate nanogels. Carbohydrate Polymers. 2020;**229**:115551

[66] Tavakoli J, Raston CL, Tang Y. Tuning surface morphology of fluorescent hydrogels using a vortex fluidic device. Molecules (Basel, Switzerland). 2020;**25**(15):3445

[67] Yue C, Ding C, Du X, Wang Y, Su J, Cheng B. Self-assembly of collagen fibrils on graphene oxide and their hybrid nanocomposite films. International Journal of Biological Macromolecules. 2021;**193**(Pt A):173-182

[68] Inam M, Cambridge G, Pitto-Barry A, Laker ZPL, Wilson NR, Mathers RT, et al. 1D vs. 2D shape selectivity in the crystallization-driven self-assembly of polylactide block copolymers. Chemical Science. 2017;**8**(6):4223-4230

[69] Liu N, Zeng X, Pidaparti R, Wang X. Tough and strong bioinspired nanocomposites with interfacial cross-links. Nanoscale. 2016;**8**(43):18531-18540

[70] Linh NVV, Thinh NT, Kien PT, Quyen TN, Phu HD. Injectable nanocomposite hydrogels and electrosprayed nano(micro)particles for biomedical applications. Advances in Experimental Medicine and Biology. 2018;**1077**:225-249

[71] Lei L, Kovacevich DA, Nitzsche MP, Ryu J, Al-Marzoki K, Rodriguez G, et al. Obtaining thickness-limited electrospray deposition for 3D coating. ACS Applied Materials & Interfaces. 2018;**10**(13):11175-11188

[72] Wahid F, Khan T, Hussain Z, Ullah H. Nanocomposite scaffolds for tissue engineering; properties, preparation and applications. In: Applications of Nanocomposite Materials in Drug Delivery. Elsevier; 2018. pp. 701-735

[73] Zhao H, Liu M, Zhang Y, Yin J, Pei R. Nanocomposite hydrogels for tissue engineering applications. Nanoscale. 2020;**12**(28):14976-14995

[74] Muniyandi P, Palaninathan V, Veeranarayanan S, Ukai T, Mohamed S. ECM mimetic electrospun porous poly (L-lactic acid) (PLLA) scaffolds as potential substrates for cardiac tissue engineering. Polymers. 2020;**12**(2):451

[75] Song X, Wang X, Zhang J, Shen S, Yin W, Ye G, et al. A tunable self-healing ionic hydrogel with microscopic homogeneous conductivity as a cardiac patch for myocardial infarction repair. Biomaterials. 2021;**273**:120811

[76] He Y, Ye G, Song C, Li C, Xiong W, Yu L, et al. Mussel-inspired conductive nanofibrous membranes repair myocardial infarction by enhancing cardiac function and revascularization. Theranostics. 2018;**8**(18):5159

[77] Wang L, Jiang J, Hua W, Darabi A, Song X, Song C, et al. Mussel-inspired conductive cryogel as cardiac tissue patch to repair myocardial infarction by migration of conductive nanoparticles. Advanced Functional Materials. 2016;**26**(24):4293-4305

[78] Nazari H, Heirani-Tabasi A, Hajiabbas M, Khalili M, Alavijeh MS, Hatamie S, et al. Incorporation of two-dimensional nanomaterials into silk fibroin nanofibers for cardiac tissue engineering. Polymers for Advanced Technologies. 2020;**31**(2):1-12

[79] Marchesan S, Bosi S, Alshatwi A, Prato M. Carbon nanotubes for organ regeneration: An electrifying performance. Nano Today. 2016;**11**(4):398-401

[80] Li H, Yu B, Yang P, Zhan J, Fan X, Chen P, et al. Injectable AuNP-HA matrix with localized stiffness enhances the formation of gap junction in engrafted

human induced pluripotent stem cell-derived cardiomyocytes and promotes cardiac repair. Biomaterials. 2021;**279**:121231

[81] Dong Y, Hong M, Dai R, Wu H, Zhu P. Engineered bioactive nanoparticles incorporated biofunctionalized ECM/ silk proteins based cardiac patches combined with MSCs for the repair of myocardial infarction: In vitro and in vivo evaluations. Science of the Total Environment. 2020;**707**:135976

[82] Liu N, Chen J, Zhuang J, Zhu P. Fabrication of engineered nanoparticles on biological macromolecular (PEGylated chitosan) composite for bio-active hydrogel system in cardiac repair applications. International Journal of Biological Macromolecules. 2018;**117**:553-558

[83] Gorustovich AA, Roether JA, Boccaccini AR. Effect of bioactive glasses on angiogenesis: A review of in vitro and in vivo evidences. Tissue Engineering Part B: Reviews. 2010;**16**(2):199-207

[84] Day RM, Boccaccini AR, Shurey S, Roether JA, Forbes A, Hench LL, et al. Assessment of polyglycolic acid mesh and bioactive glass for soft-tissue engineering scaffolds. Biomaterials. 2004;**25**(27):5857-5866

[85] Keshaw H, Forbes A, Day RM. Release of angiogenic growth factors from cells encapsulated in alginate beads with bioactive glass. Biomaterials. 2005;**26**(19):4171-4179

[86] Barabadi Z, Azami M, Sharifi E, Karimi R, Lotfibakhshaiesh N, Roozafzoon R, et al. Fabrication of hydrogel based nanocomposite scaffold containing bioactive glass nanoparticles for myocardial tissue engineering. Materials Science and Engineering: C. 2016;**69**:1137-1146

[87] Baei P, Hosseini M, Baharvand H, Pahlavan S. Electrically conductive materials for in vitro cardiac microtissue engineering. Journal of Biomedical Materials Research Part A. 2020;**108**(5):1203-1213

[88] Moore WR, Graves SE, Bain GI. Synthetic bone graft substitutes. ANZ Journal of Surgery. 2001;**71**(6):354-361

[89] Wubneh A, Tsekoura EK, Ayranci C, Uludağ H. Current state of fabrication technologies and materials for bone tissue engineering. Acta Biomaterialia. 2018;**80**:1-30

[90] Place ES, Evans ND, Stevens MM. Complexity in biomaterials for tissue engineering. Nature Materials. 2009;**8**(6):457-470

[91] Aizenberg J, Weaver JC, Thanawala MS, Sundar VC, Morse DE, Fratzl P. Skeleton of Euplectella sp.: Structural hierarchy from the nanoscale to the macroscale. Science. 2005;**309**(5732):275-278

[92] Yazaki Y, Oyane A, Sogo Y, Ito A, Yamazaki A, Tsurushima H. Control of gene transfer on a DNA–fibronectin–apatite composite layer by the incorporation of carbonate and fluoride ions. Biomaterials. 2011;**32**(21):4896-4902

[93] Langer R, Tirrell DA. Designing materials for biology and medicine. Nature. 2004;**428**(6982):487-492

[94] Thorrez L, Shansky J, Wang L, Fast L, Vanden Driessche T, Chuah M, et al. Growth, differentiation, transplantation and survival of human skeletal myofibers on biodegradable scaffolds. Biomaterials. 2008;**29**(1):75-84

[95] Oliveira JM, Rodrigues MT, Silva SS, Malafaya PB, Gomes ME, Viegas CA,

et al. Novel hydroxyapatite/chitosan bilayered scaffold for osteochondral tissue-engineering applications: Scaffold design and its performance when seeded with goat bone marrow stromal cells. Biomaterials. 2006;**27**(36):6123-6137

[96] Rakhshaei R, Namazi H, Hamishehkar H, Kafil HS, Salehi R. In situ synthesized chitosan–gelatin/ZnO nanocomposite scaffold with drug delivery properties: Higher antibacterial and lower cytotoxicity effects. Journal of Applied Polymer Science. 2019;**136**(22):47590

[97] Kordestani SS. Atlas of Wound Healing-E-Book: Atlas of Wound Healing-E-Book: Wound Healing Process. Elsevier Health Sciences; 2019:11-22

[98] Narayanan KB, Choi SM, Han SS. Biofabrication of *Lysinibacillus sphaericus*-reduced graphene oxide in three-dimensional polyacrylamide/carbon nanocomposite hydrogels for skin tissue engineering. Colloids and Surfaces B: Biointerfaces. 2019;**181**:539-548

[99] Xu T, Yang R, Ma X, Chen W, Liu S, Liu X, et al. Bionic poly(γ-glutamic acid) electrospun fibrous scaffolds for preventing hypertrophic scars, Advanced Healthcare Materials. 2019;**8**(13):1900123

[100] Dong R, Ma PX, Guo B. Conductive biomaterials for muscle tissue engineering. Biomaterials. 2020;**229**:119584

[101] Lee H, Ju YM, Kim I, Elsangeedy E, Lee JH, Yoo JJ, et al. A novel decellularized skeletal muscle-derived ECM scaffolding system for in situ muscle regeneration. Methods. 2020;**171**:77-85

[102] Patel A, Xue Y, Hartley R, Sant V, Eles JR, Cui XT, et al. Hierarchically aligned fibrous hydrogel films through microfluidic self-assembly of graphene and polysaccharides. Biotechnology and Bioengineering. 2018;**115**(10):2654-2667

[103] Tognato R, Armiento AR, Bonfrate V, Levato R, Malda J, Alini M, et al. A stimuli-responsive nanocomposite for 3D anisotropic cell-guidance and magnetic soft robotics. Advanced Functional Materials. 2019;**29**(9):1804647

[104] Vijayavenkataraman S. Nerve guide conduits for peripheral nerve injury repair: A review on design, materials and fabrication methods. Acta Biomaterialia. 2020;**106**:54-69

[105] Lee SK, Wolfe SW. Peripheral nerve injury and repair. JAAOS-Journal of the American Academy of Orthopaedic Surgeons. 2000;**8**(4):243-252

[106] Huang C-T, Shrestha LK, Ariga K, Hsu S-H. A graphene–polyurethane composite hydrogel as a potential bioink for 3D bioprinting and differentiation of neural stem cells. Journal of Materials Chemistry B. 2017;**5**(44):8854-8864

[107] Qiao K, Guo S, Zheng Y, Xu X, Meng H, Peng J, et al. Effects of graphene on the structure, properties, electro-response behaviors of GO/PAA composite hydrogels and influence of electro-mechanical coupling on BMSC differentiation. Materials Science and Engineering: C. 2018;**93**:853-863

Chapter 6

Syzygium cumini Mediated Green Synthesis of Silver Nanoparticles for Reduction of 4-Nitrophenol and Assessment of Its Antibacterial Activity

Samadhan R. Waghmode, Amol A. Dudhane and Vaibhav P. Mhaindarkar

Abstract

The biosynthesis of silver nanoparticles (AgNPs) has become more significant in the recent years owing to its applications in catalysis, imaging, drug delivery, nano-device fabrication and in medicine. We propose the synthesis of silver nanoparticles from the plant extract of *Syzygium cumini* and evaluation of its antibacterial and chemocatalytic potential. Synthesis of AgNPs carried out by using aqueous silver nitrate. The UV–Vis absorption spectrum of the synthesized AgNPs showed a broad absorption peak at 470 nm. TEM analysis shows the morphology of AgNPs as a hexagonal matrix with average particle size is about 50 nm. XRD analysis displays the crystalline structure of AgNPs. The presence of elemental silver was confirmed with EDX analysis. FTIR analysis shows that amide groups present in proteins are dominant reducing agents and play an important role in the bioreduction of Ag^+ ions to Ag^0. The bioreduced AgNPs demonstrated significant catalytic properties in a reduction reaction of 4-nitrophenol to 4-aminophenol using $NaBH_4$ in an aqueous condition. The biosynthesized AgNPs have potent antibacterial activity against common clinical pathogens. Considering the remarkable antibacterial activity against common pathogenic microorganisms, AgNPs can be used in the pharmaceutical industries.

Keywords: *Syzygium cumini*, Silver nanoparticles, Antibacterial activity, Catalytic reduction

1. Introduction

Nanobiotechnology is an economic alternative to chemical and physical methods for the synthesis of nanoparticles [1]. The nanoparticles are extensively used in cosmetics, tires, textiles, food industries and medicines [2]. Recently, nanotechnology has gained significant attention due to its unique and different properties such as catalytic, electrical, optical, magnetic and thermal, which have wide varieties of applications [3].

Numerous approaches are in practice to generate AgNPs such as chemical, electrochemical, photochemical and radiation. The significance of NPs is recognized when researchers found that size can persuade the physico-chemical properties of a substance [4]. In the past few decades, tremendous awareness and extensive research efforts were intended toward the metallic nanoparticles derived from noble metals, such as silver and gold [5]. However, there is still a need to enhance and develop high yield, low cost, non-toxic and environmentally friendly procedures. Therefore, the biological loom for the synthesis of NPs becomes crucial.

Nanotechnology has focused much more attention in recent years in several research fields. Due to their advanced optical properties, metal NPs find applications in the areas of biological, chemical and electronic sciences [5, 6]. The AgNPs have a well-built report on antimicrobial, anti-inflammatory, anti-viral, anti-angiogenesis and anti-platelet activity [7, 8]. Currently, the green synthesis of AgNPs finding medical application in the area of continued significance [9]. The new drug synthesis from AgNPs can fight against cancer and kill pathogens like bacteria, fungi, and viruses [10]. There are diverse natural sources like plants, bacteria, yeast, and fungi used to synthesize Au and AgNPs [11, 12]. The use of leaf extract for nanoparticle synthesis is low-cost and eliminates the need for culture preparations and maintenance of aseptic conditions required for microorganisms [13]. At present, plant-mediated synthesis of nanoparticles is gaining more attention due to its simplicity, rapid rate of synthesis and eco-friendliness [14]. Diverse bio-molecules such as, carbohydrates, proteins and co-enzymes are existing in plant that reduce the metallic salt into nanoparticles [1]. The phytosynthesis of nanoparticles is advantageous over chemical synthesis concerning the adverse effect of harmful chemicals on the environment [15].

Syzygium cumini is a plant recognized for its antifungal, antioxidant, anti-inflammatory, hypolipidaemic, hypoglycaemic and pharmacological properties, due to the presence of bioactive compounds in various parts of the plant [16, 17]. The fruits of *Syzygium cumini* have various medicinal purposes and currently have a huge market for treating chronic diarrhea and other enteric infections [18]. However, the remedial outcome of medicinal plants has been consistently queried due to little bio-availability of the chief constituents subsequent to metabolic conversion in the liver [19]. Therefore, the use of nanoparticles confirmed to be a valuable substitute, as they are biodegradable, biocompatible and can allow the sustained release of specific drug [20].

The antimicrobial prospective of nanoparticles is pertinent in the massive area of biology and medicine to prevent infections in burns and open wounds [21, 22]. Therefore, this manuscript describes the antibacterial activity of nanoparticles against common human pathogenic bacteria like *Pseudomonas aeruginosa*, *Serratia marcescens*, *Staphylococcus aureus*, *Salmonella typhimurium* and *Klebsiella pneumonia*. Additionally, we revealed the catalytic potential of biosynthesized AgNPs in the reduction of 4-nitrophenol to 4-aminophenol. The reduction of 4-nitrophenol to 4-aminophenol by $NaBH_4$ is one of the leading model catalytic reaction since it allows easy and reliable assessment of catalysts using the kinetic parameters [23].

Herein, we report the simple, facile, rapid and efficient process for the synthesis of AgNPs using the aqueous leaf extract of *Syzygium cumini* and their applications in antibacterial activity and catalytic reduction of 4-nitrophenol.

2. Materials and methods

2.1 Preparation of plant extract

Syzygium cumini leaves were thoroughly washed in running tap water for 15 min and then shade dried for two days at room temperature. Dry leaves were ground into fine powder in a mortar and pestle. The extract obtained was filtered through Whatman filter paper No.1. The filtrate was collected and stored at 4°C, which was further used for all experiments [24].

2.2 Synthesis of AgNPs

The well-grinded material was mixed with 100 mL of double distilled water and then transferred in 500 mL Erlenmeyer flask followed by continuous stirring on the magnetic stirrer for 10 min. The content was centrifuged at 10,000 rpm for 10 min for the removal of cell debris. 50 mL of aqueous silver nitrate (1 mM) was added to 10 mL of the leaf extract with continuous stirring. A color change from colorless to yellowish-brown, visually confirms the formation of AgNPs [25].

2.3 Characterization

The resulting solution was then diluted by using double distilled water and characterized using UV–Visible spectroscopy, X-ray diffraction, Energy dispersion spectroscopy, FT-IR and Transmission electron microscopy [26].

2.3.1 UV-visible spectroscopy

Silver nanoparticles were characterized by using Systronics UV–Vis spectrophotometer. The bio-reduction absorption spectra were monitored in 300–700 nm range.

2.3.2 X-ray diffraction spectroscopy

The biosynthesized AgNPs using *Syzygium cumini* leaf extract were lyophilized to a powder. The powdered or dried AgNPs were coated on the XRD grid, and the spectra were recorded using Rich Seifert p 300 instruments.

2.3.3 Transmission electron microscopy and energy dispersive spectroscopy

In order to know the morphology of the biosynthesized AgNPs, transmission electron microscopy (TEM) studies were carried out. The size and shape of the AgNPs were recorded by using the FEI (Netherland) model TECNAI-G2U twin operated at an accelerating voltage of 200 KV. EDS analysis was carried out at the same time by the EDS compatible with TEM.

2.3.4 Fourier transform infrared spectroscopy (FTIR) analysis

After the biosynthesis, AgNPs were centrifuged for 15 min at 10,000 rpm. The obtained pellet was re-dispersed in double distilled water to ensure the removal of any uncoordinated bio-molecules. In order to obtain the better separation of nanoparticles, the process of centrifugation was repeated twice. The purified pellet was then subjected to FTIR analysis (Shimadzu IR). AgNPs were mixed with KBr and subjected to IR source 500–4000 cm^{-1}.

2.4 Catalytic reduction of 4-nitrophenol

The catalytic reaction was studied as mentioned by Ghosh et al. with slight modification. Briefly in standard quartz cuvette 1 mL of 0.1 mM aqueous $NaBH_4$ solution mixed with 1.5 mL of 4-nitrophenol aqueous solution (0.25 mM). 100 μL of an aqueous suspension of AgNPs of *Syzygium cumini* (in double-distilled water) was added into the same and time-dependent absorption spectra were recorded every 5 min in the range of 260–520 nm at 28°C [27].

2.5 Antimicrobial activity

The antimicrobial activity of the biosynthesized AgNPs was tested against pathogenic bacteria such as *Serratia marcescens* (NCIM 2078), *Staphylococcus aureus* (NCIM 5021), *Pseudomonas aeruginosa* (NCIM 5029), *Salmonella typhimurium* (NCIM 2501) and *Klebsiella pneumonia* (NCIM 2957), etc. The organisms were collected from National Chemical Laboratory (NCL), Pune. Uniform spreading of bacterial cultures was carried out in the individual plates using a sterile glass spreader. Wells were made on the agar plates using a cork borer to about 10 mm diameter in nutrient agar medium. 100 μg of lyophilized AgNPs were added in 100 μL of distilled water. 50 μL dispersed solution was added to the well. The diameters of the inhibition zone surrounding the wells were measured in millimeters after 24 h. The antimicrobial effect of the biosynthesized AgNPs is directly proportional to the size of the spherical inhibition zone against microbial pathogens [28–30].

3. Result and discussion

3.1 Biosynthesis of AgNPs

The reduction of Ag^+ to Ag^0 NPs was carried out by using aqueous leaf extract of *Syzygium cumini*. The color change of the solution observed from colorless to yellowish-brown indicates that the synthesis of AgNPs shown in (**Figure 1(i)**). The UV–Vis absorption spectrum of the biosynthesized AgNPs demonstrated a characteristic absorption peak at 470 nm, which is a typical band for the silver shown in **Figure 1(ii)**. No other peak was observed in the spectrum, which confirmed that silver only [31]. The formation of AgNPs was further confirmed by using X-ray diffraction (XRD), FT-IR, EDS and transmission electron microscopy (TEM) analysis.

3.2 Characterization

3.2.1 Transmission electron microscopy analysis

The TEM image of the AgNPs is shown in **Figure 2**. TEM has been used to describe the size, shape and morphology of the biosynthesized AgNPs. From the figures, it is observed that the morphology of AgNPs is a hexagonal matrix. **Figure 2** shows the average particle size measured from the TEM image is 50 nm, which are in good agreement with the particle size calculated from XRD analysis.

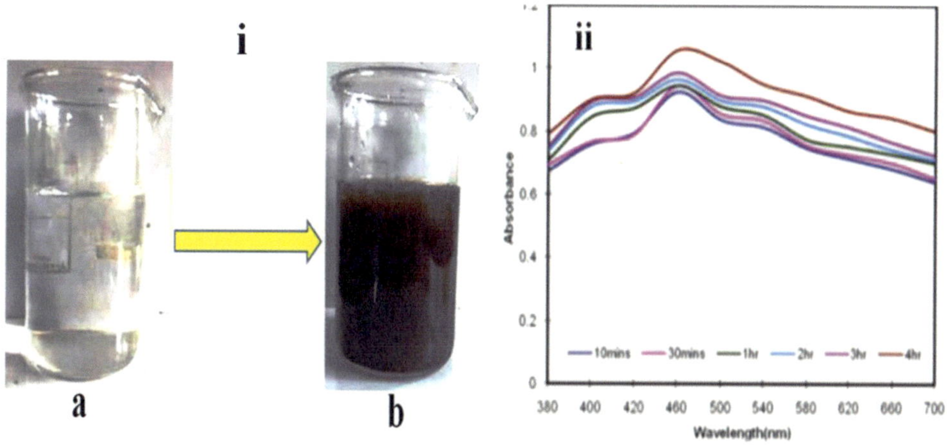

Figure 1.
(i) Color change observed (a) before and (b) after formation of AgNPs. (ii) UV–visible spectra of the synthesized AgNPs.

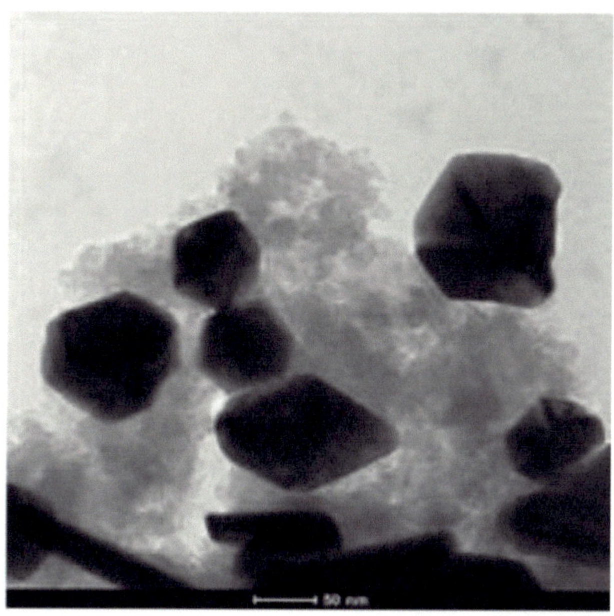

Figure 2.
TEM image of AgNPs.

3.2.2 X-ray diffraction analysis

The presence of Ag crystal in the sample was confirmed by using an X-ray diffractometer. In XRD pattern, the Braggs reflections were observed at 2θ value 38.0^0, 44.8^0, 47.5^0, 64.6^0 and 78.0^0 confirm the presence of AgNPs (**Figure 3**). A strong diffraction peak located at 38.0^0 was ascribed to the (111) facets of Ag. The XRD pattern thus clearly indicated that the AgNPs are in crystalline form. No impurities were observed in the XRD pattern.

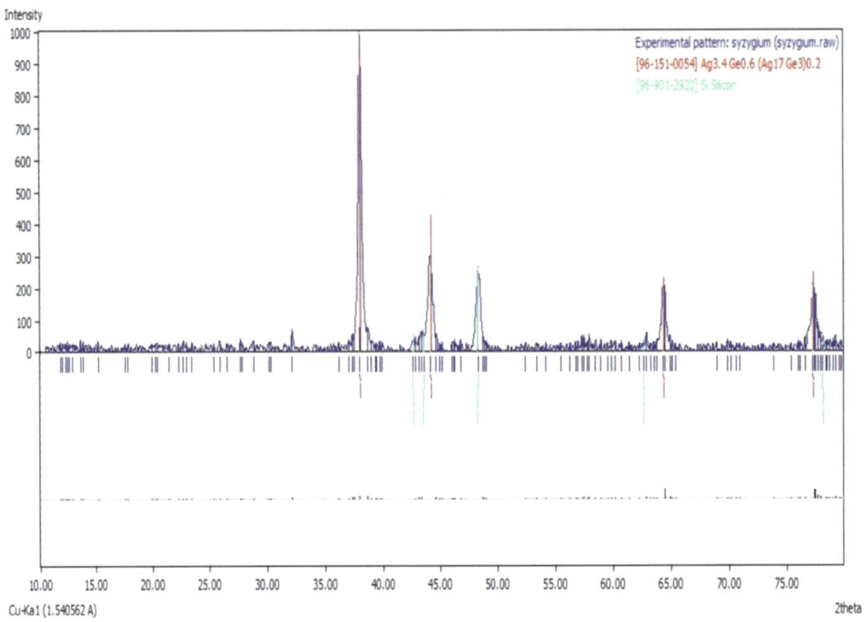

Figure 3.
XRD pattern of synthesized AgNPs.

3.2.3 Energy dispersive spectroscopy analysis

EDS analysis of synthesized particles showed the presence of elemental silver, which correlates with XRD analysis (**Figure 4**). We identified the signal energy peaks for hexagonal-shaped AgNPs produced by using *Syzygium cumini* leaf extract. The NPs showed the prominent silver energy emission peak in a range of 2–4 keV in the spectrum.

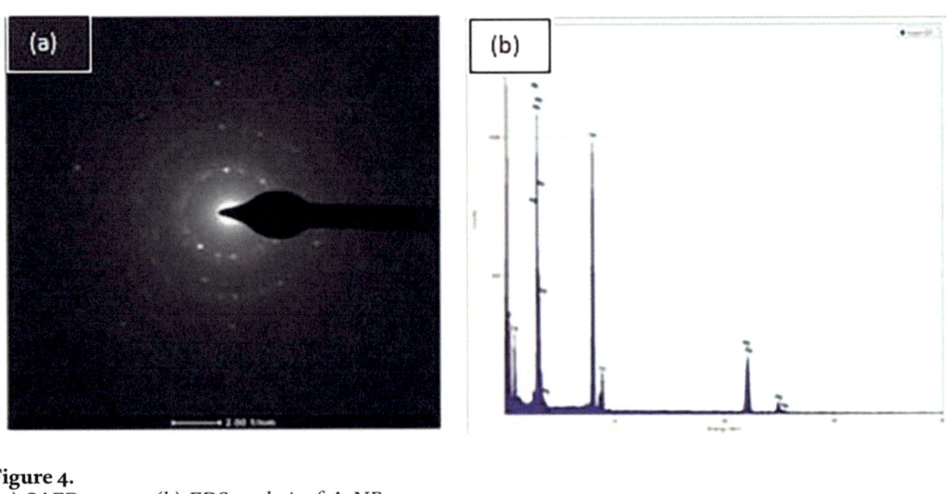

Figure 4.
(a) SAED pattern (b) EDS analysis of AgNPs.

3.2.4 Fourier transform infrared spectroscopy (FTIR) analysis

FTIR absorption spectra of AgNPs are shown in (**Figure 5**). The different possible functional groups at various positions will be determined by using FTIR analysis. The band at 1559 cm^{-1} indicates the presence of amide group [32] arises due to carbonyl stretch in proteins. It can be stated from FTIR analysis. The band at 1610 cm^{-1} is attributed to the stretching vibration of (NH) C=O group. That amide groups present in carbohydrates, proteins are dominant reducing agents and play an important role in the bio-reduction of Ag^{+} ions to Ag^{0} leads to nanoparticles synthesis.

3.3 Catalytic reduction of 4-nitrophenol

In order to study the efficiency of bio-synthesized AgNPs, the catalytic reduction of 4-nitrophenol was carried out in an aqueous medium by using $NaBH_4$ as a reductant at room temperature [33–35]. The 4-nitrophenol (0.1 mM) shows an absorption peak at 400 nm in the visible region with $NaBH_4$ (**Figure 6**). In a control experiment, it can be concluded that the reduction does not occur in the absence of AgNPs, even after the addition of excess $NaBH_4$. After the addition of AgNPs, the gradual decrease in intensity of the absorption peak at 320 nm was observed due to the formation of 4-aminophenol. The complete reduction of p-nitrophenol was also supported by a change in color from yellow to colorless.

3.4 Antibacterial activity

Antibacterial activity of biosynthesized AgNPs was investigated against human pathogens. The biosynthesized AgNPs showed a high inhibitory effect on bacteria, and it may serve as an option for decreasing bacterial infections [36]. The zone of inhibition was found to be as per **Table 1**.

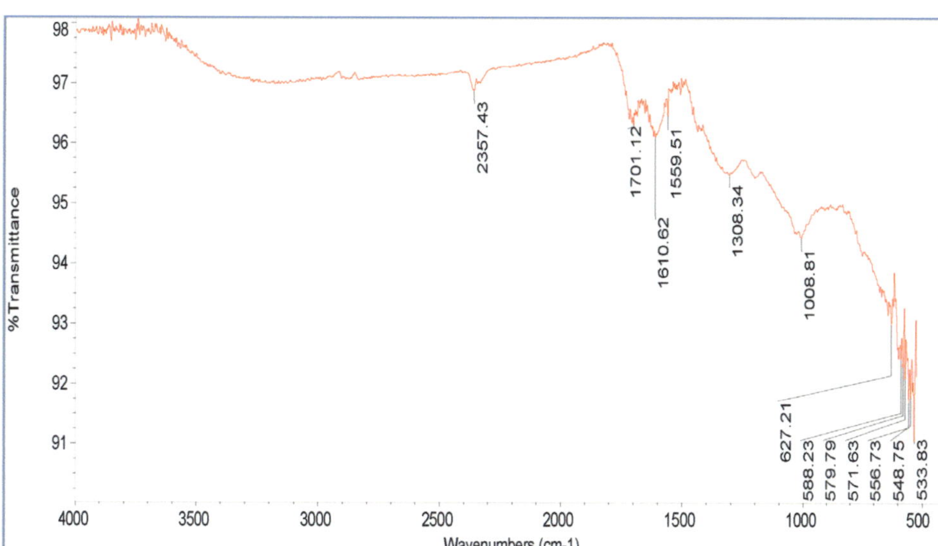

Figure 5.
FTIR image of AgNPs.

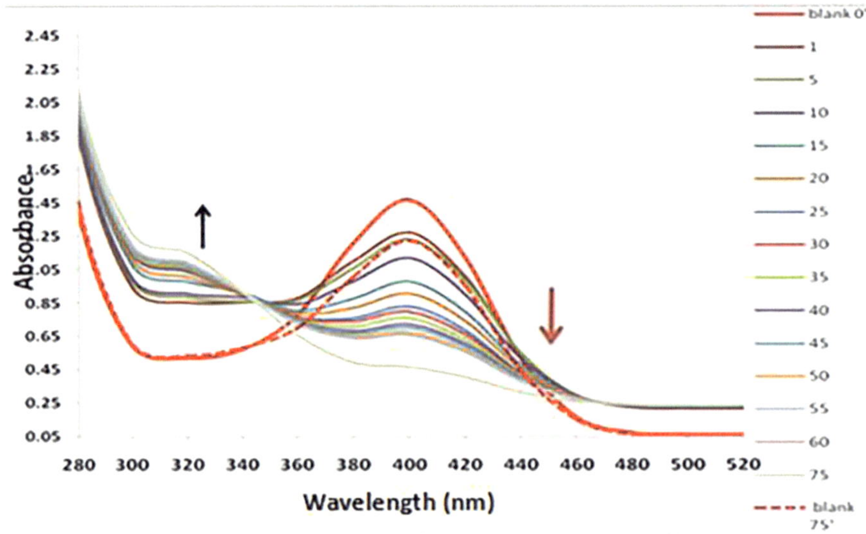

Figure 6.
The UV–vis spectrum of 4-nitrophenol reduction catalyzed by AgNPs using $NaBH_4$.

Sr. No	Name of organism	Diameter of zone of inhibition (mm)
1.	*Pseudomonas aeruginosa* NCIM 5029	14
2.	*Serratia marcescens* NCIM 2078	18
3.	*Staphylococcus aureus* NCIM 5021	22
4.	*Salmonella typhimurium* NCIM 2501	16
5.	*Klebsiella pneumonia* NCIM 2957	12

Table 1.
Antibacterial activity of Syzygium cumini *AgNPs measured as the zone of inhibition in (mm).*

4. Conclusion

The *Syzygium cumini* leaf extract reduced Ag^+ metal ions and led to the formation of AgNPs with fairly well-defined dimensions. This green approach for the synthesis of AgNPs has many advantages, such as the simplicity with which the process can be commercialized. The spherical AgNPs with unusual shapes like nano prism, hexagons and trapezoids were synthesized. The synthesized AgNPs showed excellent catalytic properties in a reduction reaction of 4-nitrophenol to 4-aminophenol by $NaBH_4$ in the aqueous phase. Thus, this rapid, eco-friendly and economical route can be used to synthesize AgNPs with wide biotechnological and chemical applications. The anti-microbial screening demonstrated that the synthesized AgNPs had a high inhibitory effect on bacteria. These observations may serve as a guide for studying the controlled release of AgNPs, in the field of controlling infectious diseases.

Conflict of interest

The authors of this have declared there is no conflict of interest.

Author details

Samadhan R. Waghmode[1*†], Amol A. Dudhane[2†] and Vaibhav P. Mhaindarkar[2]

1 Department of Microbiology, Elphinstone College, Mumbai, India

2 Department of Biotechnology, Lokmangal Biotechnology College, Solapur, India

*Address all correspondence to: samadhanwaghmode@gmail.com

† These authors have contributed equally to this work.

References

[1] Ijaz I, Gilani E, Nazir A, Bukhari A. Detail review on chemical, physical and green synthesis, classification, characterizations and applications of nanoparticles. Green Chemistry Letters and Reviews. 2020;13(3):223-245.

[2] Raj S, Jose S, Sumod US, Sabitha M. Nanotechnology in cosmetics: Opportunities and challenges. Journal of Pharmacy & Bioallied Sciences. 2012;4(3):186.

[3] Khan I, Saeed K, Khan I. Nanoparticles: Properties, applications and toxicities. Arabian Journal of Chemistry. 2019;12(7):908-931.

[4] Zhang XF, Liu ZG, Shen W, Gurunathan S. Silver nanoparticles: Synthesis, characterization, properties, applications, and therapeutic approaches. International Journal of Molecular Sciences. 2016;17(9):1534.

[5] Burdușel AC, Gherasim O, Grumezescu AM, Mogoantă L, Ficai A, Andronescu E. Biomedical applications of silver nanoparticles: An up-to-date overview. Nanomaterials. 2018;8(9):681.

[6] Thiruvengadam M, Rajakumar G, Chung IM. Nanotechnology: Current uses and future applications in the food industry. 3 Biotech. 2018;8(1):1-3.

[7] Talapko J, Matijević T, Juzbašić M, Antolović-Požgain A, Škrlec I. Antibacterial activity of silver and its application in dentistry, Cardiology and Dermatology. Microorganisms. 2020;8(9):1400.

[8] Sriramulu M, Sumathi S. Photocatalytic, antioxidant, antibacterial and anti-inflammatory activity of silver nanoparticles synthesised using forest and edible mushroom. Advances in Natural Sciences: Nanoscience and Nanotechnology. 2017;8(4):045012.

[9] Castillo-Henríquez L, Alfaro-Aguilar K, Ugalde-Álvarez J, Vega-Fernández L, Montes de Oca-Vásquez G, Vega-Baudrit JR. Green Synthesis of Gold and Silver Nanoparticles from Plant Extracts and Their Possible Applications as Antimicrobial Agents in the Agricultural Area. Nanomaterials. 2020;10(9):1763.

[10] Xu L, Wang YY, Huang J, Chen CY, Wang ZX, Xie H. Silver nanoparticles: Synthesis, medical applications and biosafety. Theranostics. 2020;10(20): 8996.

[11] Salem SS, Fouda A. Green synthesis of metallic nanoparticles and their prospective biotechnological applications: An overview. Biological Trace Element Research. 2020;6:1-27.

[12] Pantidos N, Horsfall LE. Biological synthesis of metallic nanoparticles by bacteria, fungi and plants. Journal of Nanomedicine & Nanotechnology. 2014;5(5):1.

[13] Aljabali AA, Akkam Y, Al Zoubi MS, Al-Batayneh KM, Al-Trad B, Abo Alrob O, et al. Synthesis of gold nanoparticles using leaf extract of *Ziziphus zizyphus* and their antimicrobial activity. Nanomaterials. 2018(3):174.

[14] Khandel P, Yadaw RK, Soni DK, Kanwar L, Shahi SK. Biogenesis of metal nanoparticles and their pharmacological applications: Present status and application prospects. Journal of Nanostructure in Chemistry. 2018;8(3):217-254.

[15] Husen A, Siddiqi KS. Phytosynthesis of nanoparticles: Concept, controversy and application. Nanoscale Research Letters. 2014;9(1):1-24.

[16] Ayyanar M, Subash-Babu P. *Syzygium cumini* (L.) Skeels: A review of its phytochemical constituents and traditional uses. Asian Pacific Journal of Tropical Biomedicine. 2012;2(3):240-246.

[17] Srivastava S, Chandra D. Pharmacological potentials of *Syzygium cumini*: A review. Journal of the Science of Food and Agriculture. 2013;93(9): 2084-2093.

[18] Veigas JM, Narayan MS, Laxman PM, Neelwarne B. Chemical nature, stability and bioefficacies of anthocyanins from fruit peel of *Syzygium cumini* Skeels. Food Chemistry. 2007;105(2):619-627.

[19] Ayyanar M, Subash-Babu P, Ignacimuthu S. *Syzygium cumini* (L.) Skeels., a novel therapeutic agent for diabetes: Folk medicinal and pharmacological evidences. Complementary Therapies in Medicine. 2013;21(3):232-243.

[20] Bitencourt PE, Cargnelutti LO, Stein CS, Lautenchleger R, Ferreira LM, Sangoi M, Denardi L, Borges RM, Boligon A, Moresco RN, Cruz L. Nanoparticle formulation increases *Syzygium cumini* antioxidant activity in Candida albicans-infected diabetic rats. Pharmaceutical biology. 2017;55(1): 1082-1088.

[21] Murugan K, Senthilkumar B, Senbagam D, Al-Sohaibani S. Biosynthesis of silver nanoparticles using *Acacia leucophloea* extract and their antibacterial activity. International Journal of Nanomedicine. 2014;9:2431.

[22] Keshari AK, Srivastava A, Chowdhury S, Srivastava R. Green synthesis of silver nanoparticles using *Catharanthus roseus*: Its antioxidant and antibacterial properties. Nanomedicine Research Journal. 2021;6(1):17-27.

[23] Menumerov E, Hughes RA, Neretina S. Catalytic reduction of 4-nitrophenol: A quantitative assessment of the role of dissolved oxygen in determining the induction time. Nano letters. 2016;16(12):7791-7797.

[24] Kumar V, Yadav SC, Yadav SK. *Syzygium cumini* leaf and seed extract mediated biosynthesis of silver nanoparticles and their characterization. Journal of Chemical Technology & Biotechnology. 2010;85(10):1301-1309.

[25] Asghar MA, Zahir E, Asghar MA, Iqbal J, Rehman AA (2020) Facile, one-pot biosynthesis and characterization of iron, copper and silver nanoparticles using *Syzygium cumini* leaf extract: As an effective antimicrobial and aflatoxin B1 adsorption agents. PLoS ONE.2020;15(7): e0234964.

[26] Kumar V, Yadav SC, Yadav SK. *Syzygium cumini* leaf and seed extract mediated biosynthesis of silver nanoparticles and their characterization. Journal of Chemical Technology & Biotechnology. 2010;85(10):1301-1309.

[27] Kästner C, Thünemann AF. Catalytic reduction of 4-nitrophenol using silver nanoparticles with adjustable activity. Langmuir. 2016;32(29):7383-7391.

[28] Ojo OA, Oyinloye BE, Ojo AB, Ajiboye BO, Olayide II, Idowu O, et al. Green-route mediated synthesis of silver nanoparticles (AgNPs) from *Syzygium cumini* (L.) Skeels polyphenolic-rich leaf extracts and investigation of their antimicrobial activity. IET Nanobiotechnology. 2017;12(3):305-310.

[29] Siddiqi KS, Husen A, Rao RA. A review on biosynthesis of silver nanoparticles and their biocidal properties. Journal of Nanobiotechnology. 2018;16(1):1-28.

[30] Prasad R, Swamy VS. Antibacterial activity of silver nanoparticles synthesized by bark extract of *Syzygium cumini*. Journal of Nanoparticles. 2013;2013.

[31] Loo YY, Chieng BW, Nishibuchi M, Radu S. Synthesis of silver nanoparticles by using tea leaf extract from *Camellia sinensis*. International Journal of Nanomedicine. 2012;7:4263.

[32] Kannan RR, Arumugam R, Ramya D, Manivannan K, Anantharaman P. Green synthesis of silver nanoparticles using marine macroalga *Chaetomorpha linum*. Applied Nanoscience. 2013;3:229-233

[33] Ghosh S, Patil S, Ahire M, Kitture R, Gurav DD, Jabgunde AM, et al. *Gnidia glauca* flower extract mediated synthesis of gold nanoparticles and evaluation of its chemocatalytic potential. Journal of Nanobiotechnology. 2012;10(1):1-9.

[34] Gangula A, Podila R, Karanam L, Janardhana C, Rao AM. Catalytic reduction of 4-nitrophenol using biogenic gold and silver nanoparticles derived from *Breynia rhamnoides*. Langmuir. 2011;27(24):15268-15274.

[35] Dudhane AA, Waghmode SR, Bhosale MA, Mhaindarkar VP. *Caesalpinia pulcherrima* mediated green synthesis of silver nanoparticles: Evaluation of their antimicrobial and catalytic activity. International Journal of Nanoparticles. 2017;9(3):153-165.

[36] Dudhane AA, Waghmode SR, Dama LB, Mhaindarkar VP, Sonawane A, Katariya S. Synthesis and characterization of gold nanoparticles using plant extract of *Terminalia arjuna* with antibacterial activity. International Journal of Nanoscience and Nanotechnology. 2019;15(2):75-82.

Chapter 7

Solid Lipid Nanoparticles (SLN)

Rosa-Alejandra Hernández-Esquivel, Gabriela Navarro-Tovar, Elvia Zárate-Hernández and Patricia Aguirre-Bañuelos

Abstract

Solid lipid nanoparticles (SLN) are nanocarriers in the 10–1000 nm range of a solid core, containing both hydrophilic and hydrophobic active pharmaceutical ingredients. SLNs are composed of well-tolerated and biodegradable solid lipids such as mono-, di-, and triglycerides, fatty acids, waxes, and steroids, as well as lipophilic and hydrophilic emulsifying agents. This composition of biocompatible molecules makes SLNs one of the most successful options for the administration of drugs with different routes of administration. To determine its size, morphology, and surface charge, laser diffraction spectroscopy techniques, dynamic light scattering, coulter counter, scanning ion occlusion sensing, and advanced microscopy techniques such as scanning electron microscopy, transmission electron microscopy, and atomic force microscopy are some of the most widely used methods. Surface morphology and length can be measured by electron microscopy, while dynamic light scattering and photon correlation spectroscopy determine particle size and size distribution. In addition, colloidal stability can be determined by zeta potential analysis, indirect measurement of surface charge, and differential scanning calorimetry to characterize particles and drug interactions.

Keywords: nanocarriers, solid lipid nanoparticles, components, preparation method, SLN analysis

1. Introduction

Pharmaceutical nanocarriers comprise nanoparticles, nanospheres, nanocapsules, nanoemulsions, nanoliposomes, and nanoniosomes [1]. Solid lipid nanoparticles (SLN) can be defined as solid lipid colloidal particles composed of organic matter and carbon in a range of 10–1000 nm in which the active pharmaceutical ingredients (API) are dissolved or encapsulated in lipids (**Figure 1**) [2–4]. The underlying mechanism for the formation of lipid nanocarriers is hydrophilic-hydrophobic interactions and van der Waals forces between phospholipids and water molecules [5]. In addition, the physicochemical properties of lipids, such as biocompatibility, low susceptibility to erosion phenomena, and slow water absorption, make lipids an ideal nanocarrier system to improve aqueous solubility and bioavailability of APIs [6, 7].

Some advantages of SLNs as a vehicle for APIs include: covering the bitter taste of the drug in oral administration; maintenance of therapeutic drug

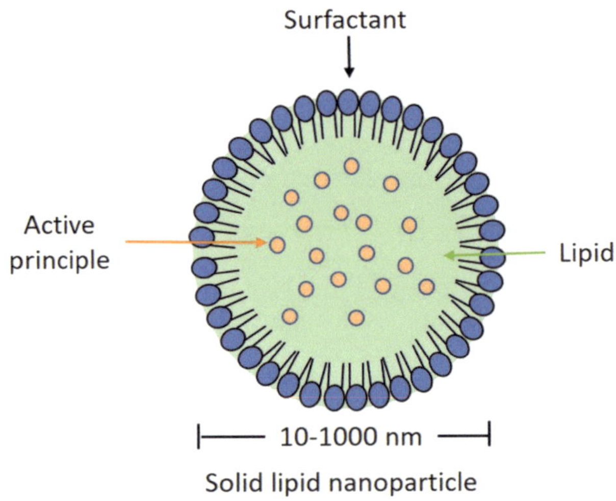

Figure 1.
General diagram of the solid lipid nanoparticles.

concentrations and circulatory time at target sites; protection against premature degradation in the gastrointestinal tract; improved pharmacokinetics, solubility, bioavailability, and stability; reduced toxicity; dose reduction and dose frequency; improvement of patient compliance; and prevention, reduction or delay of the onset of resistance, in addition to the fact that the by-products or metabolites of SLNs are significantly safe and can be easily eliminated through the normal process of excretion [8].

SLNs have been sought as a means to improve the solubility and bioavailability of many drugs, both hydrophilic and lipophilic, especially drugs belonging to class two (high permeability and low solubility drugs) and four (low permeability and low solubility drugs) of the Biopharmaceutical Classification System (BCS) [9].

SLNs have been developed for various applications, including nutraceuticals, cosmetics, pharmaceuticals, and biomedicals, as they can transport a variety of components, including small drug molecules, large biomacromolecules (polysaccharides, etc.), genetic material (DNA/RNA) [10], vaccine antigens [11], antineoplastic [12, 13], antimicrobial [14], they can also be applied in for the targeted delivery of brain medications since enhancing the ability of the drug to penetrate through the blood-brain barrier (BBB) [15].

By focusing on cellular delivery, SLNs can enhance drug delivery to target cells by various mechanisms, such as passive mechanisms that take advantage of the tumor microenvironment, active mechanisms by surface modification of the SLN, and the co-distribution mechanism. SLNs can combine many different drugs and be effective in various types of tumors (i.e., breast, lung, colon, liver, and brain), supporting their potential [16].

SLNs provide several indirect ways to address resistance problems, such as achieving a sustained release profile of a drug, maintaining concentrations within its therapeutic range, and thus avoiding potential adverse effects. Suboptimal levels can promote the selection of resistant bacteria, reduce drug toxicity by encapsulation, permitting higher doses. Promote accumulation in target cells using active targeting, and increase the inhibitory effect (i.e., decrease MIC) on bacterial strains [17].

2. Composition of SLN

SLNs are composed of lipids, surfactants, co-surfactants, and the API to be encapsulated. In addition, coating materials, antioxidants, preservatives, adhesives, viscosity enhancers, absorption enhancing agents, and other excipients can also be used.

SLNs are composed of approximately 0.1–30 (% w/w) solid fat dispersed in an aqueous phase. Surfactants are used in approximately 0aboutentrations to improve stability [18, 19].

2.1 Lipids

Lipids are the main component of the formulation and determine the stability, release, encapsulation, and loading of any API. Therefore, the selection of lipids is the key to a successful formulation, biocompatible/physiological and biodegradable lipids, as well as those that are in the classification of generally recognized as safe (GRAS), with a melting point more significant than 40°C to guarantee a solid state at room temperature and also at body temperature, are the ones of choice for the formulation of SLNs, which reduces the danger of toxicity [20].

GRAS, for its acronym in English "Generally Recognized as Safe," is a designation of the Food and Drug Administration of the United States (FDA) that a chemical or substance added to food is considered safe by experts, therefore which is exempt from the Federal Food, Drug, and Cosmetic Act (FFDCA) food additive tolerance requirements [21].

The melting point of lipids depends on their chemical structure, degree of crystallinity, and polymorphic variation. For example, the melting temperatures of saturated fatty acids and triglycerides increase in proportion to the number of C atoms. In contrast, the melting temperatures of unsaturated fatty acids decrease with increasing carbon atoms and double bonds. The polymorphic variation of lipids also determines physical characteristics such as melting and solidifying properties, morphology, and aggregation of large fat crystals and emulsions.

Selection of a solid lipid or lipid blend relevant to SLN generally depends on several factors: (i) the ability to produce particles in the submicron range, (ii) biodegradability, (iii) biocompatibility, (iv) carrying the capacity adequate drug, and (v) storage stability [22].

One of the most important considerations is the polarity of the drug to be used. Ideally, it is recommended that the lipid phase be considerably lipophilic so that lipophilic drugs can easily be incorporated and solubilized in it. Another parameter, which plays a crucial role, is the lipid's viscosity and contact angle (or a mixture of lipids) with the aqueous solution. Highly viscous lipids are challenging to work with and require higher energy for sonication, which is used to form nanoparticles [23].

The use of relatively low melting point solid lipids and the increase in the oil content in the lipophilic phase of the nanoparticle dispersion reduce the viscosity of the molten droplets during homogenization; this is how nanoparticles are formed in smaller sizes [24]. The average particle size of the SLN dispersion increases when higher melting lipids are used. The main technical point behind this phenomenon is a higher viscosity of the dispersed phase [25].

The degree of crystallinity and polymorphic modification of lipids is another factor that influences the properties of a lipid nanoparticle system because they affect the incorporation of the drug (drug loading efficiency, state, and location of the drug in the nanoparticles) and dictate its releasing properties [26, 27].

2.1.1 Polymorphism and crystallinity of lipids

Polymorphism can exist in more than one crystalline form due to the different lattice arrangements of the molecules. For example, the low-fusion lipids used in the preparation of SLN can exist in various polymorphic forms. The differences between the polymorphic forms are due to the packing of the hydrocarbon chain, the inclination of the chains against the plane of the methyl end group, and the differences in the region of the methyl end group. The predominant forms of triglycerides are polymorphs α and polymorphs β. Polymorphs α tend to convert to β' and finally to polymorphs β, with a narrow chain arrangement and packaging. Polymorph β is considered the most stable, whereas polymorph β' is metastable.

The lipids that make up the inner layer of colloidal lipid systems exhibit complex crystalline behavior. The crystallization of these materials occurs during the cooling process and can continue in the storage stage. The crystalline behavior of the solid lipid in the particles is determined by the composition and the droplet size of the molten lipids. Therefore, depending on the design and the preparation process, the lipids of the internal structure of the particles can have various conformations, such as liquid crystals, gels, or crystalline lamellar phases [28].

To avoid drug shedding due to polymorphic transitions, the use of complex lipids, such as long fatty acid chains, is the ideal choice to improve long-term stability and increase the number of drugs that can be encapsulated. These chains increase the average size of the particle, so the lipid nanoparticle formulation consists of a combination of long and short-chain fatty acids.

Among the most used lipids are triglycerides such as tristearin, tripalmitin, trilaurin; greases such as the Witepsol® series; acylglycerols such as glyceryl behenate; waxes such as cetyl palmitate; fatty acids with different hydrocarbon chain lengths such as stearic or palmitic acid. Cationic lipids, such as stearylamine, can improve drug penetration since ionic interactions are created between positively charged parts of the molecule and negatively charged cells, promoting better cellular internalization [27–29]. This increases the residence time on the surface and improves the drug's bioavailability [27]. **Table 1** shows some of the lipids used in the preparation of SLNs.

2.2 Surfactants

In general, in the preparation of lipid nanoparticles, surfactants play two critical roles: the dispersion of the lipid melt in the aqueous phase and the stabilization of the lipid nanoparticles in dispersions after cooling [30]. The main aspects to consider when using surfactants in the preparation of SLN are their safety, compatibility with other excipients, ability to produce the desired size with the minimum amount consumed, and also provide sufficient stability to the SLNs, covering their surfaces [25]. The surfactants used to produce these carriers can improve epithelial permeability (e.g., disrupt the cell membrane) and, therefore, overcome limitations in drug absorption [26].

On the other hand, the surfactants that surround the particles, in addition to guaranteeing their steric stability in aqueous dispersion, induce specific surface chemical properties and can also modulate the biopharmaceutical profile [30]. For selecting the best surfactant, several parameters must be considered: hydrophilic-lipophilic balance (HLB) values, its effect on lipid polymorphism, and particle size. HLB values for stabilizing oil dispersions in water vary between 8 and 18. The correct choice of surfactant minimizes the risk of producing particle aggregates that can compromise

Fatty acids	Lauric acid, myristic acid, palmitic acid, behenic acid, stearic acid, arachidic acid, oleic acid, linoleic acid
Waxes	Cetyl palmitate, beeswax, carnauba wax
Fatty alcohols	Dodecanol, stearyl alcohol, oleyl alcohol, cetyl alcohol, myristyl alcohol, lauryl alcohol
Triacylglycerides	Tripalmitin, trimyristin, triolein, tricaprylate, tristearin, trilaurin, medium-chain triglycerides (MCT)
Acylglycerols	Glyceryl monostearate, glyceryl distearate, glyceryl behenate (compritol), glyceryl palmitostearate, hydrogenated coco-glycerides, hydrogenated palm kernel glycerides
Cyclic complexes	Cyclodextrin, para-acyl-calix-arenes
Hard fat	Witepsol H 35, Witepsol W 35, Witepsol H 42, Witepsol E 85
Cationic lipids	Stearylamine (SA), benzalkonium chloride (alkyl dimethyl benzyl ammonium chloride, BA), cetrimide (tetradecyl trimethylammonium bromide, CTAB), cetylpyridinium chloride (hexadecylpyridinium chloride, CPC), dimethyl dioctadecyl ammonium bromide [DDAB], n-[1-(2,3-dioleoyloxy)propyl]-n,n,n-trimethylammonium chloride (DOTAP)
Others	Castor oil, hydrogenated castor oil, soybean oil, peanut oil, fully hydrogenated soybean oil, cocoa butter, anhydrous milk fat, hydrogenated palm oil, goat fat

Table 1.
Lipids used in the preparation of SLN.

the stability of the distribution in vitro and its performance in vivo [20]. **Table 2** shows some of the surfactants used in the preparation of SLNs.

Surfactants can be classified into three classes according to their electrical charge: ionic, non-ionic, and amphoteric. Ionic surfactants confer electrostatic stability, while non-ionic surfactants confer steric repulsion stability; amphoteric surfactants have negative and positively charged functional groups, so they exhibit characteristics of a cationic and anionic surfactant under low and high pH conditions, respectively [27]. The toxicity of a surfactant is an important consideration, and not all surfactants can be used to prepare all types of SLNs. The surfactants arranged in decreasing order of toxicity are: cationic, anionic, non-ionic, amphoteric.

2.3 Other agents

In addition to lipids and surfactants, lipid nanoparticle formulations may also contain other ingredients, including cryoprotectants used in SLN drying techniques such as lyophilization, spray drying, and charge modifiers and the surface. Adapting the surface of lipid nanoparticles with surface modifiers such as hydrophilic polymers can reduce their absorption by the reticuloendothelial system (RES) [27]. The rapid absorption of SLNs can be prevented by coating with a biocompatible polymer such as poly (ethylene) glycol (PEG), which can increase blood circulation time [31]. In **Table 3**, some of the charge and surface modifiers used in the preparation of lipid nanoparticles are listed.

3. Structure of solid lipid nanoparticles

As mentioned above, SLNs are composed of a lipid substrate surrounded by stabilizers and one or more active molecules incorporated into these particles.

Surfactant/emulsifiers	Amphoteric surfactants	Phospholipids: egg phosphatidylcholine, soybean phosphatidylcholine, hydrogenated egg phosphatidylcholine, hydrogenated soybean phosphatidylcholine, egg phospholipid, soybean phospholipid Steroids: cholesterol, cholesteryl oleate
	Non-ionic surfactants	Polyethylene glycol / polyoxypropylene copolymers: poloxamer 188, poloxamer 182, poloxamer 407, poloxamine 908 Polyoxyethylene sorbitan copolymers: polysorbate 20, polysorbate 60, polysorbate 80, polysorbate 85, sorbitan monooleate Polyoxyethylene alkyl/aryl ethers: tyloxapol, polyoxyethylene-20-cetyl ether, polyoxyethylene-20-isohexadecyl ether, polyoxyethylene-20-oleyl ether Others: polyglyceryl-6 distearate, polyglyceryl-3 methyl glucose distearate, caprylic/capric triglycerides of PEG, macrogol [15]—hydroxy stearate, polyoxyethylene glyceryl monostearate
	Anionic surfactants	Sodium lauryl sulfate Bile salts: sodium dihydrofolate, sodium taurocholate, sodium cholate, sodium glycocholate, sodium taurodeoxycholate
	Cationic surfactants	Dimethyldioctadecylammonium bromide, cetrimonium bromide, 1,2-dioleoyl-3-trimethylammonium-propane (DOTAP), chlorhexidine salts, DOTMA
Co-surfactant	Alcohols	Butanol, ethanol, polyvinyl alcohol (PVA)
	Others	Diethylene glycol monoethyl ether, low molecular weight PEG, propylene glycol, sorbitan monostearate, butyric acid, sodium dioctyl sulfosuccinate, sodium monooctyl phosphoric acid

Table 2.
Surfactants used in the preparation of SLN.

Preservative	**Thiomersal**
Cryoprotective	Gelatin, glucose, fructose, sorbitol, mannose, maltose, lactose, mannitol, glycine, polyvinyl alcohol, polyvinyl pyrrolidone, etc.
Load modifiers	Organic salts: mono-octyl phosphate, mono-hexadecyl phosphate, mono-decyl phosphate, sodium hexadecyl phosphate, ionic polymers, sodium dextran sulfate salt Others: dipalmitoylphosphatidylcholine (DPPC), stearylamine, diacetyl phosphate, dimyristoyl phosphatidylglycerol (DMPG)
Surface modifiers	Dipalmitoyl-phosphatidyl-ethanolamine conjugated with polyethylene glycol 2000 (DPPE-PEG2000), distearoyl-phosphatidyl-ethanolamine-n-poly (ethylene glycol) 2000 (DSPE-PEG2000), stearic acid-PEG 2000 (SA-PEG2000), α-methoxy-PEG 2000-carboxylic acid-α -lipoamino acids (mPEG2000-C-LAA18), α-methoxy-PEG 5000-carboxylic acid-α-lipoamino acids (mPEG5000-C-LAA18)

Table 3.
Miscellaneous components used in the preparation of SLN.

Thus, the structure of a particle consists of an outer layer that determines its surface properties and an inner layer or core material that determines factors such as the size, shape, and location of the active ingredient [32]. Therefore, how the lipid substrates stabilizers and active components that make up the colloidal lipid system are organized and interconnected determines their behavior as drug transporters and is strongly influenced, among other factors, and the crystalline behavior of lipids used.

It has been reported that under heat treatment or during storage, polymorphic transformations can occur, and the particles can change from spherical shapes to platelets. Platelet forms are associated with the stable polymorphic state β for the solid lipid of the matrix, and spheroidal or disc-shaped records are associated with the polymorphic state α [33].

In general, two structure models have been proposed for SLN, in which it is assumed that the particles are surrounded by the surfactant that forms the outer layer. The first is that the molten lipid droplets solidify during cooling, maintaining their spherical shape. On the other hand, the second model proposes that the cooling of molten lipids produces a flat laminar structure with surfaces structured by folds, edges, and steps that occur during the recrystallization process and where the lipid structure from polymorph α to polymorph β comes from stable [28].

3.1 API localization within SLNs

The location of the active molecule depends on the structural organization of the colloidal lipid systems; in this way, from the model corresponding to the basic spherical model for SLN, three general and commonly used classifications emerge: (i) active molecule is homogeneously distributed along the length of the particle structure (homogeneous matrix); (ii) the API is concentrated within the particle and is surrounded by the lipid matrix in a core-shell structure (enriched core); and (iii) the API is concentrated on the surface of the particle (enriched coat) (**Figure 2**).

In general, the higher temperature of the aqueous medium facilitates more excellent solubility of the drug-containing them. Considering that in most studies, the temperature used to prepare the granules is typically at least 10° above the melting point of solid lipids, this operating condition facilitates a substantial amount of drug in phase. Then during cooling, drug solubility decreases, the aqueous phase is supersaturated, and in theory, the active molecules should migrate to the lipid substrate due to their lipophilic properties.

The outer shell or enriched shell structure can be obtained when solid lipids have a high melting point and a high crystallization temperature, which leads them to crystallize before the active molecule during the nanoemulsion system O/W hot cooling process housing the active molecule on the surface of the particle. This concentrated outer shell layer shows an explosion effect on drug release, so this model is unsuitable for prolonged drug release [34].

Conversely, suppose the use of fat has a low melting point or crystallization point. In that case, high temperature may remain as a supercooled liquid or in a metastatic crystalline form where more particles are present at high temperatures. Active elements can only be contained in rotation. Instead, it can promote nucleation and thus be encapsulated within the seed (core-shell structure) or uniformly dispersed throughout the grain structure, mainly when homogenization is used at high cold pressure.

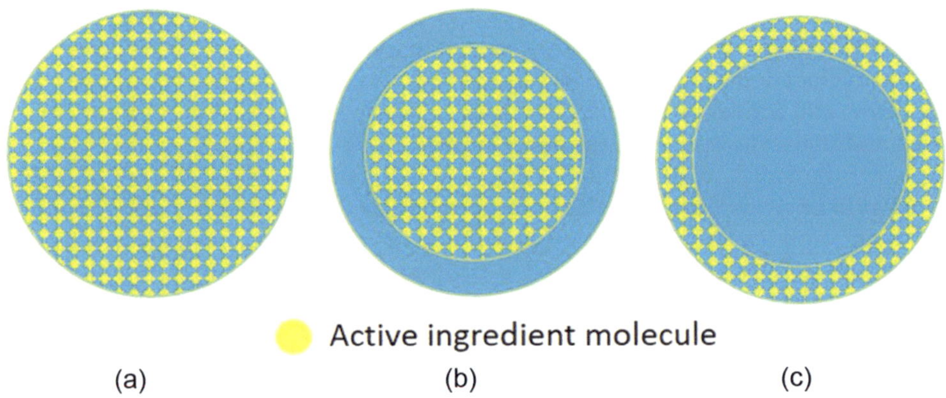

Figure 2.
Structure of solid lipid nanoparticles. Models of drug incorporation of solid lipid nanoparticles. (a) Homogeneous matrix model; (b) rich core model; (c) enriched roof model.

The homogeneous matrix model can be obtained using the cold homogenization method and by incorporating lipophilic active molecules into the SLNs with the hot homogenization method. This solidified structure relies on its solid state to provide a uniform drug distribution in SLN [35, 36].

In the core-shell or enriched core structure, the concentration of the active molecule is close to the saturation point in lipid fusion; when this highly concentrated lipid is cooled, the solubility of the active molecule in lipid fusion is reduced, and it is deposited in the center of the SLN, forming a nucleus enriched with the active molecule. This type of structure leads to a sustained release profile [37, 38].

4. SLN preparation methods

There are multiple methodologies for the preparation of SLN; the selected technique will depend on the physicochemical properties of the drug about the lipid matrix, the route of administration, among other parameters. The methods used for the preparation of SLN fall broadly into three main categories: (i) high-energy methods for dispersion of the lipid phase (such as high-pressure homogenization), (ii) low-energy methods where requires the precipitation of nanoparticles from homogeneous systems (such as microemulsions), and (iii) methods based on organic solvents.

4.1 High energy methods

High-energy-based methods are those that, in general, require the use of equipment capable of generating high shear forces, pressure distortions, or any other mechanism to achieve particle size reduction. In an emulsion, the mechanical energy required exceeds the interfacial energy by several orders of magnitude, thus requiring high energy to form submicron drops. The high-energy process uses intense mechanical force, resulting in large interfacial areas to form nanoscale emulsions [39]. The energy barrier for droplet fragmentation is highly dependent on the interfacial tension: the higher the interfacial tension, the more energy must be supplied to obtain tiny emulsion droplets. The interfacial energy and the curvature of the drop interface give rise to a pressure difference between the inside of the drop and the outside, called Laplace pressure. This pressure will act as a resistance to whatever stress is applied, so the reductions are expected to break approximately at the point when the applied stress equals or exceeds the Laplace pressure. The nanoemulsion droplets have a greater thickness due to the adsorption layer of the emulsifier concerning the radius of the droplet, which makes them more stable towards coalescence [26, 38]. In general, the high-energy process is followed by two steps: (i) the deformation and disruption of macro droplets into the smallest droplets; (ii) the adsorption of surfactant at its interface (to ensure steric stabilization) [40].

4.1.1 *High-pressure homogenization (HPH)*

Depending on the temperature used for the production of SLN, this technique can be classified into hot homogenization and cold homogenization. The advantage associated with this method is that SLNs are obtained with small particle sizes and high entrapment efficiency. High-pressure homogenization, the molten lipid is pumped through a narrow space with a 500–5000 bar pressure at high speed. Generally, 5–10% lipid content is used, but up to 40% lipid content has also been investigated. Two general approaches to HPH are hot homogenization and cold homogenization [41–43].

4.1.1.1 *Hot homogenization*

In the hot homogenization method, the procedure is carried out at temperatures above the melting temperature of the lipid. Here, the lipid and the drug are fused and combined with an aqueous surfactant at the same temperature. By using the high shear device, a hot pre-emulsion is formed. The hot colloidal emulsion droplets are recrystallized by cooling the emulsion to room temperature to generate the SLNs. In general, higher temperatures result in smaller particle sizes due to the decrease in the viscosity of the internal phase. However, high temperatures can also increase the rate of degradation of the drug and vehicle. In most cases, 3–5 homogenization cycles at 500–1500 bar are sufficient [41, 42]. Increasing the number of cycles or the homogenization pressure often increases particle size, inducing coalescence due to its high kinetic energy. Particle sizes < 500 nm can be obtained [42].

4.1.1.2 *Cold homogenization*

In this technique, the lipid melt containing the drug is cooled. Solid lipids are ground into lipid microparticles. These lipid microparticles are dispersed in a cold surfactant solution producing pre-excitation. This hypothetical process is then homogenized at room temperature or below; gravity is strong enough to reak lipid microparticles directly into SLNs [41, 42]. The particle sizes achieved by this technique are generally in the range of 50–1000 nm [41].

4.1.2 *Ultrasonic/high-speed homogenization*

This technique is based on the use of ultrasound waves which create cavitation phenomena that include the formation, growth, and implosive collapse of microbubbles/cavities in the medium; It consists of very high temperatures up to 5000 K and pressures up to 1000 bar. The particle size of the SLNs in this technique depends on the stirring speed, the emulsion time, and the cooling temperature, reaching heights less than 100 nm [44].

4.1.3 Supercritical fluid-based method

A supercritical fluid (SCF) increases the dissolving capacity of compounds, and fluid becomes a supercritical fluid when its temperature and pressure exceed its critical values. Supercritical fluid technology has the unique property of producing solids with a small size and irregular morphology. Supercritical fluids (SCF) have unique properties, such as high diffusivity, low viscosity, and high compressibility. Supercritical CO_2 (SC-CO_2) is the most common SCF because it is non-toxic, non-flammable, easy to obtain, and easily accessible. SLNs can be formulated by five main (SCF) methods: (i) rapid expansion of supercritical solutions (RESS), (ii) particles of gas-saturated solutions/suspensions (PGSS), (iii) supercritical fluid extraction of emulsions (SFEE) [45, 46].

4.2 Low energy methods

Low energy methods do not consume significant energy to achieve particle size reduction, and some even proceed spontaneously. These inferior energy methods are based on the properties of the system and its complex interfacial hydrodynamic mechanisms. The chemical energy released during emulsification is believed to occur as a consequence of the change in spontaneous curvature of surfactant molecules from negative to positive (o/w) or from positive to negative (w/o) [47].

Low energy techniques are classified into thermal or isothermal methods. Emulsion formation due to temperature-dependent changes in surfactant properties is typical of thermal processes, while emulsion formation due to continuous temperature changes is typical of the isothermal method. Spontaneous emulsion (self-emulsifying) and inverting emulsions are included among the isothermal techniques. Spontaneous emulsion involves adding an oily surfactant mixture to the water, whereas, for emulsion phase inversion, the emulsion is formed when water is added to the oily surfactant mixture. An emulsion is formed through the phase inversion temperature method when the temperature of the oil, water, and surfactant mixture rapidly drops below the phase inversion temperature under continuous mixing. These methods have the advantage of producing tiny droplets without specialized equipment. Therefore they are cost-effective and easy to use [48]. The following variables influence the spontaneity of the emulsification process: structure of the surfactant, concentration and initial location, composition of the oil phase, addition of co-surfactant and non-aqueous solvent, and salinity and temperature [49].

4.2.1 Microemulsion based method

The IUPAC defines the microemulsion as a dispersion made of water, oil, and surfactant (s) anisotropic and thermodynamically stable system with a dispersed domain diameter ranging from approximately 1–100 nm, generally 10–50 nm [50].

4.2.1.1 Hot microemulsion technique/microemulsion dilution technique

In this method, the microemulsion forms spontaneously due to the high surfactant-lipid ratio. This method is simple and includes some common steps. Initially, the fat is melted and mixed with a hot surfactant solution. Light shaking is carried out until a microemulsion is formed. In the second step, the hot microemulsion is dispersed in a large amount of cold water (2–10°C) with moderate stirring [51]. Temperature gradients facilitate the rapid crystallization of lipids and prevent aggregation. Due to the dilution step (1:25–1:50), achievable lipid contents are considerably lower than HPH-based formulations [52]. The particle size of the nanoparticles obtained varies from 50 to 800 nm [53].

4.2.1.2 Microemulsion cooling technique

The method consists of the preparation of an o/w microemulsion in which the emulsified wax melts at 37–55°C and the addition of water heated to the same temperature with minimal agitation to form a suspension homogenous milky white, after adding a specific amount of a pharmaceutically appropriate macromolecular surfactant to the water a stale and transparent microemulsion is produced in the form of a liquid matrix. This o/w microemulsion is further cooled to room temperature or 4°C to precipitate SLNs from it. This method is reproducible, simple, and easy to scale [54]. This method works at a moderate temperature, requires less time and obtains high drug entrapment. The particle size of the nanoparticles received varies from 50 to 300 nm [42, 54].

4.2.2 Double emulsion method

The double emulsion technique is one of the most used techniques to prepare nanoparticles encapsulated with hydrophilic active ingredients using stabilizers or surfactants. This method is also known as the multiple emulsion method, where it has three basic steps: (i) formation of the water-in-oil emulsion or reverse emulsion, (ii) addition of the W1/O emulsion in the aqueous surfactant solution to form a W1/O/W2 emulsion with continuous stirring (sonication or homogenization) and (iii) evaporation of the solvent or filtration of the multiple emulsion to form the nanoparticles [41].

4.2.3 Phase inversion method by temperature (PIT)

The temperature-induced phase inversion technique involves performing, using heating and cooling cycles, successive phase inversions of an O/W emulsion to a W/O emulsion until finally an emulsion is obtained. O/W correlation, where each inversion conducts to reduce the size of the droplet so that nanoparticles can be obtained with a final dilution step in cold water [36, 37]. This method brings SLNs of 30–100 nm in diameter.

4.2.4 Membrane contactor method

In this method, the lipid phase is pressed through the membrane's pores and allows the formation of tiny droplets at the outlets of the pores, which are carried away by the circulating water. SLNs are obtained after cooling the preparation to room temperature [55]. This method can achieve particle sizes of 100–200 nm.

4.2.5 Coacervation technique

The coacervation method is based on the precipitation of free fatty acids from their micelles in the presence of a surfactant. In this process, a fatty acid salt is uniformly dispersed in the stabilizer solution. The mixture was heated to the Krafft point of the fatty acid salt and stirred continuously until a clear solution was obtained. Subsequently, an ethanolic solution of the API is slowly added under continuous stirring to get a single phase. Then, a coacervation agent or an acidifying solution is added to obtain the nanoparticle suspension [54]. The particle size of SLN depends on the concentration of the micellar solution and the degree of polymer used for stabilization and can produce a particle size of 260–500 nm [56].

4.2.6 Organic solvent free dual emulsion/melt dispersion technique

In this technique, the lipid phase is heated above its melting point and dispersed in water using a low HLB surfactant. Again, an aqueous solution of surfactant with a high HLB content is added to the emulsion without thus forming; this double w/o/w emulsion is poured into cold water with gentle agitation to promote the formation of SLN [57].

4.3 Organic solvent approaches

4.3.1 Solvent evaporation emulsion method

The solvent evaporation emulsion (SEE) method has three basic steps for preparing nanoparticles. In step (I), lipid material is added to a known volume of organic solvent (immiscible in water) and suitably mixed to produce a clear homogeneous lipid solution. In step (II), the solution prepared above is added to the correct volume of a hot aqueous solution containing surfactant above the melting point of lipids to form a thick emulsion using a high-speed homogenizer. The nanoemulsion is then obtained in step (III) using a high-pressure homogenizer, which converts the coarse emulsion into a nanoemulsion due to the high pressure. Nanodispersion is formed after evaporation of the organic solvent since the lipid material will precipitate in the water. The lipids precipitated in an aqueous medium are separated by filtration through the sintered disk filter funnel. The nanoparticles prepared by this strategy are nano-sized, not flocculated (single entity), and have a high trapping efficiency [58]. This method can achieve 30–500 nm particle sizes.

4.3.2 Solvent emulsion and diffusion method

The method is based on the organic phase's initial saturation and thermodynamic equilibrium with a stabilizer containing an aqueous phase. The drug is dissolved with the help of a homogenizer in the saturated solution formed; in the next step, an o/w emulsion is produced by dispersing it in an aqueous solution with an emulsifier. Diffusion of the solvent into water is facilitated by adding more water in an appropriate ratio to the emulsion under moderate magnetic stirring, which leads to nanoprecipitation and the formation of SLNs. Particles with an average diameter of 30,100 nm can be obtained with this technique [59].

4.3.3 Solvent injection technique (or solvent displacement)

In this technique, the lipid and the active ingredient are dissolved in a water-miscible solvent. The mixture is then dispersed in an aqueous solution of a surfactant with gentle mechanical agitation, producing a suspension of lipid nanoparticles; the solvent is subsequently removed. The particle size depends on the speed of the distribution process. Higher speed makes smaller particles. More lipophilic solvents give larger particles that can become a problem. The strategy offers advantages, for example, low temperatures, low cutting pressure, easy handling, and a fast production process without really advanced equipment (for example, high-weight homogenizer) [60]. The particle size of the nanoparticles obtained varies from 100 to 500 nm [53].

5. SLN characterization

The typical composition of the SLN identifies the following characteristics of great importance: (i) particle size and particle size distribution; (ii) surface morphology, functionalization, and zeta potential; (iii) structure, depending on the degree of crystallinity, lipid variability, drug binding and carrying capacity; (iv) drug release; (v) dynamic processes and the general existence of other nano and microstructures; and (vi) toxicity assessment according to the manufacturing process, lipids, and excipients used, their effect on drug toxicity and route of administration. In **Table 4**, some parameters and techniques used to characterize SLNs will be described later.

5.1 Particle size and surface charge

The particle size range and concentrations reported for SLNs are wide. However, in vitro studies have shown that SLNs are acceptable at concentrations <1 mg/mL (total lipids). They can be less tolerated with a particle diameter > 500 nm, explained by their aggregation [20]. A decrease in the size of the nanoparticles provides easier absorption and leads to a significant increase in the rate of cellular absorption. For example, a particle size greater than 300 nm provides a sustained delivery of drugs; in this case, when the size range is 50–300 nm, rapid action is shown [24].

Particle size is a critical attribute of lipid nanocarriers, affecting stability, encapsulation efficiency, drug release profile, biodistribution, mucoadhesion, and cell uptake.

Regarding the particle size distribution characterization, the "polydispersity index" (PDI) defines the size range of lipid nanocarrier systems. The term "polydispersity" (or "dispersion" as recommended by IUPAC) is used to describe the degree of non-uniformity of a particle size distribution. Also known as the heterogeneity index, PDI is calculated from a two-parameter fit to the correlation data (cumulative analysis). This index is dimensionless and is scaled so that values less than 0.05 are mainly seen with highly monodisperse standards, while values greater than 0 [5].

The choices of a specific technique depend on several parameters, such as the expected size and distribution of the nanoparticles. Since the diameter of the particles affects the release of encapsulated agents, the smaller particles provide a larger surface area. Commonly used particle size analysis techniques include laser diffraction (LD), photo relation spectroscopy (PCS) or dynamic light scattering (DLS), colter counter, scanning ion occlusion sensing (SIOS), flow field fractionation (FFF), and nanoparticle tracking analysis (NTA).

On the other hand, the electrokinetic behavior defined by the zeta potential provides information on the magnitude of the charge on the surface of the particles in aqueous dispersion. It allows predicting the long-term physical stability of the formulations. For electrostatically stable nano-dispersion, a zeta potential value greater than 30 mV is the absolute value required. Still, if the stabilization combines steel and electrostatics, a minimum of 20 mV is enough.

The commonly reported zeta potential values for lipid particles to range from −20 to −40 mV. It should be noted that the zeta potential is determined by the nature of

Parameters	Characterization methods
Particle size	Laser diffraction spectroscopy (LD) Dynamic light scattering (DLS) Coulter counter Scanning ion occlusion sensing (SIOS) Flow field fractionation (FFF) Nanoparticle tracking analysis (NTA) Size exclusion chromatography (SEC)
Surface morphology	Scanning electron microscopy (SEM) Transmission electron microscopy (TEM). Atomic force microscopy (AFM)
Surface load	Zeta potentiometer Laser Doppler anemometry (LDA)
Determination of lipid polymorphisms and crystallinity	Differential scanning calorimetry (DSC) X-ray diffraction (XRD) Small angle X-ray scattering (SAXS) Thermal gravimetric analysis (TGA) Nuclear magnetic resonance spectroscopy (NMR) Infrared (IR) and Raman spectroscopy Electron spin resonance (ESR)
Load capacity and entrapment efficiency	High-performance liquid chromatography (HPLC) UV spectrophotometry Fluorescence spectroscopy
Release profile	In vitro release studies Fluorescence correlation spectroscopy (FCS)

Table 4.
Main parameters for the characterization of solid lipid nanoparticles.

the surface of the particles and is a function of the medium in which the nanoparticles reside and the concentration of the sample used to make the measurement. Therefore, the nature of the solvent, the pH, the ionic strength, and the nature and concentration of the electrolytes in the solution directly affect the magnitude of the zeta potential and, in many cases, the sign of the zeta potential [28].

The surface electrical potential of nanoparticles is essential, especially in formulation science, as it regulates interactions with neighboring particles (including adjacent nanoparticles) and biological systems. It is suggested that the determination of the potential of nanoparticles is carried out in suitable simulated physical solutions/environments where they interact with natural systems [61]. The zeta potential of nanoparticles influences the in vivo fate of SLNs; in general, it is seen that SLNs with positive zeta potential have a long-circulating half-life [23, 24]. In general, the zeta potential value of lipid nanoparticles is estimated from the determination of electrophoretic/electroacoustic mobility, and the LD and DLS techniques allow these determinations.

LD is a valuable technique covering a more comprehensive detection range (0.05–3500 μm) [22]. The results generated by the LD are used to estimate the respective spherical radii of the particles according to the Mie scattering solution (also known as "Mie theory") [26, 61].

The DLS is used to analyze the hydrodynamic diameter of nanoparticles with a size range of 20–600 nm [62]. However, many instruments have an operating range of

0.3 nm–10 µm [61]. The hydrodynamic or stokes diameter is the diameter of a sphere with the same translational diffusion coefficient as the particle being measured, assuming a hydration layer surrounding the particle or molecule [61]. DLS measurements also provide a polydispersity index (PDI).

SIOS is an advancement in the lattice counting method, is a recent tool to measure particle size. This methodology uses dynamically resizable nanopores to detect, quantify, and characterize individual particles in real-time. As the extent of the jam is proportional to the particle size, the exact size of the particles (40 nm–10 µm) can be evaluated after calibration with a known standard [23, 26].

FFF techniques can include simultaneous separation and measurement. These methods are based on the mobile phase's laminar flow action, ensuring separation and perpendicular force fields. FFF techniques are classified according to the type of force field applied, such as cross-flow (Fl), sedimentary (Sd), thermal (Th), electrical (El), magnetic (Mg), and dielectric (Dl). Flow field-flow fractionation (F4) is the most versatile FFF technique. This technique allows the separation of dispersed analytes over a wide range, from nano to micrometer analytes. Three variants of F4 differ in the design of the separation channel: (i) symmetric F4 (SF4), (ii) asymmetric F4 (AF4), and (iii) hollow fiber F4 (HF5). AF4 is the most used among the FFF techniques [26].

NTA is a practical, high-resolution method for measuring nanocarrier samples' size, size distribution, and concentration within a size range of 30–1000 nm. This method allows the scope of monodisperse and polydisperse pieces to be measured. Furthermore, it can calculate the surface charge of lipid-based carriers and detect their fluorescence signals [5].

5.2 Surface morphology

Electron microscopy can provide information on actual particle size, particle size distribution, surface morphology, and structure through its different techniques.

To study nanoparticles with Scanning electron microscopy (SEM), they must first be made into a dry powder, sprinkled on a sample holder, and coated with a layer of conductive metals such as gold, platinum, graphite, osmium, iridium, tungsten, osmium, chromium, or gold/palladium alloys, using a spray coating. Then, a high-energy electron beam is passed through the sample to generate various signals on the surface of the object samples [61]. The main limitation of SEM is the resolution of the images, which restricts its application to models of ~200 nm in size and does not provide any information on the internal structure of the particles [63].

Transmission electron microscopy (TEM) generally provides two-dimensional images of the internal structure of nanoparticles with a resolution of approximately 0.4 nm. The fraction of electrons transferred depends on the electron density of the sample. Therefore, components with differences in electron density appear as regions of different intensities in the image. In contrast, models are generally stained with phosphotungstic acid or uranyl acetate.

SLN formulations have low-melting lipids, and the electron gun used in electron microscopy can cause SLNs to melt, affecting their structure and integrity. These problems can be avoided using an improved technique called cryogenic field emission scanning electron microscopy (cryo-FESEM) [23].

It allows the examination of the nanomechanical properties of each molecule and particle under closer physiological conditions. Atomic Force Microscopy (AFM) is the most significant advantage over TEM/SEM in scanning nanocarriers without special sample preparation. It provides the opportunity for 3D visualization with qualitative and quantitative information on physical qualities such as the size and texture of the surface morphology and roughness. Additionally, a wide variety of particle sizes from 1 nm to 8 μm can be characterized in the same scan [61].

5.3 Determination of lipid polymorphism and crystallinity

Differential Scanning Calorimetry (DSC) is an eminent method based on the measurement of structural modifications of materials in pharmaceutical research that are accompanied by heat exchanges such as heat absorption (involved in melting) or heat emission (involved in crystallization) [4, 61]. Depending on the device's sensitivity, the DSC can scan and account for even microscopic thermal activities in the material and recognize the temperature at which these problems occur. Still, it does not immediately reveal why trouble. Therefore, the exact nature of the thermal transition must be resolved by corresponding methods, such as thermogravimetry, microscopic observations, or X-ray diffraction (XRD) to distinguish the thermal shifts, polymorphic change, melting, loss of water from the hydrate or decomposition [20, 61].

X-ray diffraction (XRD) is a technique in which the scattering of X-rays by the atoms of a crystal creates an interference-effect. The resulting diffraction pattern helps identify and determine its various polymorphic forms' crystal structure and differentiation. According to Bragg's law, it detects fluctuations in electron density on a length scale. It is possible to determine the particle size, shape, crystalline structure of lipid nanoparticles, and changes in the crystalline order of the particles during production [63, 64].

Small Angle X-ray Scattering (SAXS) provides information on nanoparticles' size, shape, and size distribution and the internal structure of disordered and partially ordered systems. It is based on the analysis of the elastic scattering behavior of X-rays when traveling through the material, recording their scattering at small angles [26, 65].

In thermal gravimetric analysis (TGA), we can analyze the melting point, crystallinity, polymorphism, and endothermic and exothermic characteristics of the sample. In this technique, samples are heated at a controlled heating rate under different atmospheres such as nitrogen, oxygen, and argon [6].

Nuclear magnetic resonance (NMR) spectroscopy provides physical, chemical, electronic, and structural information about molecules that characterize their topology, dynamics, and three-dimensional structure in solution in the solid-state [66]. Proton nuclear magnetic resonance (1H NMR) is the most widely used and is particularly suitable for the structural characterization of liquid lipid domains in SLNs. 1H NMR can provide valuable information on the distribution of APIs in GS and the mobility of API molecules incorporated into lipid transporters [26, 67].

Infrared (IR) and Raman spectroscopy are used to explore the structural properties of lipids by relating the frequency of scattered radiation to functional groups of the molecules. Raman spectroscopy allows qualitative (measuring the frequency of scattered radiation) and quantitative (measuring the intensity of scattered radiation) analysis of SLNs [26]. Fourier transform IR spectroscopy (FTIR) is a type of IR spectroscopy commonly used to characterize nanoparticles. Modification of the surface of the particles can be controlled and confirmed using FTIR based on the functional groups present in the material [68].

5.4 Load capacity and entrapment efficiency

The loading capacity (DL) and entrapment efficiency (EE) are critical parameters related to the nanoparticles' dose regimen and efficiency. API loading capacity (DL) represents the relationship between the active ingredient in the particles and the total weight of the particles, while entrainment efficiency (EE), sometimes called encapsulation efficiency, is related to the number of active ingredients. Incorporated into the granules relative to the total amount of active ingredients present in the dispersion. The loading capacity (DL%) and the encapsulation efficiency (EE%) are determined using the following equations:

$$DL(\%) = \frac{amount\ of\ API\ determined\ experimentally\,(mg)}{total\ weight\ of\ SLN\,(mg)} \times 100 \quad (1)$$

$$EE(\%) = \frac{amount\ of\ API\ determined\ experimentally\,(mg)}{amount\ of\ theoretical\ API\ in\ the\ formulation\,(mg)} \times 100 \quad (2)$$

In general, the efficiency of trapping the active ingredients in lipid nanoparticles is usually greater than 70%. Unfortunately, drug-carrying capacity is not reported in most research papers, and reported values fluctuate between 0.1% and 80% [28].

The amount of drug encapsulated in the nanocarriers influences the release kinetics; therefore, it is essential to determine the encapsulation efficiency. EE is estimated after the removal of the free API. Free API is separated by ultracentrifugation, extensive dialysis, ultrafiltration, gel filtration, or centrifugal ultrafiltration. After removing free API, the amount of drug is estimated using a standard analytical technique such as UV spectroscopy or high-performance liquid chromatography (HPLC).

The main aspects that must be taken into account in the discussion on drug entrapment within SLNs are: (i) existence of supercooled melts; (ii) presence of various lipid modifications; (iii) form of lipid nano-dispersions; and (iv) gelation phenomena [66].

5.5 Release of the active ingredient in vitro

The drug release profile is performed analogously to encapsulation efficiency determination assays measured over time intervals to discover the release mechanism.

The release of trapped API from SLNs is governed by the following principles: (i) there is an inverse relationship between API release and drug partition coefficient; (ii) a smaller particle size promotes a greater surface area, which leads to a more significant release of API; (iii) the homogeneous dispersion of the drug in the lipid matrix causes the slow release of the drug; (iv) lipid crystallinity, and high API mobility lead to rapid drug release from SLNs.

In general, the drug release study is determined under controlled shaking and centrifugation. Essentially, drug release from nanomolecular systems occurs through five possible mechanisms for (a) dissociation of the API bound to the outer layer, (b) diffusion through the polymers matrix, (c) membrane-controlled diffusion, (d) erosion of the nanoparticle matrix, or (e) a combination of diffusion and erosion processes [69]. The kinetics of the nanocarrier release pattern could be evaluated using a bi-exponential equation:

$$C = Ae - \alpha t + Be - \beta t \tag{3}$$

where C represents the drug levels within the nanocarriers at time t, A and B are constant; it depends on the properties of the matrix (A means for diffusion control and B for erosion control matrices) and, B is the rate constants that could be determined using a semi-logarithmic representation [61].

The release profile of API trapped in SLNs generally reveals a biphasic pattern with an initial burst effect followed by a prolonged release over several hours or days. Immediate removal of embedded drugs from solid lipid nanoparticles is based on diffusion from the surface of the external particles or matrix erosion induced by hydrolytic degradation. Thus, the distribution of the active substance is gradually released from the lipid core and promotes prolonged release and dissolution. The rate of freedom can be influenced by the nature and composition of the lipid substrate, the choice, and the concentration of surfactants, including technological parameters [70].

Drug release from nano-sized dosage forms can be evaluated using one of the following three categories, separate and sample (SS), continuous flow (CF), and dialysis membrane (DM).

5.6 SLN storage stability

The physical properties of SLNs during prolonged storage can be determined by monitoring changes in zeta potential, particle size, drug content, shape, and viscosity over time. External parameters such as temperature and light seem to be primary importance for long-term stability. Several factors can influence the physical strength of SLN, such as stress conditions (for example, high temperatures, exposure to light, and mechanical stress), the presence of liquid phase and electrolytes, contact with different surfaces, and high concentrations of particles. For example, storage temperatures at 4°C offer a more favorable environment; long-term storage at 20°C did not result in drug-laden SLN aggregation or drug release, while rapid particle size growth was observed at 50°C [29, 31, 71].

Spray drying and lyophilization are used in SLNs to increase or prolong stability, especially for preparations intended for intravenous administration. The aggregation of SLN can be decreased by adding cryoprotectants and obtaining a better redispersion of the dried product. Cryoprotectants favor the vitreous state of the frozen sample and reduce the osmotic activity of water and crystallization by avoiding contact between discrete lipid nanoparticles [72].

5.7 In vitro evaluation of SLN

A promising technique for in situ nanoparticle characterization is Fluorescence Correlation Spectroscopy (FCS). This technique is based on the measurement of fluctuations in the fluorescence intensity of the emissive species that diffuses through a low excitation focal volume. Autocorrelation analysis of the fluorescence intensity in the focal book provides information on the concentration, diffusion constant, and fluorescent particles' brightness. Furthermore, the fluorescence fluctuation spectroscopy analysis of fluorescence intensity fluctuations allows the quantitative analysis of the brightness distribution, making it possible to characterize heterogeneous samples containing assembled molecules. FCS serves as a tool to measure the size and polydispersity of nanoparticles and evaluate their behavior in complex biological media and their stability. Also, in many reports, FCS has been used to characterize crown protein formation on the surface of nanoparticles or the interaction of human serum albumin with liposomes. However, only a few studies have reported the use of FCS to study charge release from nanoparticles [73].

In addition to the physical and chemical characterization of the nanocarriers, their biological responses are also measured in animal cell culture studies before the start of in vivo administration. This translocation of particles is determined mainly by flow cytometry (determines the amount of translocation) and confocal microscopy (determines the location).

The toxicity of nanoparticles is highly dependent on several factors, such as surface properties, coating, structure, size, and aggregation capacity, and these factors can be altered and manipulated in the manufacturing process. Nanoparticles with poor solubility have exhibited more pronounced toxicity [74]. Toxicological studies make it possible to evaluate its toxic effects, identify routes of exposure, and predict the risks of its synthesis or use.

Various in vitro cell culture techniques have been used to analyze nanotoxicity qualitatively and to study nanoparticle uptake (cell uptake assays), localization, and biodistribution. Cell culture assays for (i) cytotoxicity (altered metabolism, decreased growth, lytic or apoptotic cell death), (ii) genotoxicity, (iii) altered gene expression, and (iv) proliferation can be performed to rule out risks associated with nanotoxicity [35].

The cell viability study is the most widely used test for in vitro cytotoxicity evaluation. Other tests also evaluate cells' effect without necessarily leading to cell death. This includes oxidative stress: increased production of reactive oxygen species (ROS), lipid peroxidation, and alteration in the oxidized/reduced glutathione pool or DNA damage [26].

5.8 In vivo evaluation of nanocarriers

After nanocarriers reveal preliminary affectivity in vitro, these carriers further evaluate their toxicity and response profile in biological species. Some in vivo evaluations that can be carried out are: (i) the therapeutic dose-response study; (ii) the biodistribution of nanocarriers among the various organs of the body; (iii) acute and multidose efficacy studies and (iv) studies on safety parameters and pharmacokinetics (ADME processes). The ultimate goal of in vitro and in vivo evaluation is to match the physicochemical aspects of nanocarriers with their biological function [61].

6. Conclusions

SLNs are the most studied lipid-based drug delivery systems, which can deliver drugs and nutrients for various routes of administration due to their biocompatibility, low toxicity, high loading capacity, slow-release rate, and high stability.

The physicochemical properties and stability of lipid nanoparticles depend on the composition of the lipid nanoparticle formulation. The lipid nature of these support systems is one of the key features that has attracted the interest of many researchers. Based on the organization of lipids and drugs in the particles, many structural models of SLNs have been described.

Lipid nanocarriers such as SLNs offer much more flexibility in drug loading, modulation of release, and improved performance in the production of final dosage forms such as creams, tablets, capsules, and injectables. The effort to develop alternative routes and treat other diseases with these systems must continue to expand their applications. Penetration through the gastrointestinal tract and BBB may be a new trend, and combining two therapeutically active agents in a single nanosystem is another consideration for future development.

In conclusion, solid lipid nanoparticles are a promising drug delivery system due to the non-toxicity aspect and a variety of drug-carrying capacity, along with the advantages of drug delivery through all routes of administration.

Author details

Rosa-Alejandra Hernández-Esquivel, Gabriela Navarro-Tovar,
Elvia Zárate-Hernández and Patricia Aguirre-Bañuelos*
Faculty of Chemistry Sciences, Autonomous University of San Luis Potosi, Mexico

*Address all correspondence to: paguirreb@uaslp.mx

References

[1] Kammari R, Das NG, Das SK. Nanoparticulate systems for therapeutic and diagnostic applications. In: Mitra AK, Cholkar K, Mandal A, editors. Emerging Nanotechnologies for Diagnostics, Drug Delivery and Medical Devices. Boston: Elsevier; 2017. pp. 105-144

[2] Sharma M, Sharma D, Sudan P, Landran M. Solid lipid nanoparticles: A Critical review. Paideuma Journal of Research. 2019;**XII**(XII 2019):532-541

[3] Teja VC, Chowdary VH, Raju YP, Surendra N, Vardhan RV, Reddy B. A glimpse on solid lipid nanoparticles as drug delivery systems. Journal of Global Trends in Pharmaceutical Sciences. 2014;**5**(2):1649-1657

[4] Omwoyo WN, Ogutu B, Oloo F, Swai H, Kalombo L, Melariri P, et al. Preparation, characterization, and optimization of primaquine-loaded solid lipid nanoparticles. International Journal of Nanomedicine. 2014;**9**:3865

[5] Danaei M, Dehghankhold M, Ataei S, Hasanzadeh Davarani F, Javanmard R, Dokhani A, et al. Impact of particle size and polydispersity index on the clinical applications of lipidic nanocarrier systems. Pharmaceutics. 2018;**10**(57):17

[6] Kumar R. Chapter 8 - Lipid-based nanoparticles for drug-delivery systems. In: Mohapatra SS, Ranjan S, Dasgupta N, Mishra RK, Thomas S, editors. Micro and Nano Technologies, Nanocarriers for Drug Delivery. Elsevier; 2019. pp. 249-284. ISBN: 9780128140338. DOI: 10.1016/B978-0-12-814033-8.00008-4

[7] Mishra V, Bansal KK, Verma A, Yadav N, Thakur S, Sudhakar K, et al. Solid lipid nanoparticles: Emerging colloidal nano drug delivery systems. Pharmaceutics. 2018;**10**(4):191

[8] Paliwal R, Paliwal SR, Kenwat R, Kurmi BD, Sahu MK. Solid lipid nanoparticles: A review on recent perspectives and patents. Expert Opinion on Therapeutic Patents. 2020;**30**(3):179-194

[9] Poovi G, Damodharan N. Lipid nanoparticles: A challenging approach for oral delivery of BCS Class-II drugs. Future Journal of Pharmaceutical Sciences. 2018;**4**(2):191-205

[10] Ezzati Nazhad Dolatabadi J, Valizadeh H, Hamishehkar H. Solid lipid nanoparticles as efficient drug and gene delivery systems: Recent breakthroughs. Advanced Pharmaceutical Bulletin. 2015;**5**(2):151-159

[11] Anderluzzi G, Lou G, Gallorini S, Brazzoli M, Johnson R, O'Hagan DT, et al. Investigating the impact of delivery system design on the efficacy of self-amplifying RNA vaccines. Vaccine. 2020;**8**(2):212

[12] Chaudhari VS, Hazam PK, Banerjee S. Lipid nanoarchitectonics for natural products delivery in cancer therapy. In: Saneja A, Panda AK, Lichtfouse E, editors. Sustainable Agriculture Reviews 44: Pharmaceutical Technology for Natural Products Delivery Vol 2 Impact of Nanotechnology. Cham: Springer International Publishing; 2020. pp. 169-203

[13] Haider M, Abdin SM, Kamal L, Orive G. Nanostructured lipid carriers for delivery of chemotherapeutics: A review. Pharmaceutics. 2020;**12**(3):288

[14] Yousefi M, Ehsani A, Jafari SM. Lipid-based nano delivery of

antimicrobials to control food-borne bacteria. Advances in Colloid and Interface Science. 2019;**270**:263-277

[15] Pottoo FH, Sharma S, Javed MN, Barkat MA, Harshita, Alam MS, et al. Lipid-based nanoformulations in the treatment of neurological disorders. Drug Metabolism Reviews. 2020;**52**(1):185-204

[16] Bayón-Cordero L, Alkorta I, Arana L. Application of solid lipid nanoparticles to improve the efficiency of anticancer drugs. Nanomaterials. 2019;**9**(3):474

[17] Scioli Montoto S, Muraca G, Ruiz ME. Solid lipid nanoparticles for drug delivery: Pharmacological and biopharmaceutical aspects. Frontiers in Molecular Biosciences. 2020;**7**(319):587997

[18] Naseri N, Valizadeh H, Zakeri-Milani P. Solid lipid nanoparticles and nanostructured lipid carriers: Structure, preparation and application. Advanced Pharmaceutical Bulletin. 2015;**5**(3):305-313

[19] Souto EB, Baldim I, Oliveira WP, Rao R, Yadav N, Gama FM, et al. SLN and NLC for topical, dermal, and transdermal drug delivery. Expert Opinion on Drug Delivery. 2020; **17**(3):357-377

[20] Campos JR, Severino P, Santini A, Silva AM, Shegokar R, Souto SB, et al. Chapter 1 - Solid lipid nanoparticles (SLN): Prediction of toxicity, metabolism, fate and physicochemical properties. In: Shegokar R, editor. Nanopharmaceuticals. Elsevier; 2020. pp. 1-15. ISBN: 9780128177785. DOI: 10.1016/B978-0-12-817778-5.00001-4

[21] U.S. Food & Drug Administration. Generally Recognized as Safe (GRAS). 2019

[22] Deshpande A, Mohamed M, Daftardar SB, Patel M, Boddu SHS, Nesamony J. Solid lipid nanoparticles in drug delivery: Opportunities and challenges. In: Mitra AK, Cholkar K, Mandal A, editors. Emerging Nanotechnologies for Diagnostics, Drug Delivery and Medical Devices. Boston: Elsevier; 2017. pp. 291-330

[23] Kathe N, Henriksen B, Chauhan H. Physicochemical characterization techniques for solid lipid nanoparticles: Principles and limitations. Drug Development and Industrial Pharmacy. 2014;**40**(12):1565-1575

[24] Üner M. Characterization and imaging of solid lipid nanoparticles and nanostructured lipid carriers. In: Aliofkhazraei M, editor. Handbook of Nanoparticles. Cham: Springer International Publishing; 2016. pp. 117-141

[25] Shah MR, Imran M, Ullah S. Chapter 1 - Solid lipid nanoparticles. In: Shah MR, Imran M, Ullah S, editors. Micro and Nano Technologies, Lipid-Based Nanocarriers for Drug Delivery and Diagnosis. William Andrew Publishing; 2017. pp. 1-35. ISBN: 9780323527293. DOI: 10.1016/B978-0-323-52729-3.00001-9

[26] Andonova V, Peneva P. Characterization methods for solid lipid nanoparticles (SLN) and nanostructured lipid carriers (NLC). Current pharmaceutical design. 2017;**23**(43): 6630-6642

[27] Balamurugan K, Chintamani P. Lipid nano particulate drug delivery: An overview of the emerging trend. The Pharma Innovation Journal. 2018;**7**(7):779-789

[28] Gordillo-Galeano A, Mora-Huertas CE. Solid lipid nanoparticles

and nanostructured lipid carriers: A review emphasizing on particle structure and drug release. European Journal of Pharmaceutics and Biopharmaceutics. 2018;**133**:285-308

[29] Surender V, Deepika M. Solid lipid nanoparticles: A comprehensive review. Journal of Chemical and Pharmaceutical Research. 2016;**8**(8):102-114

[30] Rajabi M, Mousa SA. Lipid nanoparticles and their application in nanomedicine. Current Pharmaceutical Biotechnology. 2016;**17**(8):662-672

[31] Geszke-Moritz M, Moritz M. Solid lipid nanoparticles as attractive drug vehicles: Composition, properties and therapeutic strategies. Materials Science and Engineering: C. 2016;**68**:982-994

[32] Bunjes H. Structural properties of solid lipid based colloidal drug delivery systems. Current Opinion in Colloid & Interface Science. 2011;**16**(5):405-411

[33] Schoenitz M, Joseph S, Nitz A, Bunjes H, Scholl S. Controlled polymorphic transformation of continuously crystallized solid lipid nanoparticles in a microstructured device: A feasibility study. European Journal of Pharmaceutics and Biopharmaceutics. 2014;**86**(3):324-331

[34] Shukla T, Upmanyu N, Prakash Pandey S, Gosh D. Chapter 1 - Lipid nanocarriers. In: Grumezescu AM, editor. Lipid Nanocarriers for Drug Targeting. William Andrew Publishing; 2018. pp. 1-47. ISBN 9780128136874. DOI: 10.1016/B978-0-12-813687-4.00001-3. Available from: https://www.sciencedirect.com/science/article/pii/B9780128136874000013

[35] Rawal SU, Patel MM. Chapter 2 - Lipid nanoparticulate systems: Modern versatile drug carriers. In: Grumezescu AM, editor. Lipid Nanocarriers for Drug Targeting. William Andrew Publishing; 2018. pp. 49-138. ISBN: 9780128136874. DOI: 10.1016/B978-0-12-813687-4.00002-5

[36] Shah R, Eldridge D, Palombo E, Harding I. Physicochemical stability. In: Lipid Nanoparticles: Production, Characterization and Stability. SpringerBriefs in Pharmaceutical Science & Drug Development. Springer, Cham; 2015. DOI: 10.1007/978-3-319-10711-0_5

[37] Ganesan P, Narayanasamy D. Lipid nanoparticles: Different preparation techniques, characterization, hurdles, and strategies for the production of solid lipid nanoparticles and nanostructured lipid carriers for oral drug delivery. Sustainable Chemistry and Pharmacy. 2017;**6**:37-56

[38] Ghasemiyeh P, Mohammadi-Samani S. Solid lipid nanoparticles and nanostructured lipid carriers as novel drug delivery systems: Applications, advantages and disadvantages. Research in Pharmaceutical Sciences. 2018;**13**(4):288-303

[39] Che Marzuki NH, Wahab RA, Abdul HM. An overview of nanoemulsion: Concepts of development and cosmeceutical applications. Biotechnology & Biotechnological Equipment. 2019;**33**(1):779-797

[40] Håkansson A, Rayner M. Chapter 5 - General principles of nanoemulsion formation by high-energy mechanical methods. In: Jafari SM, McClements DJ, editors. Nanoemulsions. Academic Press; 2018. pp. 103-139. ISBN: 9780128118382. DOI: 10.1016/B978-0-12-811838-2.00005-9

[41] Jain AK, Thareja S. Solid lipid nanoparticles. In: Bhushan I, Singh K, Tripathi D, editors. Nanomaterials and Environmental Biotechnology.

Nanotechnology in the Life Sciences. Springer, Cham; 2020. DOI: 10.1007/978-3-030-34544-0_13

[42] Singh R. Preparation of solid lipid nanoparticles through various methods using different precursors. Journal of Drug Delivery Therapeutic. 2019;**9**(2):415-419

[43] Yadav KS, Kale K. High pressure homogenizer in pharmaceuticals: Understanding its critical processing parameters and applications. Journal of Pharmaceutical Innovation. 2020;**15**:690-701. DOI: 10.1007/s12247-019-09413-4

[44] Cheaburu-Yilmaz CN, Karasulu HY, Yilmaz O. Chapter 13 - Nanoscaled dispersed systems used in drug-delivery applications. In: Vasile C, editor. Micro and Nano Technologies, Polymeric Nanomaterials in Nanotherapeutics. Elsevier; 2019. pp. 437-468. ISBN: 9780128139325. DOI: 10.1016/B978-0-12-813932-5.00013-3

[45] Akbari Z, Amanlou M, Karimi-Sabet J, Golestani A, Shariaty NM. Application of supercritical fluid technology for preparation of drug loaded solid lipid nanoparticles. International Journal of Nanoscience and Nanotechnology. 2020;**16**(1):13-33

[46] Trucillo P, Campardelli R. Production of solid lipid nanoparticles with a supercritical fluid assisted process. The Journal of Supercritical Fluids. 2019;**143**:16-23

[47] Safaya M, Rotliwala YC. Nanoemulsions: A review on low energy formulation methods, characterization, applications and optimization technique. Materials Today: Proceedings. 2020;**27**:454-459

[48] Chircov C, Grumezescu AM. Chapter 6 - Nanoemulsion preparation, characterization, and application in the field of biomedicine. In: Grumezescu AM, editor. Nanoarchitectonics in Biomedicine. William Andrew Publishing; 2019. pp. 169-188. ISBN: 9780128162002. DOI: 10.1016/B978-0-12-816200-2.00019-0

[49] Li Z, Xu D, Yuan Y, Wu H, Hou J, Kang W, et al. Advances of spontaneous emulsification and its important applications in enhanced oil recovery process. Advances in Colloid and Interface Science. 2020;**277**:102119

[50] Kale SN, Deore SL. Emulsion micro emulsion and nano emulsion: A review. Systematic Reviews in Pharmacy. 2017;**8**(1):39

[51] Battaglia L, Ugazio E. Lipid nano- and microparticles: An overview of patent-related research. Journal of Nanomaterials. 2019;**2019**:2834941

[52] Chandana KV, Gupta NV, Kanna S. Nanostructured lipid carriers: The frontiers in drug delivery. Asian Journal of Pharmaceutical and Clinical Research. 2019;**12**(7). DOI: 10.22159/ajpcr.2019.v12i7.33595

[53] Mahajan PS, Mahajan KB, Darekar A. A review on solid lipid nanoparticle (SLN): An advanced treatment modality. International Journal of Pharmaceutical Sciences Research. 2015;**6**(9):3698-3612

[54] Battaglia L, Muntoni E, Chirio D, Peira E, Annovazzi L, Schiffer D, et al. Solid lipid nanoparticles by coacervation loaded with a methotrexate prodrug: Preliminary study for glioma treatment. Nanomedicine (London, England). 2017;**12**(6):639-656

[55] Bhatt S, Sharma J, Singh M, Saini V. Solid lipid nanoparticles: A promising technology for delivery of poorly water-soluble drugs. Acta

Pharmaceutica Sciencia. 2018;**56**(3). DOI: 10.23893/1307-2080

[56] Blasi P, Giovagnoli S, Schoubben A, Ricci M, Rossi C. Solid lipid nanoparticles for targeted brain drug delivery. Adv Drug Deliv Rev. 2007;**59**(6):454-477. DOI: 10.1016/j.addr.2007.04.011. Epub [22 May 2007]. PMID: 17570559

[57] Becker Peres L, Becker Peres L, de Araújo PHH, Sayer C. Solid lipid nanoparticles for encapsulation of hydrophilic drugs by an organic solvent free double emulsion technique. Colloids and Surfaces B: Biointerfaces. 2016;**140**:317-323

[58] Wang Y, Li P, Truong-Dinh Tran T, Zhang J, Kong L. Manufacturing techniques and surface engineering of polymer based nanoparticles for targeted drug delivery to cancer. Nanomaterials (Basel, Switzerland). 2016;**6**(2):26

[59] Patravale VB, Mirani AG. Preparation and characterization of solid lipid nanoparticles-based gel for topical delivery. In: Weissig V, Elbayoumi T, editors. Pharmaceutical Nanotechnology: Basic Protocols. New York, NY: Springer New York; 2019. pp. 293-302

[60] Qushawy M, Nasr A. Solid lipid nanoparticles (SLNs) as nano drug delivery carriers: Preparation, characterization and application. International Journal of Applied Pharmaceutics. 2020;**12**(1). DOI: 10.22159/ijap.2020v12i1.35312

[61] Jain AK, Thareja S. In vitro and in vivo characterization of pharmaceutical nanocarriers used for drug delivery. Artificial Cells, Nanomedicine, and Biotechnology. 2019;**47**(1):524-539

[62] Langevin D, Raspaud E, Mariot S, Knyazev A, Stocco A, Salonen A, et al. Towards reproducible measurement of nanoparticle size using dynamic light scattering: Important controls and considerations. NanoImpact. 2018;**10**:161-167

[63] Manaia EB, Abuçafy MP, Chiari-Andréo BG, Silva BL, Oshiro Junior JA, Chiavacci LA. Physicochemical characterization of drug nanocarriers. International Journal of Nanomedicine. 2017;**12**:4991-5011

[64] Khosa A, Reddi S, Saha RN. Nanostructured lipid carriers for site-specific drug delivery. Biomedicine & Pharmacotherapy. 2018;**103**:598-613

[65] Pauw BR, Kastner C, Thunemann AF. Nanoparticle size distribution quantification: Results of a small-angle X-ray scattering inter-laboratory comparison. Journal of Applied Crystallography. 2017;**50**(5):1280-1288

[66] Tekade RK, Maheshwari R, Tekade M, Chougule MB. Chapter 8 - Solid lipid nanoparticles for targeting and delivery of drugs and genes. In: Mishra V, Kesharwani P, Mohd Amin MCI, Iyer A, editors. Nanotechnology-Based Approaches for Targeting and Delivery of Drugs and Genes. Academic Press; 2017. pp. 256-286. ISBN: 9780128097175. DOI: 10.1016/B978-0-12-809717-5.00010-5

[67] Lin P-C, Lin S, Wang PC, Sridhar R. Techniques for physicochemical characterization of nanomaterials. Biotechnology Advances. 2014;**32**(4):711-726

[68] Singer A, Barakat Z, Mohapatra S, Mohapatra SS. Chapter 13 - Nanoscale drug-delivery systems: In vitro and in vivo characterization. In: Mohapatra SS, Ranjan S, Dasgupta N, Mishra RK, Thomas S, editors. Micro and Nano Technologies, Nanocarriers for Drug Delivery. Elsevier; 2019. pp. 395-419.

ISBN: 9780128140338. DOI: 10.1016/B978-0-12-814033-8.00013-8

[69] Jug M, Hafner A, Lovrić J, Kregar ML, Pepić I, Vanić Ž, et al. An overview of in vitro dissolution/release methods for novel mucosal drug delivery systems. Journal of Pharmaceutical and Biomedical Analysis. 2018;**147**:350-366

[70] Pandey A. Solid lipid nanoparticles: A multidimensional drug delivery system. In: Daima H, Navya PN, Ranjan S, Dasgupta N, Lichtfouse E, editors. Nanoscience in Medicine Vol 1 Environmental Chemistry for a Sustainable World. Vol. 39. Cham: Springer; 2020. pp. 249-295

[71] Samimi S, Maghsoudnia N, Eftekhari RB, Dorkoosh F. Chapter 3 - Lipid-based nanoparticles for drug delivery systems. In: Mohapatra SS, Ranjan S, Dasgupta N, Mishra RK, Thomas S, editors. Micro and Nano Technologies, Characterization and Biology of Nanomaterials for Drug Delivery. Elsevier; 2019. pp. 47-76. ISBN: 9780128140314. DOI: 10.1016/B978-0-12-814031-4.00003-9

[72] Mehnert W, Mäder K. Solid lipid nanoparticles: Production, characterization and applications. Advanced Drug Delivery Reviews. 2012;**64**:83-101

[73] Bouchaala R, Richert L, Anton N, Vandamme TF, Djabi S, Mély Y, et al. Quantifying release from lipid nanocarriers by fluorescence correlation spectroscopy. ACS Omega. 2018;**3**(10):14333-14340

[74] Kaul S, Gulati N, Verma D, Mukherjee S, Nagaich U. Role of nanotechnology in cosmeceuticals: A review of recent advances. Journal of Pharmaceutics. 2018;**2018**:3420204

Section 3

Advanced Nanocomposites for Energy Storage and Conversion

Chapter 8

Graphene Related Materials and Composites: Strategies and Their Photocatalytic Applications in Environmental Remediation

Santosh S. Patil, Lakshmana Reddy Nagappagari, Ganesh Kamble, Diksha E. Shinde and Kiyoung Lee

Abstract

Photochemical reactions hold great promise in solving energy and environment related problems and likely contribute towards development of sustainable society. Despite of recent advancements, the inherent catalytic efficiency of conventional photocatalyst has been severely limited by myriad complexity associated with (i) ineffective light absorption in visible region, (ii) unproductive recombination's of e^{-}/h^{+} pair in excited state, and (iii) low chemical stability. Contemporary researches on photocatalysts that can be viable for commercial applications has yet to be realized. Graphene has attracted an immense research interests to enhancing the photocatalysts efficiency endowing from their unique optical and electronic properties and salient features such as surface area, mechanical strength and photochemical stability. In this book chapter, we discussed graphene related material (GRMs) to produce hybrid architectures or nanocomposites that can be used as efficient photocatalysts for the degradation of organic pollutants (dyes, pharmaceutical wastes, pesticides etc.) in wastewater. Lastly, we summarize the key insights in photocatalytic electron transfer mechanism, challenges and future perspective which help understand the rationale of GRMs in this field.

Keywords: graphene-related materials, degradation, environmental pollution, hybrid photocatalysts, nanocomposites

1. Introduction

The world is facing serious water scarcity problem and thus protection of fresh water resources is critical in sustainable development of the society. Water pollution causing severe health issues because of waterborne diseases such as cholera, diarrhea, typhoid, and hepatitis leading to human sickness and deaths of (> 14,000) people/day globally [1]. Major water contaminants stemmed from pesticides, textile dyes and a bunch of different chemicals often end up in water bodies or rivers [2, 3]. Textile industries generally uses many colored dye effluents, 65–75% of them belong to azo dyes and annually ~12% of these dyes were lost during manufacturing and the processing operations [4]. The direct discharge of pollutant dyes or pharmaceuticals release toxic or carcinogenic substances into the aqueous environment leading to severe dangers and environmental disasters. To overcome these issues, globally various physical, chemical and biological processes have been used such as precipitation, adsorption (activated carbon), air stripping, coagulation, reverse osmosis, and membrane ultrafiltration, however, these conventional techniques are non-destructive, or often transfer the organic compounds from one phase to another phase triggering secondary pollution [5–7].

Ever since the discovery of water splitting via semiconductor photocatalysis by Fujishima and Honda [8], this technology has gained intense research interests due to potential applications in various fields including sustainable energy conversion, degradation of organic pollutants, bacterial elimination, CO_2 reduction, air purification, antibacterial, organic reactions and self-cleaning etc. (**Figure 1**) [9–11]. Visible light driven (VLD) photocatalyst which can directly harvest solar energy to remove various toxic organic pollutants from water through the advanced oxidation processes is a relatively new and active research area in this field. Till date, titanium dioxide (TiO_2) is the mostly used photocatalysts because of low cost, superior physico-chemical properties and environmental sustainability. However, TiO_2 has one major drawback of wide band gap (only absorb ultraviolet light <4% of solar spectrum) and limited its use under direct sunlight. In this regard, various other semiconductor nanomaterials have been exemplified as photocatalyst systems such as SnO_2, ZnO, WO_3, Fe_2O_3, $BiVO_4$, Ag_3PO_4, BiOCl and BiOBr, etc. for photocatalytic water

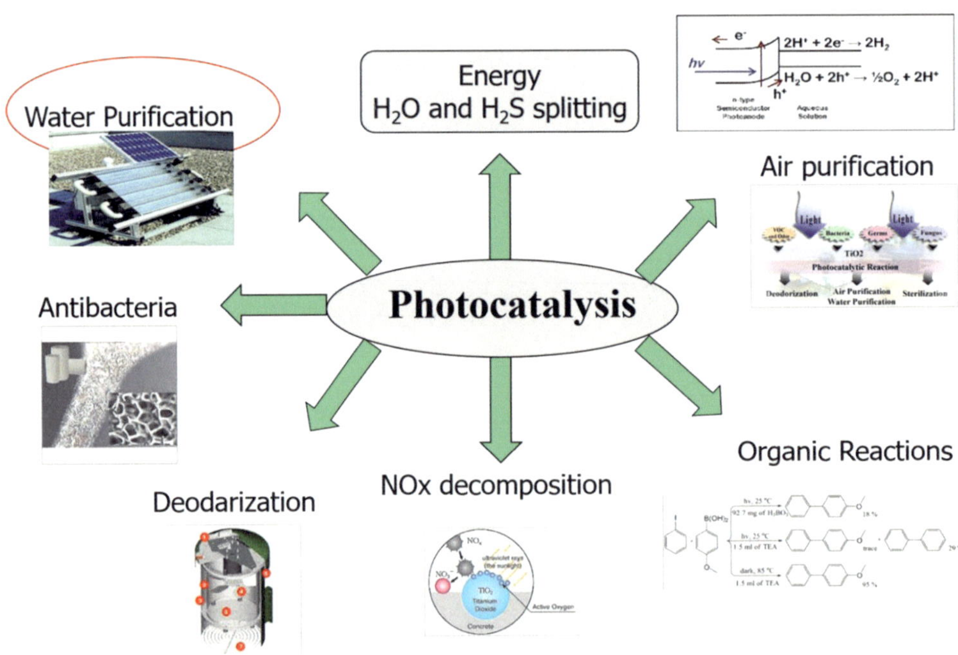

Figure 1.
The diverse applications of photocatalysis.

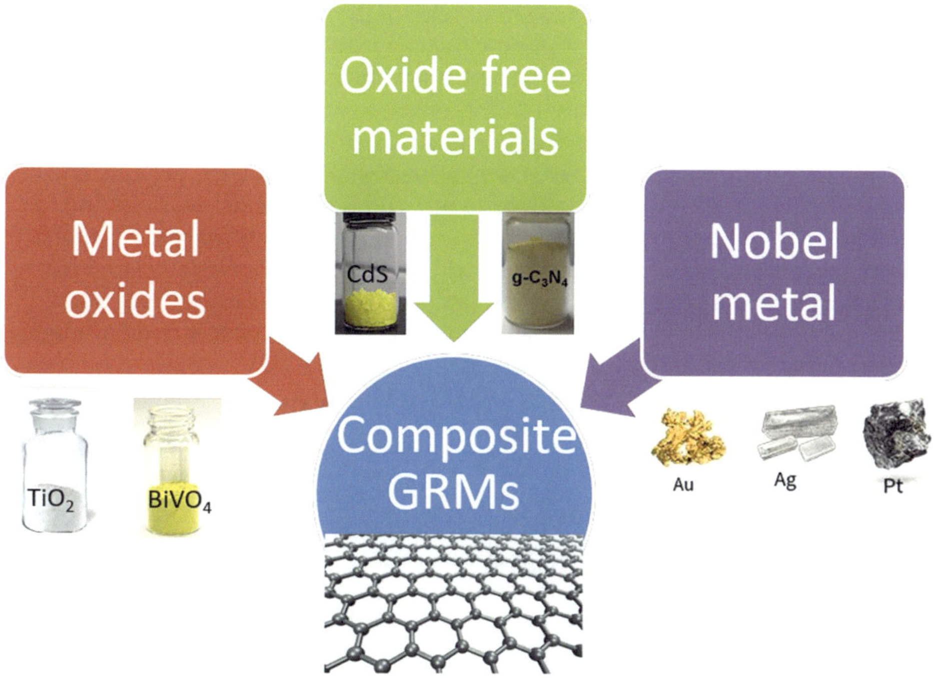

Figure 2.
Development of graphene based composites with different materials.

decontamination [11–13]. The practical utility of these semiconductors catalysts is restricted by low its absorption coefficient, high rate of recombination of electron–hole pairs and variance with the solar spectrum [14, 15]. In order to overcome these issues, a different strategy including Z-Scheme semiconductor photocatalyst [16], metal doping in semiconductor [15], semiconductor-heterostructures or nanocomposites [16–18] have been proposed. In the past decade, metal-doped-graphene or related materials have fascinated intense interests because of its potential for environmental purification and converting photon energy into chemical energy. This process obeys one of the 'Green Chemistry Principle', and widely applied for the degradation of hazardous pollutants. However, in the practical applications was restricted due to the failure to absorb visible light. Therefore, development of novel catalysts that can meet these technical needs is still a daunting challenge. It is in no doubt that the innovations of graphene-based material (GRM) will brings about massive opportunities in nurturing science and technology (**Figure 2**). Due to its exceptional atom-thick 2D structure containing sp^2 bonded arrangement of carbon atoms in a hexagonal lattice, the high specific surface area, conjugated aromatic system, admirable mechanical properties, and robust physicochemical stability and greater electron mobility, graphene is regarded as an ideal high-performance candidate for many photocatalyst systems [19, 20]. A series of different carbon materials, such as activated carbon, porous carbon, carbon nanotubes, graphene and graphitic-carbon nitride can be used as catalyst supports.

In this book chapter we focus on graphene and related materials (graphene oxide and reduced graphene oxide and derivatives) which has emerged to be excellent promoters in photocatalytic reactions because of their low cost, unique physicochemical properties, surface area, electronic conductivity, mechanical flexibility and ionic mobility. These significant features together with high stability enabling them to produce hybrids or nanocomposites photocatalysts with enhanced performances for degradation of organic pollutants (dyes, pharmaceutical wastes, pesticides etc.) in water. Advanced oxidation processes (AOP) for waste water treatment mainly relies on generation of photo-induced-charge carriers (electron and holes), and utilizing them to produce highly active reactive oxygen species (ROS). The as produced ROS radicals generally function as ultimate oxidants in photochemical reactions and subsequently degrade the organic pollutants in water. This book chapter highlights historical background, synthesis strategies and overview of graphene-related materials to produce composite photocatalysts, enabling application in environmental remediation. Finally, the remarks pertaining to state-of-art advancements, challenges and future perspectives have been discussed.

2. General principle of photocatalysis

The term "Photocatalysis" refers to change in the rate of a chemical reaction or its initiation under the action of light in the presence of a catalyst. This process works just like natural photosynthesis, takes place in presence of photosynthetic organisms that converts carbon dioxide into sugars (chemical fuels) using the light energy from sunlight. It is a green process and mainly divided into two types: (i) Homogeneous photocatalysis, and (ii) heterogeneous photocatalysis. In homogeneous photocatalysis, the reactants and the photocatalysts exist in the same phase (e.g., photocatalysis of ozone where all reactants come under one phase, i.e., gas phase). Whereas in heterogeneous photocatalysis, the reactants and the photocatalysts exist in different phases (e.g., semiconductor photocatalysis.). As shown in **Figure 3**, when a photocatalyst absorbs light irradiation from sunlight or an illuminated light source, the electrons in the valence band of semiconductor are excited to the conduction band, whereas the holes are left in the valence band. This creates electron (e^-) and hole (h^+) pairs called as semiconductor's "photo-excited" state and the energy difference between the

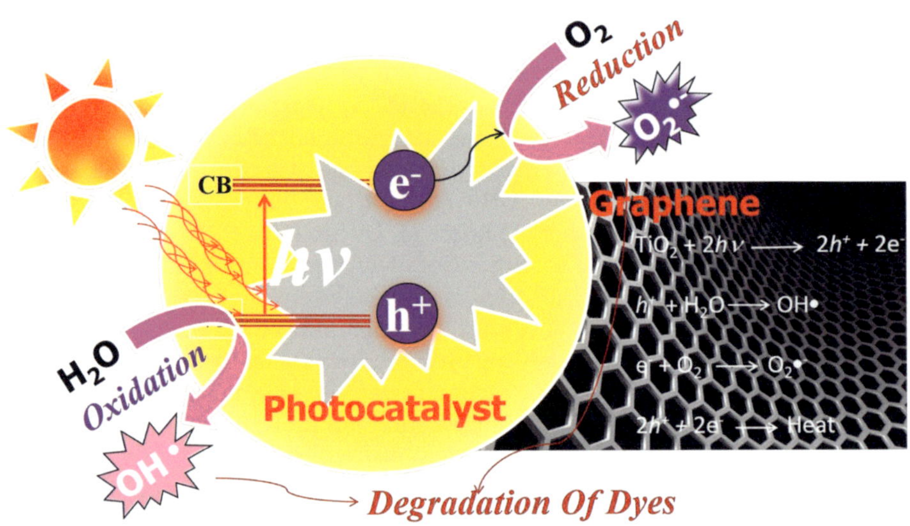

Figure 3.
Schematic illustrating principle of photocatalysis for degradation of pollutant dyes.

valence band and conduction band is referred as "band gap". After photoexcitation, the excited electrons and holes migrate to the surface of photocatalyst to carry out photochemical reactions. For example, in photocatalytic removal of organic pollutants, the photogenerated e^-/h^+ take part in oxidation/reduction process and produces reactive oxygen species (OH, O_2^- and H_2O_2), which can eventually decompose organic pollutants [21, 22]. The ideal photocatalyst requires several key factors such as (i) a suitable band gap to allow the utilization of a significant fraction of the solar spectrum; (ii) optimal band edges relative to the water redox levels; (iii) high mobilities of electrons and holes thereby to reach the surface and reduce/oxidize the targeted molecules before recombining; and (iv) chemical/structural stability.

3. Graphene related materials (GRMs): Principles and strategies

Graphene has attracted immense attention ever since 2010 Nobel prize in Physics was awarded to Andre Geim and Konstantin Novoselov for their pioneering work. Owing to exotic physicochemical properties, and wide industry application prospects including catalysis, electronics, sensing, energy conversion-storage and environmental remediation, constant attempts have been made in its synthesis, investigations and innovations. Basically, graphene is an allotrope of carbon (linked by sp^2 bonds), a one-atom-thick layer arranged in two-dimensional honeycomb network exhibiting unique properties such as high thermal conductivity (5000 W m^{-1} K^{-1}), large specific surface area (2630 m^2 g^{-1}) and high intrinsic electron mobility (200,000 cm^2 V^{-1} s^{-1}) [23, 24]. GRMs include graphene oxide (GO), reduced graphene oxide (rGO), and their derivatives (e.g., functionalized graphene or composites which can be used as building blocks

Figure 4.
General synthesis strategies for preparation of graphene related materials. Reproduced with permission from Ambrosi et al. [25].

to develop series of nanocomposites or hybrid photocatalysts via Vander Walls interaction and inherent surface active O^{2-}/OH groups and oxo ligands chemistry.

3.1 Preparation of graphene and related materials

A wide range of synthesis techniques have been developed to yield graphene and related materials. It can be broadly classified into two distinct approaches: (i) top-down, and (ii) bottom-enabling different scale-up capability and variations in the properties (**Figure 4**). Top-down synthesis strategy relies on simple exfoliation of graphite via mechanical means (e.g., Scotch tape), chemical (e.g., solution-processed, graphite oxide exfoliation/reduction), and electrochemical (oxidation/reduction and exfoliation) methods and allows weakening the van der Waals forces between the graphene layers to form the graphene with single or few atom thick layers. A special graphene nanoribbons with tuneable band gaps and edge shapes have been achieved via opening of carbon nanotubes through chemical or thermal routes [26]. On the other hand, bottom-up strategies which rely on assembly of small molecular building blocks into few layer graphene nanostructures have been achieved through different chemical routes such as catalytic (e.g., CVD), thermal (e.g., SiC decomposition), or chemical (organic synthesis) processes. The readers are directed to several recent reviews for the details of synthesis of GRMs [25, 26].

4. Graphene-related derivatives/composites

4.1 GRM/transition metal oxides composite photocatalysts

Transition metal oxides has been extensively used in photocatalytic environment remediation due to their exotic low cost, high catalytic activity and good stability [8, 9]. In semiconductor photocatalysis, the electrons are excited from the valence band to the conduction band, and electron–hole pairs are generated. These electron–hole pairs are either reunite or transfer to the surface to initiate a series of redox reactions, and generate highly reactive oxidative species (ROS), such as·OH,·O^{-2}, and H_2O_2 which ultimately participate in the degrading of the organic pollutants [27]. A number of strategies have been used to improve the photocatalytic activities of metal oxides: (i) doping of photocatalysts either by anions or cations [10, 11], (ii) coupling of surfaces with metals or semiconductors [12, 28], and (iii) increasing the surface area, reactive facets of photocatalysts [29, 30].

Benefitting from distinct properties and structures, the carbonaceous nanomaterials has attracted immense attention to produce highly active photocatalysts [16]. Westwood et al. [31] and Sigmund et al. [32] reported the combination of carbon nanotubes (CNTs), graphene and other novel carbonaceous nanomaterials with TiO_2, which lead to the dormant recombination of photogenerated electron–hole pairs. Additionally, graphene incorporation tends to offer unprecedented properties due to its unique sp^2 hybrid carbon tightly packed into a two-dimensional honeycomb structure [18, 33]. Based on the formation sequence of the graphene and semiconductor, various graphene-related nanomaterials have been synthesized [34]. The most common strategies for the fabrication of graphene-based photocatalysts are shown in (**Figure 5**) [35]. In addition, different conventional reactions including hydrothermal reaction, thermal irradiation, the adoption of reductants (hydrazine, $NaBH_4$, etc.) have been used to construct graphene-based composites [36–38].

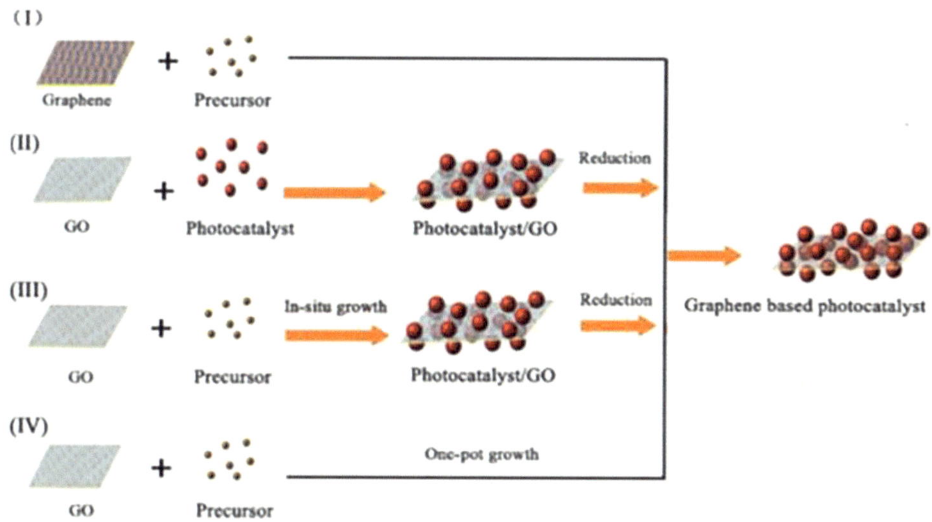

Figure 5.
Synthetic strategies used to fabricate the graphene-based photocatalysts. Reproduced with permission from An and Yu [35].

Owing to high efficiency, good stability and low cost the TiO_2 is mostly preferred for graphene-based nanocomposites [39–42]. Graphene based TiO_2 photocatalytic films also have been developed and used for photocatalytic applications owing to their salient features of easily fixing, recycling and restoring. Because of the efficient charge separation and transportation among the giant p-conjugation and planar structure, the speedy degradation of dye pollutants was achieved by coating TiO_2 films with GO [43, 44]. Apart from TiO_2, many other metal oxides based photocatalysts have been reported such as ZnO, SnO_2, WO_3 and Fe_2O_3 which showed similar photocatalytic applications [36, 45]. Among them, ZnO is often considered as a favorable alternative to TiO_2 for photocatalytic applications [46, 47]. Additionally, visible light active nanocomposites based on graphene have been constructed exhibiting significantly improved the photodegradation activities for the removal of organic pollutants dyes [48]. For instance, Patil et al. reported unique wet chemical synthesis approach for graphene-wrapped Ag_3PO_4/$LaCO_3OH$ heterostructures (**Figure 6A**) [49]. First, $LaCO_3OH$ microspheres were obtained by facile hydrothermal method (**Figure 6A-ii**). Next, an appropriate amount of $LaCO_3OH$ and graphene were dispersed in distilled water and an aqueous solution of $NH_4H_2PO_4$ and $AgNO_3$ was subsequently added dropwise under magnetic stirring to form nanocomposites (**Figure 6A-iii**).

From pre-screening of photocatalysts with Ag_3PO_4/(x wt% $LaCO_3OH$) with mass ratios of x = 5, 10, 15, 20, 25 and 30 wt% we found that Ag_3PO_4/(20 wt% $LaCO_3OH$) exhibits highest photocatalytic degradation performance. On the basis of control experiments and physicochemical characterization., we observed that enhanced photoactivity is attributed to the co-catalytic effect of $LaCO_3OH$, accelerates charge separation due to creation of heterojunction interface. Thus, incorporation of graphene can effectively avoid the disintegration of Ag_3PO_4 into metallic Ag (photocorrosion), featuring excellent photoactivity and stability (**Figure 6A-iv**). Likewise, $BiVO_4$/graphene nanocomposites reported to have intrinsic visible-light driven performance among the Bi^{3+} containing oxides materials [51]. A remarkably high photocatalytic reaction activity of $BiVO_4$ was found, when graphene was incorporated which is attributed to electronic charge

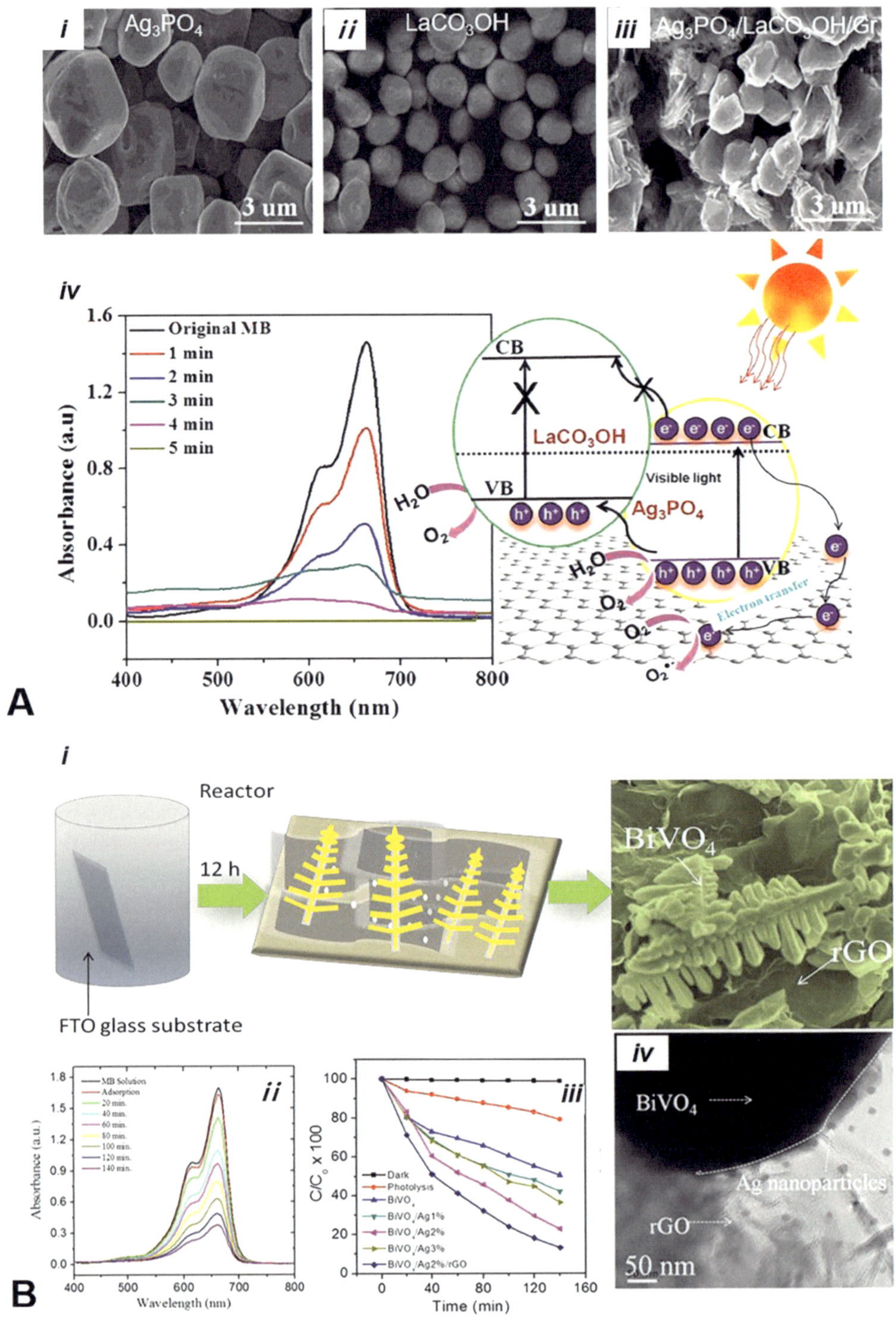

Figure 6.

(A) Graphene wrapped $Ag_3PO_4/LaCO_3OH$ heterostructures. SEM images of Ag_3PO_4 (i), $LaCO_3OH$ (ii) and $Ag_3PO_4/LaCO_3OH$/graphene (iii). Schematic representation of photocatalytic degradation of MB and electron transfer process. Reprinted with permission from Patil et al. [49]. (B) One-pot synthesis of $BiVO_4$/Ag/rGO nanocomposite (i) schematic illustration of hydrothermal synthesis, (ii and iii) photodegradation of MB dye. (iv) corresponding TEM image of composite showing ternary composite of $BiVO_4$, Ag and rGO nanostructures. Adopted from Patil et al. [50].

equilibration between $BiVO_4$ and graphene, and likely contribute to shift of the Fermi level and reduction in conduction band potential [52]. Recently, Patil et al. demonstrated one pot in-situ preparation of ternary $BiVO_4$/Ag/rGO hybrid heterostructures via simple hydrothermal method and tested for photoelectrochemical (PEC) water splitting and photocatalytic MB degradation (**Figure 6B**) [50]. These results revealed that the combined effects of the incorporated Ag and rGO nanostructures lead to interface creation wherein visible light absorption, charge separation-transfer and superior surface characteristics was greatly improved. An excellent degradation rate (k_{app}) of 1.29×10^{-2} and 0.192×10^{-2} was obtained for the photocatalytic MB and phenol degradation, respectively using ternary $BiVO_4$/Ag/rGO hybrid nanostructures which is three times higher than pristine $BiVO_4$ photocatalyst (**Figure 6b-ii** and **iv**). Moreover, the synthesis of magnetically separable α-Fe_3O_4 [53] and $ZnFe_2O_4$ [54] on graphene support has also been reported, these nanocomposites exhibited an excellent organic pollutant removal efficiency from wastewater. These photocatalysts can be easily separated from aqueous solution by applying an external magnetic field enabling well retained photoactivity even after repeated use. **Table 1** summarizes the recent reports on the degradation of organic contaminants by using GRM-based photocatalysts.

Composite photocatalyst/ amount	Mass ratio of GRM	Dye concentration	Degradation percentage (%)	Irradiation time (hours)	Light source	Reference
P25 TiO_2-graphene	1 wt. %	MB (0.01 g/L, 2.7 x 10 M^{-5} M)	65	1	500 W Xe lamp (λ > 420 nm)	[55]
TiO_2-graphne (1 mg/mL)	30 mg	MB (1 x 10 M^{-5} M)	75	3	Sunlight	[56]
TiO_2–GO (0.05 g)	0.14 wt. %	Methyl orange (MO; 12 mg/L)	35	3	1000 W Xe lamp	[57]
SnO_2-graphene (6.3 mg)	15 wt. %	Rhodamine B (RhB; 5.3 x 10 mM^{-5} mM)	~ 80	~ 2	350 W Xe lamp	[36]
ZnO-graphene (50 mg)	2 wt. %	MB (1 x 10 M^{-5} M)	90	0.4	UV light	[58]
$ZnFe_2O_4$/graphene	20 wt. %	MB (20 mg/L)	88	1.30	450 W Xe lamp	[54]
$BiVO_4$/rGO (0.02 g)	No data	MB (3 mg/L)	90	2.0	54 W halogen lamp	[59]
CdS/graphene (0.050 g)	5 wt. %	MO (10 mg/L)	95	1	200 W Xe lamp	[60]
CdS/CNTs (0.050 g)	88 wt. %	MO (10 mg/L)	88	1		[60]
g-C_3N_4/rGO	No data	RhB (10 mg/L)	94.2	2.30	150 W Xe lamp	[61]
Ag/AgCl/Graphene	0.6 wt. %	MO (15 mg/L)	71	~ 0.5	500 W Xe lamp	[62]
Ag/Ag_2S/TiO_2/GO	No data	Crystal violet	28.92	2	Xe lamp	[63]

Table 1.
Summary of recent GRM-metal oxide nanocomposite photocatalysts degradation of organic pollutants.

4.2 GRM/oxide free semiconductor photocatalysts

Graphene based-nanocomposite with oxide free materials (metal sulphide, metal nitrides, graphitic carbon nitride (g-C_3N_4) and Bi-oxyhalides) has gained increasing attention in environmental remediation [37, 38, 40, 64]. Metal sulfides such as CdS, MoS_2, SnS_2, Sn_2S_3, CuS, and ZnS generally hold narrow energy band gaps and negative conduction band (CB) edge positions [65]. Many metal sulphides (mono, binary and ternary) have been developed with tunable band structures and successfully employed in photocatalytic dye degradation [65, 66]. Among them, CdS most studied photocatalyst that can directly absorb sunlight at wavelengths under 550 nm [41]. Furthermore, graphene/CdS composite reported to have excellent visible-light-driven photocatalytic activity for organic pollutant degradation. Ma et al. reported that optimal weight percentage of graphene in the CdS clusters/graphene nanocomposites was found to be 1.0 wt%, which resulted in a high photocatalytic degradation of methyl 3,5-dichloro-4-hydroxybenzoate (MDHB) [40]. Heterojunction construction-based copper sulfide nanostructures was observed to be an effective strategy for environmental applications. Andronic et al. demonstrated copper sulfide/graphene heterojunction photocatalysts for dye photodegradation enabling relatively large surface area, porous morphology, the ability to photogenerated electrons across the composite interface, and high adsorption capacity for organic molecules [42]. Graphene-CuS composites with different surface morphologies were prepared via different synthesis strategies such as CuS-GO/TiO_2 composites were synthesized by sol-gel method [43], flower-like CuS/rGO composites synthesized by a facile one-step solvothermal procedure [44]. The dye degradation efficiency of GR/CuS composite was observed to be 30% higher than crystalline pure phase CuS. Thus, enhanced photoactivity was attributed to the not only high electronic conductivity of graphene but also its significant influence on the morphology of the CuS/Gr nanocomposite [44].

Most recently, an innovative metal free polymeric photocatalysts—graphitic carbon nitride (g-C_3N_4) has been developed representing low cost, easy scalable synthesis and superior photoactivity [67]. It is composed of C, O, N and some contamination of H atoms, coordinated by tris-triazine-based patterns. It is a highly stable both thermally (up to 600°C) and chemically due to covalent C − N bonds. The state-of-the-art catalytic and optoelectronic properties are ascribed to sheet-like structure of g-C_3N_4 with appropriate band gap energy (Eg-2.7 eV), metal-free nature, and tunable electronic band structure and stability [68]. The existence of primary surface sites of g-C_3N_4 are striking for several optoelectronic applications including photocatalysis [69]. g-C_3N_4 contains of C–N bonds deprived of electron localization in the π state and the number of surface defects are found due to presence of hydrogen specifies, which could be beneficial in catalysis [67]. g-C_3N_4 has been synthesized via different chemical routes can act as efficient photocatalyst for photodegradation of organic dyes [70, 71].

Furthermore, g-C_3N_4 was reported to be doped with metallic impurities, in which band gap energy was reduced enhancing photo response and photocatalytic properties [72]. In order to dope metal ions, a salt of equivalent solubility were mixed with the g-C_3N_4 precursor [73]. Various transition metals such as Pd, Cu, Fe, W, Zr has been doped to endow remarkable photocatalytic activities due to alteration of the electronic and atomic structure of g-C_3N_4 [74]. Specifically, the light absorption and mobility of charge carriers can be increased which are essential prerequisites for better photocatalytic performance. Metal cations and the negatively charged atoms of nitrogen attributed to the lone pairs of electrons on the nitrogen edges of g-C_3N_4 [73].

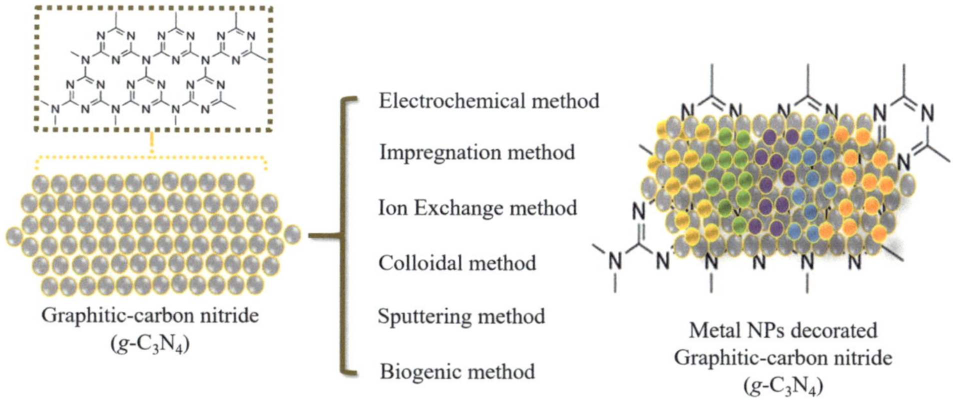

Figure 7.
Schematic illustrating the decoration of graphitic-carbon nitride (g-C_3N_4) with metal nanoparticles. Reprinted with permission from Khan [71].

Noble metals also have been used as a doping metal in the g-C_3N_4, such as platinum and palladium have been utilized which could result in enhanced transporter mobility, improved separation of electron hole pairs, and narrowing of the band gap values [39, 75]. In the recent years several strategies have been adopted for the incorporation of metal or transition metal-into g-C_3N_4, as shown in **Figure** 7.

4.3 GRM/noble metals nanocomposite photocatalyst

Graphene-based noble metal composites have been fabricated by introduction or mixing of noble metal precursor into the graphene solution via relatively simplistic methods such as in situ growth or wet chemical methods. Metals like Au, Ag, Pd and Pt, were incorporated into GRMs to exploit high efficiency composite photocatalysts [76, 77]. The noble metal nanoparticles could act as an electron trap, increases the mobility of charge carriers and likely suppresses the rate of recombination. Therefore, it can be considered as appealing platforms in the design of high-performance visible light driven photocatalysts. Among them, Ag and Au NPs exhibit unique optical properties due to the collective oscillation of free electrons on their surfaces while being interaction with incoming electromagnetic radiation [78]. In the photocatalysis process, noble metal nanomaterials have been extensively studied to enhance the light absorption capability of metal oxide-noble metal photocatalysts often called as plasmonic photocatalysts [79]. Apparently, doping of metals could decrease the band gap and accelerate the interfacial electron transfer. It has been reported that Au and Ag nanoparticles could allow capturing and scattering photons with a relatively high excitation wavelength in the visible light region [80]. Many noble metal-graphene-hybrid nanostructures have been prepared and tested for the degradation of organic pollutant dyes in water [81–83].

The incorporation of metal oxide photocatalysts with graphene and noble metal nanostructures could provide synergic effect to boost the photocatalytic efficiency. Interactions of graphene with metal nanostructures could facilitate electrons transport together with plasmonic effects. Fabricating of as such multifunctional heterogeneous systems may provide several pathways for electron transport stemmed from noble metal–metal oxide, metal oxide–graphene, and noble metal–graphene interfaces [84]. For example, Wen et al. [85] synthesized the graphene based TiO_2-Ag doped photocatalyst and indicated that strong absorption in the visible light region can be realized increasing the photocatalytic efficiency for the degradation of methyl orange (MO) dye. This is attributed to the combined effects of the surface plasmon resonance (SPR) properties of the Ag NPs and the strong interaction of graphene assembled on the TiO_2 surface. Furthermore, at the Ag/TiO_2 interface, formed Schottky junctions could enhance the ability of electron transfer with graphene and resulted in increased photocatalytic efficiency [83]. Similarly, Huang et al. reported a layered structure of Ag NPs deposited on graphene-TiO_2 nanorods exhibiting a greater efficiency of photocatalytic degradation of MB [86]. Jafari et al. [87] and Vasilaki et al. [88] fabricated a ternary-graphene-Ag/TiO_2 nanocomposite by hydrothermal method and demonstrated that composite system can efficiently degrade MB dye related to pristine TiO_2. Similarly, Ghasemi et al. evaluated the degradation of acid blue 92 under UV and visible light irradiation using Au–TiO_2–graphene nanocomposite photocatalyst [89]. In this case, Au NPs help extend the visible light absorption ability by decreasing the band gap energy of the system TiO_2–Pt/Pd–graphene nanocomposite photocatalysts were investigated for the degradation of Reactive Red 195 and 2,4-dichlorophenoxyacetic acid under UV and visible light. Due to greater photonic efficiency, Pt metal showed the higher degradation rate of pollutants [90]. Graphene-Ag-ZnO nanocomposite have been used as a high performance photocatalysts for the degradation of RhB dye. The rate of degradation is 13.80 times higher than bare ZnO photocatalysts [91]. Correspondingly, Au/rGO/ZnO nanocomposite found to be an efficient photocatalysts for degradation MB dye in water [92, 93]. Zhang et al. decorated palladium (Pd) on ZnO–graphene nanostructures in which photocatalytic activity was dramatically enhanced as a result of charge separation at the Pd–ZnO interface [94].

Doping of noble metals in graphitic carbon nitride (g-C_3N_4) could endow interesting physicochemical properties owing to two-dimensional layered structure and likely contribute visible-light photo response. For example, Au/Pd/g-C_3N_4 nanocomposites were realized by loading of Au and Pd nanoparticles on the surface of g-C_3N_4 sheets [95]. This nanocatalyst demonstrated >90% degradation efficacy for degradation of tetracycline hydrochloride. Similarly, Au/g-C_3N_4 nanosheet/reduced graphene oxide (Au/CNNS/rGO) nanocomposite was produced by inducing simple thermal oxidation exfoliation combined with in-situ photoreduction reactions leading to significantly enhanced photocatalytic activities for MB degradation and H_2 production reaction [96]. As shown in **Figure 8a–c**, Au nanoparticles were uniformly decorated on the surface of the thin carbon nitride nanosheets (CNNS) indicating intimate interaction between them and can be beneficial for improving the plasmonic characteristics of the nanocomposite. **Figure 8c** shows size distribution of Au nanoparticles with an average diameter of 5.5 nm. Additionally, the rGO was found to be integrated on the opposite surface of CNNSs, which resulted in bidirectional nanostructure to promote the electron transfer. According to high resolution transmission electron microscopy (HRTEM) analysis (**Figure 8d**), the lattice spacing of 0.203 nm was confirmed and in agreement with (200) lattice planes of metallic Au. The authors detected a dramatic improvement in H_2 production reaction and methylene blue degradation which was approximate ly 9.6X and 6 X fold higher than pure g-C_3N_4 under visible light irradiation. Kim et al. reported a a strategy to incorporate different noble metal nanoparticles (Pd, Pt, Au, and Ag) into GO nanosheets whereby noble metals and GO was reduced simultaneously using ascorbic acid as a reductant [97]. Wang et al. reported Ag NPs@GO nanocomposite through light-induced synthesis method in which size dependent extremely

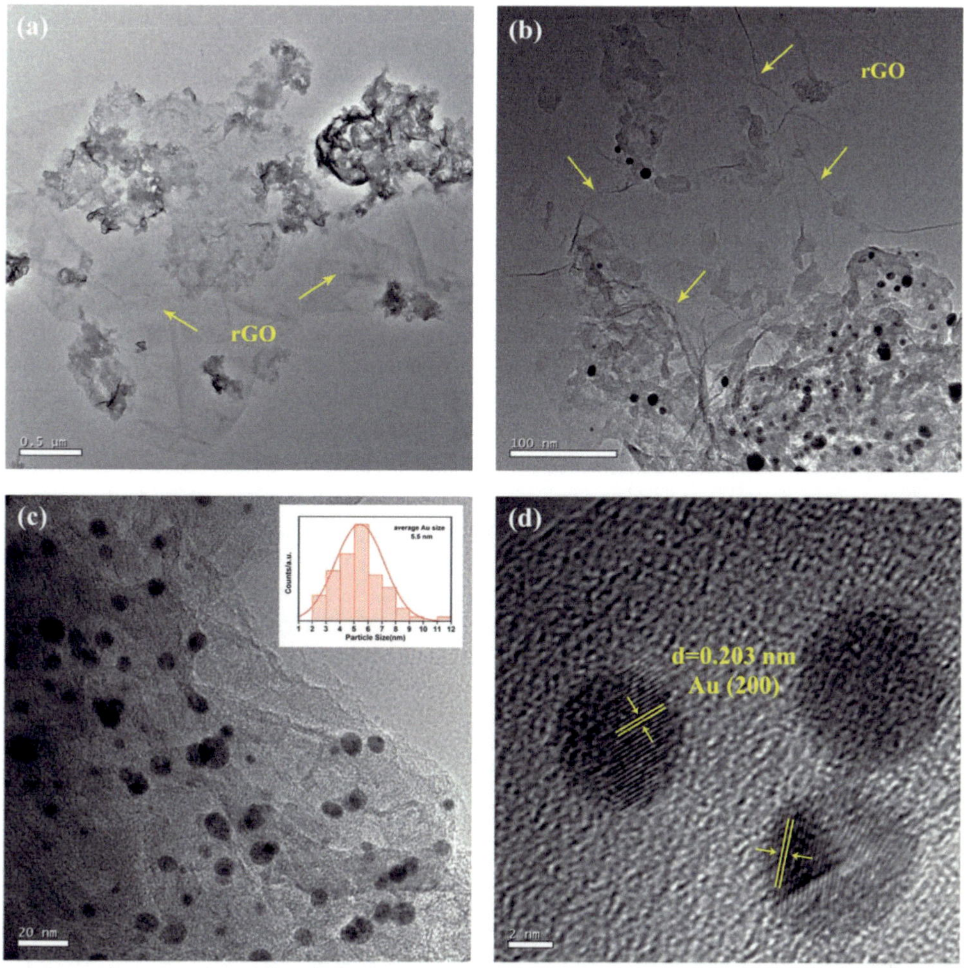

Figure 8.
TEM images of 5% Au/CNNS/rGO composites at low-magnification (a) and high-magnification (b) and (c). Inset in (c) represents particle size distribution curve of Au nanoparticles. HRTEM image of only Au nanoparticles (d). Reprinted with permission from Li et al. [96].

high catalytic activity was investigated towards degradation of 4-nitrophenol [98]. Similarly, Ji et al. obtained Ag NPs@GO nanocomposite by in-situ reduction of Ag^+ ions into on GO nanosheets exhibiting efficient catalytic activity for the reduction of 4-nitrophenol into 4-aminophenol [99]. Rajesh et al. reported anchoring of Ag NPs and Au NPs onto chitosan grafted GO via $NaBH_4$ reduction method which displayed superior photocatalytic activity towards degradation of aromatic nitroarenes and azo dyes [100]. Gu et al. was chosen functionalized graphene/Fe_3O_4 hybrid as nanocarrier to deposit AuPt alloy NPs via controlled self-assembly strategy. The nanocomposite displays magnetic features which is beneficial to recover the catalysts easily and repeated use endowing superior catalytic activity for the reduction of 4-nitrophenol [45]. Similarly, Islam et al. obtained magnetically recyclable carbon nanotube-rGO-Fe_3O_4-Ag NPs nanocomposite by combining in-situ reduction and hydrothermal methods which was demonstrated high catalytic efficiency for the removal of toxic dyes-MB and 4-nitrophenol [101].

5. Overview of charge-transfer mechanisms in emerging GRM composites

In this section, we particularly focus on GRMs composites for environmental remediation, with special emphasis on charge-transfer mechanisms. For this purpose, the composites of GRMs have been taken in account, since the charge transfer mechanism was mainly influenced by the composite systems, or heterojunctions formation between the two semiconductors which is together called as nanocomposite system. GRMs in association with several other semiconductor materials for example GO, rGO@MO, (MO = TiO_2, WO_3, $BiVO_4$, $AgPO_4$, AgCl, Fe_2O_3, Bi_2O_3, etc.) has been widely studied for environmental applications [35, 102–104]. Several charge transfer mechanisms have elucidated to understand the boosts in photocatalytic degradation efficiency. This charge transfer mechanism mainly depends on the type of heterojunction formed with other metal oxide semiconductors with GRMs. Especially the type-I, type-II and Z-scheme heterojunctions are the most important ones that influence the charge transfer mechanism during the photocatalytic process and eventually improve the photocatalytic efficiency. Gao et al. [105] studied the charge transfer mechanism of graphene-Bi_2WO_6 (G-BWO) for the photocatalytic degradation of Rhodamine B (RhB). Here, authors elucidated the type-1 mechanism based in the electrochemical studied and photocatalytic activity results (**Figure 9a**). In another study the Simsek et al. [106], demonstrated the charge transfer mechanism of ternary heterojunction system

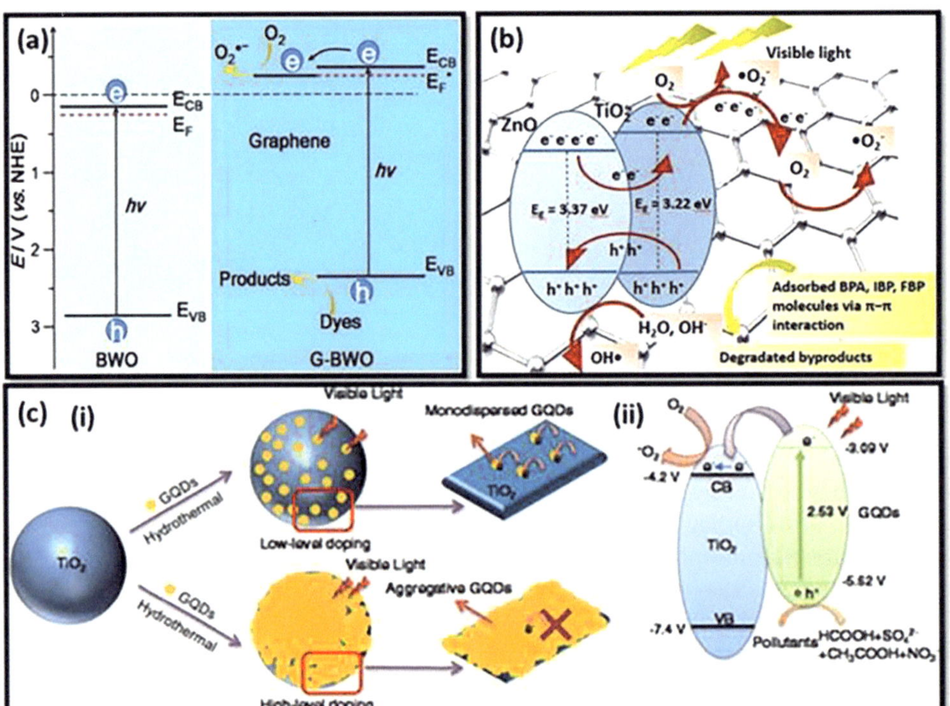

Figure 9.
(a) Scheme of energy band diagram and photocatalytic degradation at BWO (Bi_2WO_6) and G-BWO (Graphene-Bi_2WO_6), reproduced with permission from Gao et al. [105], (b) mechanism of photocatalytic degradation over RGO/TiO_2/ZnO hybrid structure. Reproduced with permission from Bilgin Simsek et al. [106]. Copyright @ Elsevier 2018, (c) schematic representation of (i) photo-generated electron transfer of GQD on the surface of TiO_2 under visible-light irradiation and (ii) energy position of GQD/TiO_2 under visible-light irradiation, reproduced with permission from Pan et al. [107]. Copyright @ ACS 2015.

consists of RGO/TiO_2/ZnO (**Figure 9b**) towards photocatalytic degradation of estrogen (bisphenol-A) and pharmaceuticals (ibuprofen, flurbiprofen). When RGO/TiO_2/ZnO catalyst is irradiated with light, the electrons are excited from conduction band of ZnO to the conduction band of TiO_2 structure. These photoinduced electrons can react with O_2 to generate O_2 radicals, and the holes on the surface captured by OH^- or H_2O to form OH. Recently, Ran and co-workers studied the graphene quantum dots (GQDs) decorated by TiO_2 as heterojunction composite for degradation of methyl orange (MO) under visible-light irradiation. The reaction mechanism follows the type II heterojunction (**Figure 9c**) [107]. Therefore, from these studies it can be understood that the GRMs or its composites likely contribute for charge transfer mechanisms for removal or pollutants through photocatalytic process. The reaction mechanism varies based on the engineered material that is deposited on graphene and its related material.

5.1 Photocatalytic degradation of organic dye pollutants/water decontamination

GRMs possesses several properties that make it attractive for environmental applications. The most studied aspect of graphene & GO is probably its electronic properties [108, 109]. It has extremely high electron mobility, reaching 10,000 $cm^2 V^{-1} s^{-1}$ to 50,000 $cm^2 V^{-1} s^{-1}$ at room temperature, with an intrinsic mobility limit of >200,000 $cm^2 V^{-1} s^{-1}$ [110]. More importantly the graphene can sustain current densities up to six orders of magnitude higher than copper [108]. These remarkable electronic properties of graphene, however, were obtained under ideal conditions, with mechanically exfoliated graphene under vacuum. Nonetheless, the promising electronic properties of graphene have triggered research and development for its extreme use in photocatalytic materials for degradation of various pollutants in water [104]. **Table 2** summarizes the most common adsorption mechanisms as well as advantages and disadvantage of using graphene related materials and their derivatives as adsorbents for environmental remediation and sequestration of metal ions from aqueous solutions.

The GRMs provide lot of scope for photocatalytic degradation of organic dyes in combination with other metal oxide semiconductors. Especially due to its low cost and strong oxidizing activity, TiO_2 is the most commonly used semiconductor for forming graphene based photocatalytic nanocomposites for the photodegradation of organic and biological contaminants [131, 132]. The rGO-TNT (TiO_2 nanotubes) composites were prepared with different concentrations of rGO and the photocatalytic degradation of malachite green (dye) was found to be influenced by the rGO/TNT ratio. The rGO-TNT containing 10% rGO showed the highest photocatalytic degradation activity against malachite green, a performance of three times higher compared to neat TiO_2 nanotubes [133]. As dyes are aromatic molecules, their adsorption on graphene is endorsed by p-p stacking interactions between the sp^2 domains from both systems. Therefore, the adsorption capacity of graphene-TiO_2 composites for organic dyes can be higher than bare TiO_2 nanomaterials [55]. After interaction with graphene sheets the oxidative species nearby the catalyst can readily access the adsorbed dye, making the photocatalytic degradation process more efficient. In another study, Zhang et al. [36], studied the photocatalytic degradation of rhodamine B (RhB) with reduced graphene oxide (rGO) which was modified with TiO_2 and SnO_2 to form rGO-TiO_2 and rGO-SnO_2 respectively. The composite materials performed very interesting photocatalytic properties for degradation of RhB under visible light irradiation. First, their photocatalytic activities were higher than that of P25 (a commercial TiO_2 as a benchmark photocatalyst). Second, the reaction mechanism catalyzed by the composite

Graphene related materials	Mechanisms involved in the adsorption of metal ions	Advantages	Disadvantages	References
Graphene oxide (GO)	Electrostatic interactions Ion exchange	High dispersibility in water; good colloidal stability; abundant presence of oxygenated functional groups	Limited amount of sorption sites	[111–115]
Reduced GO (rGO) Pristine graphene	Electrostatic interactions Lewis-base-acid mechanism	Reestablishment of sp^2 domains; better electron-transport property	Low density of oxygen-containing functional groups; lower colloidal stability	[116–120]
Magnetic graphene nanocomposites	Electrostatic interactions with graphene Interactions with the surface of the particles Magnetic properties of the nanoparticles	Larger surface area compared to the pristine forms; increased number of binding sites compared to pristine graphene; easy recovery from aqueous solutions	Co-reduction of GO during the attachment of the particles reduces the colloidal stability	[121–126]
Graphene materials modified with organic molecules	Electrostatic interactions Complexation with organic molecules	Larger surface area compared to pristine forms; good colloidal stability; improved amount of functional groups ($-NH^2$, $-OH$)	The stability of the loaded molecules varies according to the modification strategy	[117, 119, 127–130]

Table 2.
The mechanisms of interaction and possible advantages and disadvantages of using graphene related materials as adsorbents to remove metal ions from aqueous solutions as part of environmental remediation.

materials was different from that of semiconductor photocatalysis (**Figure 10**). The careful characterization showed that the excellent photocatalytic performance of the composite materials was associated with the good electrical conductivity and effective charge separation because of the presence of RGO. In presence of catalysts RGO–SnO_2 and RGO–TiO_2, the degradation was remarkably enhanced with rate constants of about 6.2×10^{-3} min^{-1} for RGO–SnO_2 and 3.6×10^{-3} min^{-1} for RGO–TiO_2, significantly higher than that of RGO (3.9×10^{-4} min^{-1}) as shown in **Figure 10a**. In addition, photocatalyst RGO–SnO_2 performed better than RGO–TiO_2 and commercial product P25. The adsorption of RhB on the catalysts could be accredited to two parts, the adsorption of RhB on the surface of the RGO and the surface of the metal oxides. Although the surface area of RGO–TiO_2 (341 m^2 g^{-1}) is much larger than that of RGO–SnO_2 (241 m^2 g^{-1}), benefiting from the uptake of RhB from water, RGO–SnO_2 exhibited higher photocatalytic activity compared to RGO–TiO_2. Apparently, the enhanced photocatalytic degradation of RhB over RGO–SnO_2 is not mainly due to the adsorption of dyes on the RGO sheet. Considering the work function of RhB (−5.45 eV), excited RhB (−3.08 eV) [134], graphene (−4.42 eV) and the conduction bands of SnO_2 (−4.5 eV) and TiO_2 (−4.4 eV) in **Figure 10b**, the different locations of adsorbed RhB molecules on the catalyst surface would lead to different degradation efficiencies. It was demonstrated that excited RhB can efficiently inject electrons into the graphene

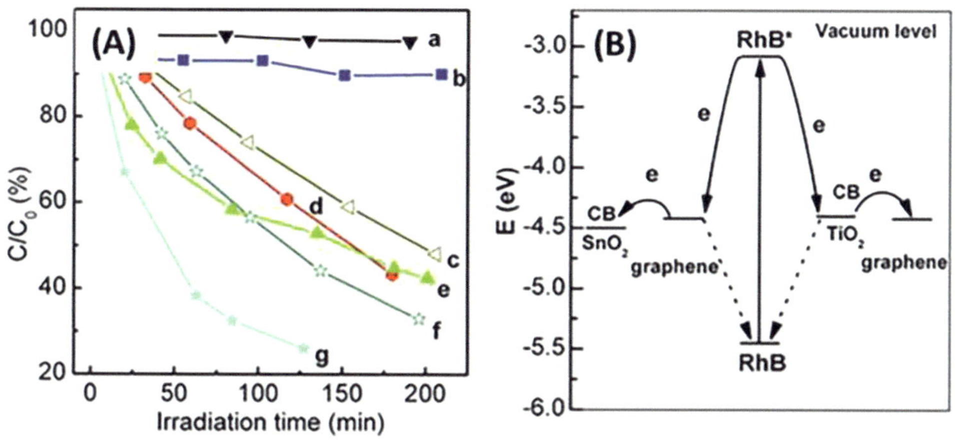

Figure 10.
(A) Photocatalytic degradation of RhB (a) without catalyst, with (b) RGO, (c) RGO–TiO$_2$, (d) P25, (e) RGO–SnO$_2$, (f) RGO–TiO$_2$-400, and (g) RGO–SnO$_2$-400. (B) the energy diagrams of RhB, graphene, TiO$_2$ and SnO$_2$. Reproduced with permission from Zhang et al. [36]. Copyright @ RSC 2011.

plane, and the degradation rate was even faster than that of TiO_2 [134]. However, due to the electron recombination between the injected electron and the surface adsorbed RhB^+ (dotted line in **Figure 10b**), the degradation of RhB over the RGO was excruciatingly slow [134]. After the RGO was loaded with SnO_2, the injected electron could further move to the conduction band of SnO_2 due to the higher work function of SnO_2 than RGO. The electron on the SnO_2 surface could also be trapped by dissolved oxygen to form various reactive oxygen species (ROSs), thus greatly enhancing the degradation of RhB. Here, the RGO acted as an electron mediator, facilitating the electron transfer from RhB^* to SnO_2. However, when RGO was loaded with TiO_2, since the work function of RGO is higher than that of TiO_2, the electron on the RGO surface cannot further transfer to the TiO_2 nanoparticles. Therefore, from this study it was concluded that the work function of the metal oxides plays a crucial role in activating the degradation efficiency of the dye molecules.

5.2 Photocatalytic degradation of pesticides, pharmaceutical wastes etc.

GRMs were found to be quite effective in the degradation of pesticides and pharmaceutical wastes as similar to degradation of organic dye pollutant molecules. In this section we discussed some of degradation mechanisms of GRMs for this important topic of research. Chlorpyrifos is widely used to control pest insects in residential, agricultural, and commercial applications. Its common use has led to the release of chlorpyrifos into sediments, wastewater and water sources. The presence of chlorpyrifos in wastewaters and water sources may affect ecosystem and human health due to its chronicle toxicity to aquatic organisms. In this regard, Vinod Kumar and co-workers studied the magnetic recoverable $CoFe_2O_4@TiO_2$ decorated reduced graphene oxide nanocomposite for investigating the photocatalytic degradation of chlorpyrifos (**Figure 11**) [135]. The effect of initial concentration of CP on the degradation rate was studied within the range of 1 to 40 mg/L at a catalyst amount of 0.4 g/L. The nanocomposite was separated easily from the solution within 12 s by using a magnet without filtering and it was stabilized into solution for the next cycle. The stability experiments of $CoFe_2O_4@TiO_2$/rGO nanocomposite showed no loss in

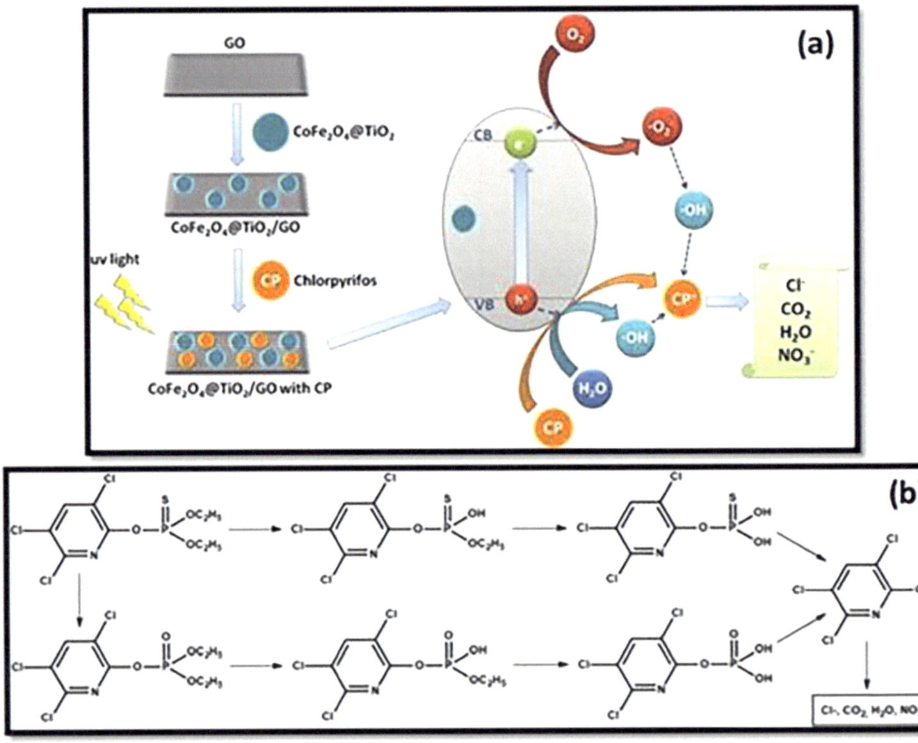

Figure 11.
(a) Photocatalytic degradation mechanism for the chlorpyrifos by using $CoFe_2O_4@TiO_2/rGO$ nanocomposite, and (b) degradation pathways showing various intermediate by products. Reproduced with permission from Gupta et al. [135]. Copyright @ Elsevier 2015.

the efficiency after 8 cycles [135]. The photochemical mechanism for the degradation of the chlorpyrifos by using $CoFe_2O_4@TiO_2/rGO$ nanocomposite is shown in **Figure 11a**. Similarly, the photocatalytic degradation pathways of chlorpyrifos has been shown in **Figure 11b** which generally undergone various intermediate byproducts on part of photocatalytic mechanism.

Recently Cruz et al. [136] studied the photodegradation of a mixture of four pesticides classified by the European Union as priority pollutants: diuron, alachlor, isoproturon and atrazine by using GO-TiO_2. The influence of two water matrices (ultrapure or natural water) was also investigated. The photocatalytic activity of GO-TiO_2 composite under visible light was remarkably higher if compared to commercial TiO_2 P25, shorter reaction times to photo-degrade 50% of pesticides as well as faster chloride formation rate being obtained with GO-TiO_2.

In the similar manner, GRMs-composites have been extensively applied for degradation of pharmaceutical wastes through photo degradation process [137, 138]. Since the water pollutants emerging from pharmaceutical, cosmetics, heavy metals, pesticides, industrial additives, and solvents are becoming new global water quality threats. The presence of pharmaceuticals in municipal wastewater, hospital wastes, and industrial effluents are the major sources of contaminants in drinking water [139]. In this connection the Sravya et al. reported the photocatalytic degradation of pharmaceutical wastes specifically the paracetamol [138]. Highly efficient visible light active polyaniline (PANI)/Ag nanocomposites grafted reduced graphene oxide (rGO–Ag/PANI) was prepared and tested for the photocatalytic degradation of paracetamol under visible light radiation. The results reveled that ~99.6% degradation of paracetamol in the acidic medium (pH 5) and 75.76% in the basic medium (pH 9), respectively. The enhanced degradation efficiency is attributed to the synergetic effect of rGO, PANI, and Ag NPs in the nanocomposites. Many other researchers widely studied the degradation of pharmaceutical wastes by using graphene and its related materials by making good composites [140–142].

6. Concluding remark and perspectives

In summary, GRMs (GO, rGO, graphene and derivatives) can be incorporated into different semiconductors or metals to construct GRM-based composite photocatalysts. An integration of GRMs into these composites has proven to be effective approach to endow unprecedented properties and concurrently improves the high dye adsorption capability, light absorption, and also facilitate charge separation and transfer properties, enabling advancement in the overall photocatalytic activity. Despite of decades intensive research, the scalable synthesis of high-quality graphene is in unmatured state and several issues need to be addressed. We discussed simple, cost-effective synthetic strategies such as in situ reduction, coprecipitation, hydrothermal or solvothermal reactions etc. to develop emerging GRM based composites and applicability towards photocatalytic degradation of pollutant dyes, organics and pharmaceutical wastes in water.

With regard to breakthrough in photocatalysis technology, the studies are in early stage and further specific advancements are required. The following insights may be useful.

1. Although conventional photocatalysts such as titanium dioxide (TiO_2) and graphitic carbon nitride (g-C_3N_4), and graphene have proven their fitness to work as effective supports, the coordinating sites for metal or single atoms are limited which indeed succumbed to inefficient solar light harvesting due to their wide band gaps and no band gap (graphene). Therefore, finding a low-cost new transition metal based nanosheet materials and single atom photocatalysts is challenging. Like GRMs, the Bi-containing oxyhalide nanosheets or combinations thereof could be possible choice materials for the development of new generation photocatalysts.

2. Additionally, although present GRM/Nobel metal nanocomposites have shown increasing plasmonic photocatalytic activity and promises for the potential applications still they have one drawback. Metal atoms used in the photocatalysts design are rare and expensive. Therefore, from economical or practical perspective exploring new low-cost transition metals (Cu, Mo, Bi, Co, Ni etc.) as cocatalysts together with inexpensive light harvester nanosheets support photocatalysts (narrow band gap) could be possible means to produce feasible high efficiency photocatalysts. Altogether there is plenty of room towards developing noble metal free single atoms/GRM nanosheet photocatalysts for sustainable energy conversion and environmental remediation.

3. Importantly, most of the researches in this field are mainly focused on material design, without much understanding about what is going on in the process, and photoreactor/devices system. It is anticipated that the knowledge of understanding all three factors is equally important which may lead to significant opportunities in solving real world problems.

Author details

Santosh S. Patil[1*], Lakshmana Reddy Nagappagari[2], Ganesh Kamble[3], Diksha E. Shinde[4] and Kiyoung Lee[1*]

1 Department of Chemistry and Chemical Engineering, Inha University, Incheon, South Korea

2 Division of Advanced Materials Engineering, Jeonbuk National University, Jeonju, South Korea

3 Department of Engineering Chemistry, Kolhapur Institute of Technology's College of Engineering (Autonomous), Kolhapur, India

4 Modern College of Arts, Science and Commerce (Autonomous) Shivajinagar, Pune, India

*Address all correspondence to: santoshkumar.patil.19@gmail.com; kiyoung@inha.ac.kr

References

[1] Agoramoorthy G. India's pollution nightmare: Can it be tackled? Environmental Science & Technology. 2012;**46**:1305-1306

[2] Comstock MJ. Evaluation of pesticides in ground water, copyright, ACS symposium series, foreword. In: Comstock MJ, editor. Evaluation of Pesticides in Ground Water. Vol. 315. Washington, DC: American Chemical Society; 1986

[3] Joo DJ et al. Decolorization of reactive dyes using inorganic coagulants and synthetic polymer. Dyes and Pigments. 2007;**73**:59-64

[4] Zhang D, Zeng F. Visible light-activated cadmium-doped ZnO nanostructured photocatalyst for the treatment of methylene blue dye. Journal of Materials Science. 2012;**47**:2155-2161

[5] Patil SS et al. Green approach for hierarchical nanostructured Ag-ZnO and their photocatalytic performance under sunlight. Catalysis Today. 2016;**260**:126-134

[6] Zhang S et al. Visible-light photocatalytic degradation of methylene blue using SnO_2/α-Fe2O3 hierarchical nanoheterostructures. ChemPlusChem. 2013;**78**:192-199

[7] Loeb SK et al. The technology horizon for photocatalytic water treatment: Sunrise or sunset? Environmental Science & Technology. 2019;**53**:2937-2947

[8] Fujishima A, Honda K. Electrochemical photolysis of water at a semiconductor electrode. Nature. 1972;**238**:37-38

[9] Fujishima A, Zhang X, Tryk DA. Heterogeneous photocatalysis: From water photolysis to applications in environmental cleanup. International Journal of Hydrogen Energy. 2007;**32**: 2664-2672

[10] Ni M, Leung MKH, Leung DYC, Sumathy K. A review and recent developments in photocatalytic water-splitting using TiO_2 for hydrogen production. Renewable and Sustainable Energy Reviews. 2007;**11**:401-425

[11] Kubacka A, Fernández-García M, Colón G. Advanced nanoarchitectures for solar photocatalytic applications. Chemical Reviews. 2012;**112**:1555-1614

[12] Sivula K, Le Formal F, Grätzel M. Solar water splitting: Progress using hematite (α-Fe2O3) photoelectrodes. ChemSusChem. 2011;**4**:432-449

[13] Reimer T et al. Single step integration of ZnO nano- and microneedles in Si trenches by novel flame transport approach: Whispering gallery modes and photocatalytic properties. ACS Applied Materials & Interfaces. 2014;**6**:7806-7815

[14] Fanourakis SK, Peña-Bahamonde J, Bandara PC, Rodrigues DF. Nano-based adsorbent and photocatalyst use for pharmaceutical contaminant removal during indirect potable water reuse. NPJ Clean Water. 2020;**3**:1

[15] Luo J et al. A critical review on energy conversion and environmental remediation of photocatalysts with remodeling crystal lattice, surface, and interface. ACS Nano. 2019;**13**:9811-9840

[16] Chnadel N et al. Z-scheme photocatalytic dye degradation on AgBr/ Zn(Co)Fe_2O_4 photocatalysts supported on nitrogen-doped graphene. Materials Today Sustainability. 2020;**9**:100043

[17] Mafa PJ, Ntsendwana B, Mamba BB, Kuvarega AT. Visible light driven $ZnMoO_4/BiFeWO_6/rGO$ Z-scheme photocatalyst for the degradation of anthraquinonic dye. Journal of Physical Chemistry C. 2019;**123**:20605-20616

[18] Khan F et al. Recent advances in graphene oxide and reduced graphene oxide based nanocomposites for the photodegradation of dyes. Journal of Materials Chemistry C. 2020;**8**:15940-15955

[19] Panneri S et al. C3N4 anchored ZIF 8 composites: Photo-regenerable, high capacity sorbents as adsorptive photocatalysts for the effective removal of tetracycline from water. Catalysis Science & Technology. 2017;7:2118-2128

[20] Kamble GSKE-G. Introductory Chapter: Graphene Oxide: Applications and Opportunities. Rijeka: IntechOpen; 2018. DOI: 10.5772/intechopen.79640

[21] Zhang N, Zhang Y, Pan X, Yang M-Q, Xu Y-J. Constructing ternary CdS–graphene–TiO_2 hybrids on the flatland of graphene oxide with enhanced visible-light photoactivity for selective transformation. Journal of Physical Chemistry C. 2012;**116**:18023-18031

[22] Hoffmann MR, Martin ST, Choi W, Bahnemann DW. Environmental applications of semiconductor photocatalysis. Chemical Reviews. 1995;**95**:69-96

[23] Giovannetti R, Rommozzi E, Zannotti M, D'Amato CA. Recent advances in graphene based TiO_2 nanocomposites ($GTiO_2Ns$) for photocatalytic degradation of synthetic dyes. Catalysts. 2017;7:305

[24] Nurazzi NM et al. The frontiers of functionalized graphene-based nanocomposites as chemical sensors. Nanotechnology Reviews. 2021;**10**:330-369

[25] Ambrosi A, Chua CK, Bonanni A, Pumera M. Electrochemistry of graphene and related materials. Chemical Reviews. 2014;**114**:7150-7188

[26] Backes C et al. Production and processing of graphene and related materials. 2D Materials. 2020;**7**:22001

[27] Mecklenburg M et al. Aerographite: Ultra lightweight, flexible nanowall, carbon microtube material with outstanding mechanical performance. Advanced Materials. 2012;**24**:3486-3490

[28] Lee SY, Kang D, Jeong S, Do HT, Kim JH. Photocatalytic Degradation of rhodamine B dye by TiO_2 and gold nanoparticles supported on a floating porous polydimethylsiloxane sponge under ultraviolet and visible light irradiation. ACS Omega. 2020;**5**:4233-4241

[29] Rajendrachari S et al. Photocatalytic degradation of Rhodamine B (RhB) dye in waste water and enzymatic inhibition study using cauliflower shaped ZnO nanoparticles synthesized by a novel One-pot green synthesis method. Arabian Journal of Chemistry. 2021;**14**:103180

[30] Wetchakun N, Chainet S, Phanichphant S, Wetchakun K. Efficient photocatalytic degradation of methylene blue over $BiVO_4/TiO_2$ nanocomposites. Ceramics International. 2015;**41**:5999-6004

[31] Leary R, Westwood A. Carbonaceous nanomaterials for the enhancement of TiO2 photocatalysis. Carbon. 2011;**49**:741-772

[32] Woan K, Pyrgiotakis G, Sigmund W. Photocatalytic carbon-nanotube–TiO2

composites. Advanced Materials. 2009;**21**:2233-2239

[33] Singh S, Faraz M, Khare N. Recent advances in semiconductor–graphene and semiconductor–ferroelectric/ferromagnetic nanoheterostructures for efficient hydrogen generation and environmental remediation. ACS Omega. 2020;**5**:11874-11882

[34] Peter CN et al. N-doped ZnO/graphene oxide: A photostable photocatalyst for improved mineralization and photodegradation of organic dye under visible light. Ionics (Kiel). 2019;**25**:327-339

[35] An X, Yu JC. Graphene-based photocatalytic composites. RSC Advances. 2011;**1**:1426-1434

[36] Zhang J, Xiong Z, Zhao XS. Graphene–metal–oxide composites for the degradation of dyes under visible light irradiation. Journal of Materials Chemistry. 2011;**21**:3634-3640

[37] Lü W et al. Graphene-enhanced visible-light photocatalysis of large-sized CdS particles for wastewater treatment. Nanoscale Research Letters. 2014;**9**:148

[38] Wang C et al. Facile synthesis of nanoporpous graphitic carbon nitride photocatalyst coupled with N-doped graphene quantum dots for efficient photo-degradation dyes under nature solar light radiation. Diamond and Related Materials. 2018;**89**:197-205

[39] Tiwari B, Ram S. Biogenic synthesis of graphitic carbon nitride for photocatalytic degradation of organic dyes. ACS Omega. 2019;**4**:10263-10272

[40] Ma N, Chen A, Bian Z, Yang Y, Wang H. In situ synthesis of a cadmium sulfide/reduced graphene oxide/bismuth Z-scheme oxyiodide system for enhanced photocatalytic performance in chlorinated paraben degradation. Chemical Engineering Journal. 2019;**359**:530-541

[41] Wang H et al. Photodeposition of metal sulfides on titanium metal–organic frameworks for excellent visible-light-driven photocatalytic Cr(vi) reduction. RSC Advances. 2015;**5**:32531-32535

[42] Isac L, Cazan C, Enesca A, Andronic L. Copper sulfide based heterojunctions as photocatalysts for dyes photodegradation. Frontiers in Chemistry. 2019;**7**:694

[43] Park C et al. Preparation of CuS-graphene oxide/TiO2 composites designed for high photonic effect and photocatalytic activity under visible light. Chinese Journal of Catalysis. 2013;**34**:711-717

[44] Hu X-S, Shen Y, Zhang Y-T, Nie J-J. Preparation of flower-like CuS/reduced graphene oxide(RGO) photocatalysts for enhanced photocatalytic activity. Journal of Physics and Chemistry of Solids. 2017;**103**:201-208

[45] Gu W, Deng X, Jia X, Li J, Wang E. Functionalized graphene/Fe3O4 supported AuPt alloy as a magnetic, stable and recyclable catalyst for a catalytic reduction reaction. Journal of Materials Chemistry A. 2015;**3**:8793-8799

[46] Sharma M et al. Assessment of GO/ZnO nanocomposite for solar-assisted photocatalytic degradation of industrial dye and textile effluent. Environmental Science and Pollution Research. 2020;**27**:32076-32087

[47] Al-Rawashdeh NAF, Allabadi O, Aljarrah MT. Photocatalytic activity of graphene oxide/zinc oxide nanocomposites with embedded metal nanoparticles for the degradation

of organic dyes. ACS Omega. 2020;**5**:28046-28055

[48] Mandal S et al. An overview on graphene-metal oxide semiconductor nanocomposite: A promising platform for visible light photocatalytic activity for the treatment of various pollutants in aqueous medium. Molecules. 2020;**25**:5380

[49] Patil SS et al. Graphene-wrapped $Ag_3PO_4/LaCO_3OH$ heterostructures for water purification under visible light. Journal of Energy Chemistry. 2016;**25**:845-853

[50] Patil SS et al. One-pot in situ hydrothermal growth of $BiVO_4$/Ag/rGO hybrid architectures for solar water splitting and environmental remediation. Scientific Reports. 2017;**7**:8404

[51] Yang R, He J, Niu M, Fan Y, Zhu R. The photocatalytic activity of GO-modified BiVO4 for the degradation of phenol under visible light irradiation. Chemical Physics Letters. 2019;**735**:136770

[52] Biswas MRUD, Oh W-C. Synthesis of $BiVO_4$-GO-PVDF nanocomposite: An excellent, newly designed material for high photocatalytic activity towards organic dye degradation by tuning band gap energies. Solid State Sciences. 2018;**80**:22-30

[53] Chong S, Zhang G, Tian H, Zhao H. Rapid degradation of dyes in water by magnetic $Fe0/Fe_3O_4$/graphene composites. Journal of Environmental Sciences. 2016;**44**:148-157

[54] Fu Y, Wang X. Magnetically separable $ZnFe_2O_4$–graphene catalyst and its high photocatalytic performance under visible light irradiation. Industrial and Engineering Chemistry Research. 2011;**50**:7210-7218

[55] Zhang H, Lv X, Li Y, Wang Y, Li J. P25-graphene composite as a high performance photocatalyst. ACS Nano. 2010;**4**:380-386

[56] Zhou K, Zhu Y, Yang X, Jiang X, Li C. Preparation of graphene–TiO_2 composites with enhanced photocatalytic activity. New Journal of Chemistry. 2011;**35**:353-359

[57] Chen C et al. Synthesis of visible-light responsive graphene Oxide/TiO_2 composites with p/n heterojunction. ACS Nano. 2010;**4**:6425-6432

[58] Xu T, Zhang L, Cheng H, Zhu Y. Significantly enhanced photocatalytic performance of ZnO via graphene hybridization and the mechanism study. Applied Catalysis B: Environmental. 2011;**101**:382-387

[59] Phanichphant S, Nakaruk A, Chansaenpak K, Channei D. Evaluating the photocatalytic efficiency of the $BiVO_4$/rGO photocatalyst. Scientific Reports. 2019;**9**:16091

[60] Ye A, Fan W, Zhang Q, Deng W, Wang Y. CdS–graphene and CdS–CNT nanocomposites as visible-light photocatalysts for hydrogen evolution and organic dye degradation. Catalysis Science & Technology. 2012;**2**:969-978

[61] Liao G, Chen S, Quan X, Yu H, Zhao H. Graphene oxide modified g-C_3N_4 hybrid with enhanced photocatalytic capability under visible light irradiation. Journal of Materials Chemistry. 2012;**22**:2721-2726

[62] Zhu M, Chen P, Liu M. Graphene oxide enwrapped Ag/AgX (X = Br, Cl) nanocomposite as a highly efficient visible-light plasmonic photocatalyst. ACS Nano. 2011;**5**:4529-4536

[63] Shuang S et al. Efficient photocatalysis with graphene

oxide/Ag/Ag_2S–TiO_2 nanocomposites under visible light irradiation. RSC Advances. 2018;**8**:5784-5791

[64] Stobel Christy EJ, Pius A. Performance of metal free g-C3N4 reinforced graphene oxide bio-composite for the removal of persistent dyes. Environmental Chemistry and Ecotoxicology. 2021;**3**:220-233

[65] Shamraiz U, Hussain RA, Badshah A, Raza B, Saba S. Functional metal sulfides and selenides for the removal of hazardous dyes from Water. Journal of Photochemistry and Photobiology B: Biology. 2016;**159**:33-41

[66] Chen X et al. Fabricate globular flower-like CuS/$CdIn_2S_4$/$ZnIn_2S_4$ with high visible light response via microwave-assisted one–step method and its multipathway photoelectron migration properties for hydrogen evolution and pollutant degradation. ACS Sustainable Chemistry & Engineering. 2016;**4**:6680-6688

[67] Wang X et al. A metal-free polymeric photocatalyst for hydrogen production from water under visible light. Nature Materials. 2009;**8**:76-80

[68] Khan ME, Khan MM, Cho MH. Environmentally sustainable biogenic fabrication of AuNP decorated-graphitic g-C_3N_4 nanostructures towards improved photoelectrochemical performances. RSC Advances. 2018;**8**:13898-13909

[69] Ding F et al. Graphitic carbon nitride-based nanocomposites as visible-light driven photocatalysts for environmental purification. Environmental Science: Nano. 2017;**4**:1455-1469

[70] Bai X, Wang L, Zong R, Zhu Y. Photocatalytic activity enhanced via g-C3N4 nanoplates to nanorods. Journal of Physical Chemistry C. 2013;**117**:9952-9961

[71] Khan ME. State-of-the-art developments in carbon-based metal nanocomposites as a catalyst: Photocatalysis. Nanoscale Advances. 2021;**3**:1887-1900

[72] Tonda S, Kumar S, Kandula S, Shanker V. Fe-doped and -mediated graphitic carbon nitride nanosheets for enhanced photocatalytic performance under natural sunlight. Journal of Materials Chemistry A. 2014;**2**:6772-6780

[73] Gao H, Yan S, Wang J, Zou Z. Ion coordination significantly enhances the photocatalytic activity of graphitic-phase carbon nitride. Dalton Transactions. 2014;**43**:8178-8183

[74] Wang Y et al. Simple synthesis of Zr-doped graphitic carbon nitride towards enhanced photocatalytic performance under simulated solar light irradiation. Catalysis Communications. 2015;**72**:24-28

[75] Pan H, Zhang Y-W, Shenoy VB, Gao H. Ab initio study on a novel photocatalyst: Functionalized graphitic carbon nitride nanotube. ACS Catalysis. 2011;**1**:99-104

[76] Zheng C, Chen W, Huang Y, Xiao X, Ye X. Graphene oxide–noble metal (Au, Pt, and Pd) nanoparticle composites as optical limiters. RSC Advances. 2014;**4**:39697-39703

[77] Qiu B, Xing M, Zhang J. Recent advances in three-dimensional graphene based materials for catalysis applications. Chemical Society Reviews. 2018;**47**:2165-2216

[78] Prakash J, Harris RA, Swart HC. Embedded plasmonic nanostructures:

Synthesis, fundamental aspects and their surface enhanced Raman scattering applications. International Reviews in Physical Chemistry. 2016;**35**:353-398

[79] Zhang X, Chen YL, Liu R-S, Tsai DP. Plasmonic photocatalysis. Reports on Progress in Physics. 2013;**76**:46401

[80] Khan ME, Khan MM, Cho MH. Recent progress of metal–graphene nanostructures in photocatalysis. Nanoscale. 2018;**10**:9427-9440

[81] Kurt BZ, Durmus Z, Durmus A. Preparation and characterization of platinum (Pt) and palladium (Pd) nanoparticle decorated graphene sheets and their utilization for the elimination of basic fuchsin and indigo carmine dyes. Solid State Sciences. 2016;**51**:51-58

[82] Koklioti MA et al. Mechanistic insights into the photocatalytic properties of metal nanocluster/graphene ensembles. Examining the role of visible light in the reduction of 4-nitrophenol. Nanoscale. 2017;**9**:9685-9692

[83] Hareesh K, Joshi RP, Sunitha DV, Bhoraskar VN, Dhole SD. Anchoring of Ag-Au alloy nanoparticles on reduced graphene oxide sheets for the reduction of 4-nitrophenol. Applied Surface Science. 2016;**389**:1050-1055

[84] Singh N, Prakash J, Gupta RK. Design and engineering of high-performance photocatalytic systems based on metal oxide–graphene–noble metal nanocomposites. Molecular Systems Design and Engineering. 2017;**2**:422-439

[85] Wen Y, Ding H, Shan Y. Preparation and visible light photocatalytic activity of Ag/TiO2/graphene nanocomposite. Nanoscale. 2011;**3**:4411-4417

[86] Huang HJ, Zhen S-Y, Li P-Y, Tzeng S-D, Chiang H-P. Confined migration of induced hot electrons in Ag/graphene/TiO_2 composite nanorods for plasmonic photocatalytic reaction. Optics Express. 2016;**24**:15603-15608

[87] Jafari Z et al. Ag/TiO2/freeze-dried graphene nanocomposite as a high performance photocatalyst under visible light irradiation. Journal of Energy Chemistry. 2016;**25**:393-402

[88] Vasilaki E, Georgaki I, Vernardou D, Vamvakaki M, Katsarakis N. Ag-loaded TiO2/reduced graphene oxide nanocomposites for enhanced visible-light photocatalytic activity. Applied Surface Science. 2015;**353**:865-872

[89] Ghasemi S, Hashemian SJ, Alamolhoda AA, Gocheva I, Rahman Setayesh S. Plasmon enhanced photocatalytic activity of Au@TiO_2-graphene nanocomposite under visible light for degradation of pollutants. Materials Research Bulletin. 2017;**87**: 40-47

[90] Ghasemi S et al. Synthesis and characterization of TiO_2–graphene nanocomposites modified with noble metals as a photocatalyst for degradation of pollutants. Applied Catalysis A: General. 2013;**462-463**:82-90

[91] Yang T-H et al. Fully integrated Ag nanoparticles/ZnO nanorods/graphene heterostructured photocatalysts for efficient conversion of solar to chemical energy. Journal of Catalysis. 2015;**329**:167-176

[92] Khoa NT et al. Fast and effective electron transport in a Au–graphene–ZnO hybrid for enhanced photocurrent and photocatalysis. RSC Advances. 2015;**5**:63964-63969

[93] Khoa NT et al. Fabrication of Au/graphene-wrapped ZnO-nanoparticle-assembled hollow spheres with effective

photoinduced charge transfer for photocatalysis. ACS Applied Materials & Interfaces. 2015;7:3524-3531

[94] Zhang L et al. Significantly enhanced photocatalytic activities and charge separation mechanism of Pd-decorated ZnO–graphene oxide nanocomposites. ACS Applied Materials & Interfaces. 2014;**6**:3623-3629

[95] Zhao R et al. Au/Pd/g-C_3N_4 nanocomposites for photocatalytic degradation of tetracycline hydrochloride. Journal of Materials Science. 2019;**54**:5445-5456

[96] Li H, Zhao F, Liu T, Zhang N, Wang Y. Design of novel structured Au/g-C3N4 nanosheet/reduced graphene oxide nanocomposites for enhanced visible light photocatalytic activities. Sustainable Energy and Fuels. 2020;**4**:4086-4095

[97] Kim S-H et al. Synthesis of noble metal/graphene nanocomposites without surfactants by one-step reduction of metal salt and graphene oxide. Journal of Colloid and Interface Science. 2013;**389**:85-90

[98] Wang N et al. Sub-10 nm Ag nanoparticles/graphene oxide: Controllable synthesis, size-dependent and extremely ultrahigh catalytic activity. Small. 2019;**15**:1901701

[99] Ji Z et al. Facile synthesis of magnetically separable reduced graphene oxide/magnetite/silver nanocomposites with enhanced catalytic activity. Journal of Colloid and Interface Science. 2015;**459**:79-85

[100] Rajesh R, Sujanthi E, Senthil Kumar S, Venkatesan R. Designing versatile heterogeneous catalysts based on Ag and Au nanoparticles decorated on chitosan functionalized graphene oxide. Physical Chemistry Chemical Physics. 2015;**17**:11329-11340

[101] Islam MR et al. Recyclable Ag-decorated highly carbonaceous magnetic nanocomposites for the removal of organic pollutants. Journal of Colloid and Interface Science. 2020;**562**:52-62

[102] Perreault F, de Faria A, Elimelech M. Environmental applications of graphene-based nanomaterials. Chemical Society Reviews. 2015;**44**:5861-5896

[103] Ali I et al. Graphene based adsorbents for remediation of noxious pollutants from wastewater. Environment International. 2019;**127**:160-180

[104] Xiang Q, Yu J, Jaroniec M. Graphene-based semiconductor photocatalysts. Chemical Society Reviews. 2012;**41**:782-796

[105] Gao E, Wang W, Shang M, Xu J. Synthesis and enhanced photocatalytic performance of graphene-Bi_2WO_6 composite. Physical Chemistry Chemical Physics. 2011;**13**:2887-2893

[106] Bilgin Simsek E, Kilic B, Asgin M, Akan A. Graphene oxide based heterojunction TiO2–ZnO catalysts with outstanding photocatalytic performance for bisphenol-A, ibuprofen and flurbiprofen. Journal of Industrial and Engineering Chemistry. 2018;**59**:115-126

[107] Pan D et al. Efficient separation of electron–hole pairs in graphene quantum dots by TiO2 heterojunctions for dye degradation. ACS Sustainable Chemistry & Engineering. 2015;**3**:2405-2413

[108] Geim A, Graphene K. Status and Prospects. Science. 2009;**324**: 1530-1534

[109] Bolotin KI et al. Ultrahigh electron mobility in suspended graphene. Solid State Communications. 2008;**146**: 351-355

[110] Sreeprasad TS, Berry V. How do the electrical properties of graphene change with its functionalization? Small. 2013;**9**:341-350

[111] Zhao G, Li J, Ren X, Chen C, Wang X. Few-layered graphene oxide nanosheets as superior sorbents for heavy metal ion pollution management. Environmental Science & Technology. 2011;**45**:10454-10462

[112] Wu W et al. Highly efficient removal of Cu(II) from aqueous solution by using graphene oxide. Water, Air, and Soil Pollution. 2012;**224**:1372

[113] Sitko R et al. Adsorption of divalent metal ions from aqueous solutions using graphene oxide. Dalton Transations. 2013;**42**:5682-5689

[114] Wang H et al. Adsorption characteristics and behaviors of graphene oxide for Zn(II) removal from aqueous solution. Applied Surface Science. 2013;**279**:432-440

[115] Mi X et al. Preparation of graphene oxide aerogel and its adsorption for Cu^{2+} ions. Carbon. 2012;**50**:4856-4864

[116] Huang Z-H et al. Adsorption of lead(II) ions from aqueous solution on low-temperature exfoliated graphene nanosheets. Langmuir. 2011;**27**:7558-7562

[117] Yang Y et al. Preparation of reduced graphene oxide/poly(acrylamide) nanocomposite and its adsorption of Pb(II) and methylene blue. Langmuir. 2013;**29**:10727-10736

[118] Sreeprasad TS, Maliyekkal SM, Lisha KP, Pradeep T. Reduced graphene oxide–metal/metal oxide composites: Facile synthesis and application in water purification. Journal of Hazardous Materials. 2011;**186**:921-931

[119] Deng X, Lü L, Li H, Luo F. The adsorption properties of Pb(II) and Cd(II) on functionalized graphene prepared by electrolysis method. Journal of Hazardous Materials. 2010;**183**:923-930

[120] Ma H-L et al. Chemical reduction and removal of Cr(vi) from acidic aqueous solution by ethylenediamine-reduced graphene oxide. Journal of Materials Chemistry. 2012;**22**:5914-5916

[121] Yao Y et al. Magnetic CoFe2O4–graphene hybrids: Facile synthesis, characterization, and catalytic properties. Industrial and Engineering Chemistry Research. 2012;**51**:6044-6051

[122] Gollavelli G, Chang C-C, Ling Y-C. Facile synthesis of smart magnetic graphene for safe drinking water: Heavy metal removal and disinfection control. ACS Sustainable Chemistry & Engineering. 2013;**1**:462-472

[123] Zhu J et al. One-pot synthesis of magnetic graphene nanocomposites decorated with core@double-shell nanoparticles for fast chromium removal. Environmental Science & Technology. 2012;**46**:977-985

[124] Li J et al. Removal of Cu(II) and fulvic acid by graphene oxide nanosheets decorated with Fe3O4 nanoparticles. ACS Applied Materials & Interfaces. 2012;**4**:4991-5000

[125] Chandra V et al. Water-dispersible magnetite-reduced graphene oxide composites for arsenic removal. ACS Nano. 2010;**4**:3979-3986

[126] Jabeen H et al. Enhanced Cr(vi) removal using iron nanoparticle

decorated graphene. Nanoscale. 2011;**3**:3583-3585

[127] Cao C-Y, Qu J, Wei F, Liu H, Song W-G. Superb adsorption capacity and mechanism of flowerlike magnesium oxide nanostructures for lead and cadmium ions. ACS Applied Materials & Interfaces. 2012;**4**:4283-4287

[128] Liu L et al. Preparation and characterization of chitosan/graphene oxide composites for the adsorption of Au(III) and Pd(II). Talanta. 2012;**93**:350-357

[129] Zhang S, Shao Y, Liu J, Aksay IA, Lin Y. Graphene–polypyrrole nanocomposite as a highly efficient and low cost electrically switched ion exchanger for removing ClO4–from wastewater. ACS Applied Materials & Interfaces. 2011;**3**:3633-3637

[130] Dinda D, Gupta A, Saha SK. Removal of toxic Cr(vi) by UV-active functionalized graphene oxide for water purification. Journal of Materials Chemistry A. 2013;**1**:11221-11228

[131] Liu J et al. Self-assembling TiO_2 nanorods on large graphene oxide sheets at a two-phase interface and their anti-recombination in photocatalytic applications. Advanced Functional Materials. 2010;**20**:4175-4181

[132] Liang Y, Wang H, Sanchez Casalongue H, Chen Z, Dai H. TiO_2 nanocrystals grown on graphene as advanced photocatalytic hybrid materials. Nano Research. 2010;**3**:701-705

[133] Perera SD et al. Hydrothermal synthesis of graphene-TiO_2 nanotube composites with enhanced photocatalytic activity. ACS Catalysis. 2012;**2**:949-956

[134] Xiong Z, Zhang LL, Ma J, Zhao XS. Photocatalytic degradation of dyes over graphene–gold nanocomposites under visible light irradiation. Chemical Communications. 2010;**46**:6099-6101

[135] Gupta VK et al. $CoFe_2O_4$@TiO_2 decorated reduced graphene oxide nanocomposite for photocatalytic degradation of chlorpyrifos. Journal of Molecular Liquids. 2015;**208**:122-129

[136] Cruz M et al. Bare TiO2 and graphene oxide TiO_2 photocatalysts on the degradation of selected pesticides and influence of the water matrix. Applied Surface Science. 2017;**416**:1013-1021

[137] Thakur K, Kandasubramanian B. Graphene and graphene oxide-based composites for removal of organic pollutants: A review. Journal of Chemical & Engineering Data. 2019;**64**:833-867

[138] Sravya S, RamaDevi D, Belachew N, Rao KE, Basavaiah K. Vitamin C assisted synthesis of rGO–Ag/PANI nanocomposites for improved photocatalytic degradation of pharmaceutical wastes. RSC Advances. 2021;**11**:12030-12035

[139] He Y, Sutton NB, Rijnaarts HHM, Langenhoff AAM. Corrigendum to Degradation of pharmaceuticals in wastewater using immobilized TiO_2 photocatalysis under simulated solar irradiation. Applied Catalysis B: Environmental. 2016;**189**:283

[140] Anku WW et al. Photocatalytic degradation of pharmaceuticals using graphene based materials. In: Naushad M, editor. A New Generation Material Graphene: Applications in Water Technology. Berlin: Springer International Publishing; 2019. pp. 187-208. DOI: 10.1007/978-3-319-75484-0_7

[141] Majumder S, Chatterjee S, Basnet P, Mukherjee J. ZnO based nanomaterials for photocatalytic degradation of aqueous

pharmaceutical waste solutions – A contemporary review. Environmental Nanotechnology, Monitoring and Management. 2020;**14**:100386

[142] Anirudhan TS, Deepa JR, Nair AS. Fabrication of chemically modified graphene oxide/nano hydroxyapatite composite for adsorption and subsequent photocatalytic degradation of aureomycine hydrochloride. Journal of Industrial and Engineering Chemistry. 2017;**47**:415-430

Chapter 9

Processing of Graphene/Elastomer Nanocomposites: A Minireview

Mohammed A. Sharaf and Andrzej Kloczkowski

Abstract

Since the isolation and identification of graphene, the academic and industrial communities are utilizing its superior properties. This minireview deals with the processing of graphene-based fillers/elastomer nanocomposites. The incorporation of graphene in an elastomeric matrices has significant effects on the properties of nanocomposites. The dispersion of graphene in elastomers is discussed. The processing of graphene/elastomer nanocomposites is discussed. The mechanical properties of the elastomeric matrix can be enhanced due to the presence of graphene. In this review and due to space limitations, we will present an example of improvements in the mechanical characteristics of graphene/styrene-butadiene (SBR) elastomer nanocomposites.

Keywords: elastomers; graphene-based filers; dispersion, interfacial interactions, mechanical characteristics, mechanical percolation

1. Introduction

1.1 Elastomers

Elastomers exhibit rubber-like behavior [1–4]. They are characterized by weak intermolecular forces, ease of deformation at ambient temperatures, low modulus of elasticity, and can stretch to high ultimate strains. Also, elastomers manifest good heat resistance. To sum up, elastomers are characterized by: (i) their amorphous high molecular mass molecular chains that are randomly coiled; (ii) bonds in the elastomer molecules freely rotate or extend in response to any applied strain; (iii) glass transition temperature for elastomers ought to be lower than their service temperatures; and (iv) a low degree of vulcanization to control shape [1–4].

Elastomers have low mechanical, thermal, and electrical conductivity properties. Compounding elastomeric materials with fillers results in composites with an average mix of properties of the original constituents. Owing to their properties and their low cost, they have established their niche in several technological applications that include automotive, aerospace, military, conveyor belts, packaging, healthcare, and numerous others.

Elastomers constitute the polymeric matrix in the production of elastomer-based nanocomposites that possess improved properties [5–12]. On the other hand, elastomeric nanocomposites demand different preparative and processing procedures as compared with polymer-based nanocomposites. The process of their cross-linking is different and requires the use of an activator.

The incorporation of inorganic fillers such as silica nanoparticles, layered silicates (clay), carbon black, carbon nanotubes, and other nanomaterials results in the production of high-performance elastomeric nanocomposites [6, 13–18]. The final properties of the nanocomposite are affected by the type of filler involved due to differences in their structural and geometrical characteristics.

1.2 Graphene

Graphene (GE) is 2D a carbon allotrope consisting of sp^2 hybridized carbons, which are arranged in a honeycomb lattice [19, 20]. It is a single-atom-thick nano-structured sheet that is considered the basic building block for graphene-based fillers [21–23]. Because of its unique exceptional properties, graphene has emerged as a very promising nanomaterial. It is one of the strongest materials and a good conductor of heat and electricity; it is optically transparent and impermeable to gases [24–26]. In **Figure 1**, the graphene hexagonal honeycomb chemical structure is represented [27]. Some notable properties of graphene are listed in **Table 1** [28].

Once the mass-produced graphene has qualities comparable to those produced in research laboratories, it will be of greater interest in applications [33–35]. However, there are still various barriers and risks to overcome [36].

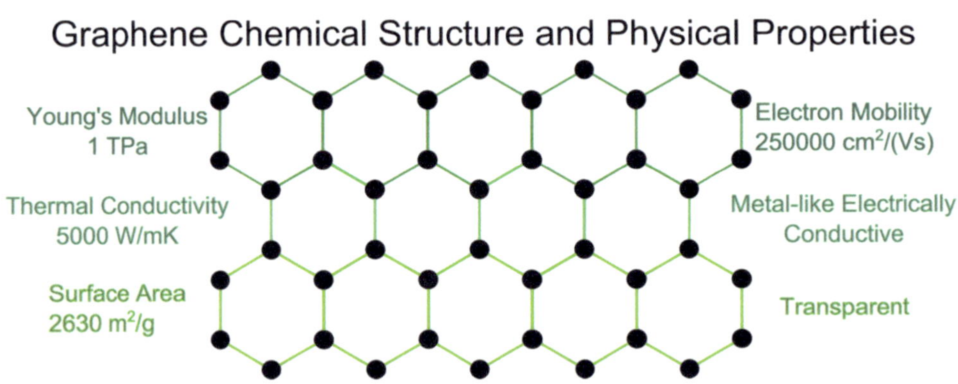

Figure 1.
Graphene hexagonal honeycomb chemical structure and its remarkable physical properties. The black dots are carbon atoms [27].

Property	Experimental values	Refs.
Elastic modulus	~1 TPa	[29]
Tensile strength	~130 GPa	
Fatigue resistance	10^9 cycles at mean stress of 71 GPa	[30]
Thermal conductivity	~5300 W m^{-1} K^{-1}	[31]
Thermal stability	~2600 K	[32]
Electrical conductivity	6000 S cm^{-1}	[33]

Table 1.
Some notable properties of graphene [28].

Polymer nanocomposites reinforced by a single- and few-layer graphene have demonstrated significant improvements in various properties [37]. Another encouraging factor in the use of graphene in polymer nanocomposites is that it has been realized that a minute quantity of can leads to significant enhancements in properties, Undoubtedly, graphene is a favorite candidate for the use in nanocomposites. Emphasis will be given here to the techniques used in the preparation of graphene, elastomers, and nanocomposites, along with the physicochemical aspects and attributes reported and the applications of the final materials.

This chapter will be concerned with the processing and properties of graphene/elastomer nanocomposites.

2. Some considerations concerning the processing graphene-based/elastomer nanocomposites

2.1 Filler dispersion

The efficient preparation of an elastomeric nanocomposite requires appropriate dispersion of the filler in the elastomeric matrix. Whence, there is a need for the filler and matrix to be compatible, physically and/or chemically. Most properties of nanocomposites depend on the nanofillers structure in the matrix. Therefore, such intimate nanofiller/elastomeric matrix interactions are dependent on the filler size, morphology, surface treatment, and activity. Undoubtedly, the fine dispersion of nanosized particles would lead to a very large interfacial area. Undoubtedly, the reinforcement changes with the specific interfacial area, for fillers with the same chemical nature [38–40]. As a result, the filler-matrix interactions and bonding will increase, in the meantime, the filler-filler interactions will decrease, and/or the platelet interlayer distance will increase [41–45]. They permit lower loadings, weight reduction of the fillers, and allow for the nanoparticles of ≤ 1 μm size the formation of a sort of colloidal suspensions. In the specific case of graphite-based fillers, because of the positive effect of aromatic rings, they would allow for the establishment of an intimate interaction between the nanofiller and the rubber matrix [40, 46–48]. One should note that elastomeric polymers possess high viscosity and thus necessitate intensive mixing and blending to achieve an even dispersion and distribution.

During the dispersive step, in the mixing process, large components undergo size reduction through erosion and rupture into smaller fragments that are then separated into the matrix [49–51]. These two last phases most likely happen rather simultaneously [49, 52]. Ideal dispersion of platelets-like (lamellar) nanofillers requires that all layers should be separated from each other, i.e., complete exfoliation. However, the strong interlayer interactions need to be circumvented, more often through intercalations to facilitate exfoliation [53–55].

Graphite is like montmorillonite clay (MMT), and researchers applied similar procedures for the dispersion of graphene-based fillers [56–58]. Intercalated and exfoliated compounds are in common usage. To reduce the strong p-p interactions between graphene platelets, graphene sheets are customarily surface-modified [56–58]. For the classification of graphene-based fillers, for example, a stack of few-layer graphene is called graphene nanoplatelets (GNP) [27, 59, 60].

Figure 2b is a representation of stacked, expanded, and exfoliated graphite platelets in an elastomeric matrix. Exfoliation of GNPs into an elastomer is desired to enhance the properties of the final material. Most exfoliation processes are performed in suspension rather than during bulk processing, during which, high shear is reported to break the platelets rather than exfoliate them [27, 59, 60].

2.2 Surface functionalization

For surface functionalization of graphene sheets, there is firstly a *non-covalent* and secondly a *covalent* approach. Non-covalent functionalization is primarily based on hydrophobic, van der Waals, and electrostatic interactions [61–65]. Covalent functionalization is dependent on the oxygen functional groups on the graphene surface. They can be utilized to change the surface functionality of graphene. The main functional groups are the carboxylic acid at the edges and epoxy and hydroxyl groups on the basal plane [66–69].

A comprehensive summary of the functionalization of graphene has been reviewed in the literature [70–74].

2.3 Graphene-based fillers in the reinforcement of elastomer nanocomposites

Fillers have been employed as reinforcing agents for ages. Fillers, depending on their activity, can be categorized as extenders, which increase the volume of materials. On the one hand, reinforcing fillers leads to improve certain physical and mechanical properties. The reinforcing ability of fillers is dependent on the particle shape, size, agglomeration (dispersion), surface properties, and most importantly, the degree of interaction between the filler and the matrix.

2.3.1 Graphene oxide (GO) as a reinforcing filler

The research community focused attention on the field of polymer nanocomposites on graphene oxide (GO). This could be attributed to the variety of chemical functional groups available on the surface of GO that make it tunable, hydrophilic, and relatively low-cost as compared with pristine graphene. Its interaction with polar polymeric matrices is improved by the presence of such hydrophilic functional groups.

Improving the interfacial adhesion of GO and the host polymers through covalent bonding has been the subject of several studies [75–77]. Thus, GO becomes intertwined as a single phase with the host chains. In consequence, the superior properties of GO are transferred to the matrix, and thus improvements in the final properties of the composites will ensue. Various approaches have been explored for producing GO-polymer nanocomposites through covalent bonding [78–83]. These methods normally involve the incorporation of thermally reduced GO (TrGO) [84, 85] or in situ chemically reduced GO (CrGO) in the rubber matrix [86, 87]. Homogeneity of TOGO in the rubber matrix is assisted by ultrasonication or high shear mixing. A strong polymer-filler interaction is critical for GO to act as a successful reinforcement agent [88–92]. For GO, the change in the degree of oxidation may significantly impact the physicochemical structure of the GO surface.

2.3.2 Other reinforcing fillers

Conventionally, the most widely used and effective conventional reinforcing filler for rubber is carbon black (CB). It provides a notable improvement in the properties of elastomers, in general. The drawback, in some cases, is when colored rubber compositions are needed. Due to its high specific surface area, silicon dioxide (SiO_2) is known to be the most effective reinforcing non-black fillers. it is reported that about 90% of the worldwide production of CB is employed and applied in the tire industry, enhancing tear strength, modulus, and wear characteristics of the tires [93]. Recently, nanofillers have gained momentum both in fundamental and industrial fields. This is mainly due to the high specific surface area and quite often to their high aspect ratio that can be dispersed in the elastomeric matrix. There exist interactions at the molecular level between the elastomeric matrix and the nanofiller. As a result, extraordinary properties are achieved compared with conventional filler materials. Based on their dimensions in host matrices, nanofillers are classified into three categories: spherical nanoparticle fillers, elongated structure with only two dimensions is in the nanometer scale, and the third is larger, e.g., nanotubes and whiskers, and layered structures with only one dimension of the filler is in the nanometer scale, such as graphene that is the subject of our study. One notes that these nanofillers have a strong tendency to form aggregates and agglomerates, owing to their high surface energies.

3. Mechanical characteristics

3.1 Effects of a covalent interphase

Many graphene-based polymer composites with significantly improved properties have been prepared and tested. Graphene with its possession of remarkable physical properties have been proposed as an efficient reinforcement agent for elastomers [94–100]. To make the best out of graphene in reinforcement, improved dispersion of graphene and higher interfacial interactions are important [101–104].

Graphene oxide is efficiently reduced by tannin derivatives. At the same time, *ortho*-quinone derivatives produced during the reduction of graphene oxide are adsorbed on the graphene surface [105, 106]. The quinones are reactive toward thiols via Michael addition [107, 108]. Polythiols are usually produced during the sulfur-cure of an elastomer [108, 109]. The *ortho*-quinone derivatives are quite suitable for constructing covalent interphase cross-linking between graphene and the rubber matrix.

In a study by Yang et al. [106], graphene/styrene-butadiene rubber (SBR) nanocomposites were prepared by latex co-coagulation of SBR latex with a suspension of graphene. They pursued a strategy to construct a strong interface in graphene/SBR nanocomposites. The *ortho*-quinone covalent interphase was realized in the graphene/rubber system. The main feature was superior dispersion and controlled reinforcement properties with the quinone-modified graphene in the rubber matrix. In this nanocomposites system, the energy loss by a tire was lower compared with those for carbon black-filled elastomers. **Figure 2** represents the manufacture of graphene/SBR nanocomposite and the mechanism of formation of covalent interphase between graphene and the SBR Matrix. The developed nanocomposites were applied to a dynamic elastomeric product, the auto tires. The very low rolling resistance coefficient is considered a great advancement in the production of tires.

3.2 Mechanical behavior of a graphene/SBR nanocomposite

With the introduction of covalent interphase, the dispersion status of graphene in the rubber matrix and the interfacial interaction between graphene and SBR were remarkably enhanced. The stress-strain isotherms of SBR and SBR nanocomposites are illustrated in **Figure 3**. The effect of graphene on the mechanical properties of

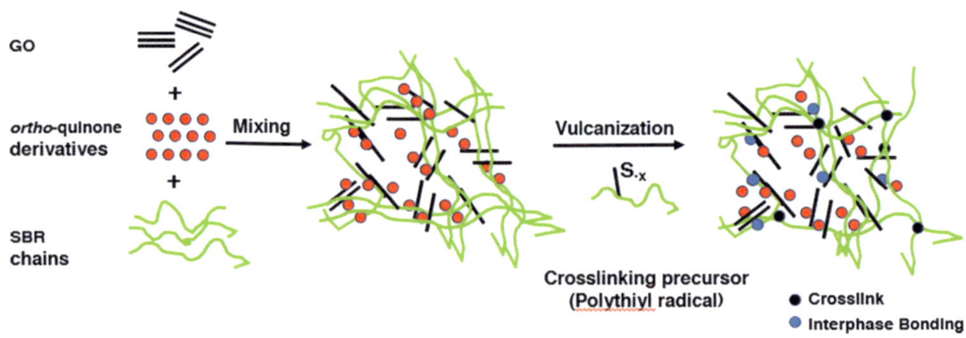

Figure 2.
Fabrication of graphene/SBR nanocomposites and a schematic illustration of proposed formation mechanism of covalent interface between graphene and rubber matrix [109].

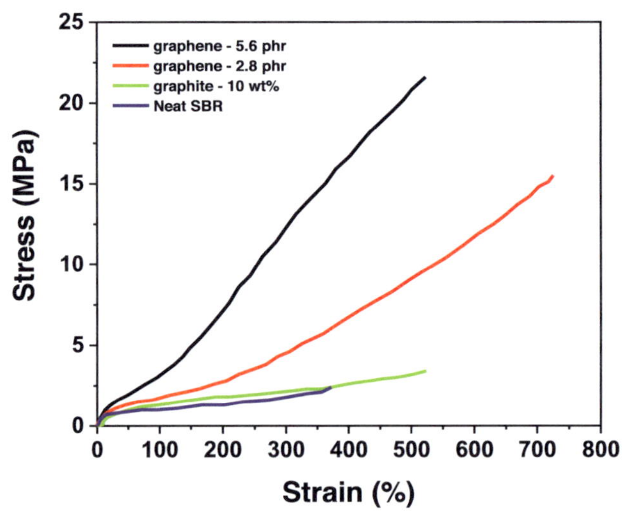

Figure 3.
Stress-strain curves of SBR/graphene nanocomposites [107].

graphene/SBR nanocomposites is apparent. For purposes of comparison, a stress-strain isotherm of 10% graphite-filled SBR composite was also made. The presented results indicate much higher efficiency in the reinforcement of SBR. In **Figure 4**, the effect of graphene loading on the mechanical properties of the nanocomposites is demonstrated. Compared with the neat SBR, the tensile strength of graphene/SBR nanocomposites with only 1.1 phr of graphene has increased by 223%. With the incorporation of 5.6 phr of graphene, the tensile strength of the nanocomposite can reach 21.5 MPa, which is over ninefold that of the neat SBR. This is a marked enhancement in reinforcement compared with other 2D reinforcing fillers such as clay [110–112].

As an additional confirmation of reinforcement, a percolation phenomenon is observed in **Figure 5** in the tensile strength with graphene loading, as has been reported earlier [106, 113–118]. Although for CNT-filled composites, 0.001 wt% percolation concentrations have been reported. In other studies, 2 wt% concentrations are reported that depend on the processing, alignment, chemistry of the CNTs, and matrix compatibility [116–118].

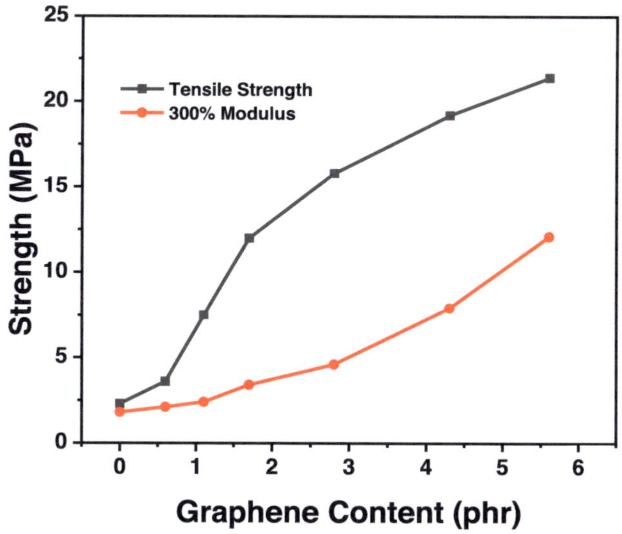

Figure 4.
Tensile modulus and strength of graphene/SBR nanocomposites [107].

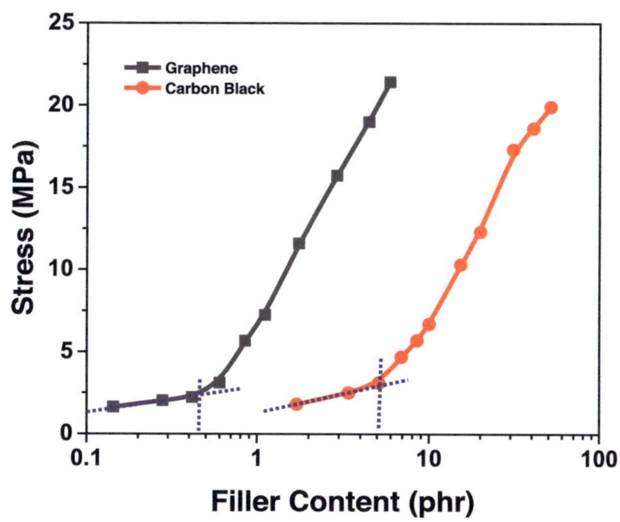

Figure 5.
Percolation phenomena of SBR nanocomposites, the intersections of the dash lines represent the percolation point [107].

In what would be considered the first stage, below the percolation threshold volume fraction of graphene, the phenomenon could be explained by the rubber strengthening mechanism that graphite nanofillers induce the formation of straightened polymer chains during the stretching process [119]. In the second stage, when the distance between the fillers decreases to a specific threshold, chains attached to the adjacent fillers will form straightened rubber chains that will enhance and strengthen the rubber [119]. In **Figure 5**, the tensile stress of the composites could be observed to initially increase slightly and then shows an abrupt increase. The percolation point of graphene/SBR nanocomposite appears surprisingly at as low as 0.42 phr, which is far below that of CB/SBR composite (5.1 phr), as seen in **Figure 5**. The ultralow percolation threshold of graphene/SBR composites might be ascribed to the excellent dispersion of ultrathin graphene layers and the strong covalent interface. The different result obtained for graphene and carbon black (CB) in the reinforcement of SBR composites is evidence that graphene possesses significantly higher reinforcing efficiency toward SBR [106].

The effect of the presence of the nanostructures on the thermal stability of natural rubber was evaluated using thermogravimetric analysis (TGA). The results indicated that the presence of GO does not affect the thermal stability of the rubber [106, 119].

4. Conclusions

Graphene-based fillers have features such as relatively high mechanical properties. They are incorporated as fillers in polymeric materials such as elastomers nanocomposites due to their pronounced differences in properties. Several parameters such as processing conditions that affect these enhancements were highlighted. Morphologies of the corresponding nanocomposites were observed to be affected by processing. One of the major challenges concerning graphene is the industrial-scale production of inexpensive graphene-based nanocomposites. This would require newer processing techniques able to prepare low-thickness stacks with higher specific areas. A crucial aspect that we addressed is the better dispersion of graphene-based fillers in an elastomeric matrix. Covalent functionalization of graphene-based sheets proved to be indeed significant in the improvement of the interfacial filler dispersion and its exfoliation in the matrix. That is why other efforts should be made to exploit graphene-based fillers at the nanoscale by creating a nanostructural organization of graphene sheets that would ensure a lower percolation threshold and consequently more pronounced enhancement in reinforcement.

Author details

Mohammed A. Sharaf[1*] and Andrzej Kloczkowski[2,3]

1 Department of Maritime Transportation Management Engineering, Istanbul University-Cerrahpasa, Istanbul, Turkey

2 The Steve and Cindy Rasmussen Institute for Genomic Medicine at Abigail Wexner Research Institute, Nationwide Children's Hospital, Columbus, OH, USA

3 Department of Pediatrics, The Ohio State University, Columbus, OH, USA

*Address all correspondence to: sharafma@gmail.com

References

[1] James HM, Guth E. Statistical thermodynamics of rubber elasticity. The Journal of Chemical Physics. 1953;**21**(6):1039-1049

[2] Treloar LR. The mechanics of rubber elasticity. Proceedings of the Royal Society of London A Mathematical and Physical Sciences. 1976;**351**(1666):301-330

[3] Mark JE. The rubber elastic state. Mark JE, et al. editors. Physical properties of polymers, Am. Chem. Soc. Publications. 1984:1-54

[4] Ogden RW. Recent advances in the phenomenological theory of rubber elasticity. Rubber Chemistry and Technology. 1986;**59**(3):361-383

[5] Špitalský Z, Kratochvíla J, Csomorová K, Krupa I, Graça MP, Costa LC. Mechanical and electrical properties of styrene-isoprene-styrene copolymer doped with expanded graphite nanoplatelets. Journal of Nanomaterials. 2015;**2015**:Article ID 168485; https://doi.org/10.1155/2015/168485

[6] Araby S, Zaman I, Meng Q, Kawashima N, Michelmore A, Kuan H-C, et al. Melt compounding with graphene to develop functional, high-performance elastomers. Nanotechnology. 2013;**24**(16):165601

[7] Papageorgiou DG, Kinloch IA, Young RJ. Graphene/elastomer nanocomposites. Carbon. 2015;**95**:460-484

[8] Mazhar S, Lawson BP, Stein BD, Pink M, Carini J, Polezhaev A, et al. Elastomer based nanocomposites with reduced graphene oxide nanofillers allow for enhanced tensile and electrical properties. Journal of Polymer Research. 2020;**27**(5):1-10

[9] Wang R, Zhang J, Kang H, Zhang L. Design, preparation and properties of bio-based elastomer composites aiming at engineering applications. Composites Science and Technology. 2016;**133**:136-156

[10] Basu D, Das A, Stöckelhuber KW, Wagenknecht U, Heinrich G. Advances in layered double hydroxide (LDH)-based elastomer composites. Progress in Polymer Science. 2014;**39**(3):594-626

[11] Yang J, Tian M, Jia Q-X, Shi J-H, Zhang L-Q, Lim S-H, et al. Improved mechanical and functional properties of elastomer/graphite nanocomposites prepared by latex compounding. Acta Materialia. 2007;**55**(18):6372-6382

[12] Das A, Kasaliwal GR, Jurk R, Boldt R, Fischer D, Stöckelhuber KW, et al. Rubber composites based on graphene nanoplatelets, expanded graphite, carbon nanotubes and their combination: A comparative study. Composites Science and Technology. 2012;**72**(16):1961-1967

[13] Maiti M, Bhattacharya M, Bhowmick AK. Elastomer nanocomposites. Rubber Chemistry and Technology. 2008;**81**(3):384-469

[14] Song SH. The effect of clay/multiwall carbon nanotube hybrid fillers on the properties of elastomer nanocomposites. International Journal of Polymer Science. 2018;**2018**:Article ID 5295973; https://doi.org/10.1155/2018/5295973

[15] Tjong SC. Structural and mechanical properties of polymer nanocomposites. Materials Science and Engineering: R: Reports. 2006;**53**(3-4):73-197

[16] Bokobza L. Mechanical and electrical properties of elastomer nanocomposites based on different carbon nanomaterials. C. 2017;**3**(2):10

[17] Paul DR, Mark JE. Fillers for polysiloxane ("silicone") elastomers. Progress in Polymer Science. 2010;**35**(7):893-901

[18] Chrissafis K, Bikiaris D. Can nanoparticles really enhance thermal stability of polymers? Part I: An overview on thermal decomposition of addition polymers. Thermochimica Acta. 2011;**523**(1-2):1-24

[19] Zhao J, Wei N, Fan Z, Jiang J-W, Rabczuk T. The mechanical properties of three types of carbon allotropes. Nanotechnology. 2013;**24**(9):095702

[20] Low I-M, Albetran HM, Degiorgio M. Structural characterization of commercial graphite and graphene materials. Journal of Nanotechnology and Nanomaterials. 2020;**1**(1):23-30

[21] Singh V, Joung D, Zhai L, Das S, Khondaker SI, Seal S. Graphene based materials: Past, present and future. Progress in Materials Science. 2011;**56**(8):1178-1271

[22] Chen D, Tang L, Li J. Graphene-based materials in electrochemistry. Chemical Society Reviews. 2010;**39**(8):3157-3180

[23] Yavari F, Koratkar N. Graphene-based chemical sensors. The Journal of Physical Chemistry Letters. 2012;**3**(13):1746-1753

[24] Edwards RS, Coleman KS. Graphene synthesis: Relationship to applications. Nanoscale. 2013;**5**(1):38-51

[25] Farooqui U, Ahmad A, Hamid N. Graphene oxide: A promising membrane material for fuel cells. Renewable and Sustainable Energy Reviews. 2018;**82**:714-733

[26] Ray SC. Application and uses of graphene oxide and reduced graphene oxide. In Ray SC, Applications of Graphene and Graphene-Oxide Based Nanomaterials, Macro & Nano Technologies Series, Elsevier; 2015. pp. 39-55

[27] Young RJ, Liu M, Kinloch IA, Li S, Zhao X, Vallés C, et al.. The mechanics of reinforcement of polymers by graphene nanoplatelets. Composites Science and Technology. 2018;**154**:110-116

[28] Nwosu CN, Iliut M, Vijayaraghavan A. Graphene and water-based elastomer nanocomposites–a review. Nanoscale. 2021;**13**(21):9505-9540

[29] Cui T, Mukherjee S, Sudeep PM, Colas G, Najafi F, Tam J, et al. Fatigue of graphene. Nature Materials. 2020;**19**(4):405-411

[30] Balandin AA, Ghosh S, Bao W, Calizo I, Teweldebrhan D, Miao F, et al. Superior thermal conductivity of single-layer graphene. Nano Letters. 2008;**8**(3):902-907

[31] Kim K, Regan W, Geng B, Alemán B, Kessler B, Wang F, et al. High-temperature stability of suspended single-layer graphene. Physica Status Solidi (RRL)–Rapid Research Letters. 2010;**4**(11):302-304

[32] Du X, Skachko I, Barker A, Andrei EY. Approaching ballistic transport in suspended graphene. Nature Nanotechnology. 2008;**3**(8):491-495

[33] Novoselov KS, Colombo L, Gellert P, Schwab M, Kim K. A roadmap for graphene. Nature. 2012;**490**(7419):192-200

[34] Wick P, Louw-Gaume AE, Kucki M, Krug HF, Kostarelos K, Fadeel B, et al. Classification framework for graphene-based materials. Angewandte Chemie International Edition. 2014;**53**(30):7714-7718

[35] Zhang Y, Xu Y. Simultaneous electrochemical dual-electrode exfoliation of graphite toward scalable production of high-quality graphene. Advanced Functional Materials. 2019;**29**(37):1902171

[36] Johnston H, Pojana G, Zuin S, Jacobsen NR, Møller P, Loft S, et al. Engineered nanomaterial risk. Lessons learnt from completed nanotoxicology studies: Potential solutions to current and future challenges. Critical Reviews in Toxicology. 2013;**43**(1):1-20

[37] Legge EJ, Paton KR, Wywijas M, McMahon G, Pemberton R, Kumar N, et al. Determining the level and location of functional groups on few-layer graphene and their effect on the mechanical properties of nanocomposites. ACS Applied Materials & Interfaces. 2020;**12**(11):13481-13493

[38] Jancar J, Douglas J, Starr FW, Kumar S, Cassagnau P, Lesser A, et al. Current issues in research on structure–property relationships in polymer nanocomposites. Polymer. 2010;**51**(15):3321-3343

[39] Medalia A. Effect of carbon black on dynamic properties of rubber vulcanizates. Rubber Chemistry and Technology. 1978;**51**(3):437-523

[40] Galimberti M, Infortuna G, Guerra S, Barbera V, Agnelli S, Pandini S. sp2 carbon allotropes in elastomer matrix: From master curves for the mechanical reinforcement to lightweight materials. Express Polymer Letters. 2018;**12**(3): 265-283

[41] Kango S, Kalia S, Celli A, Njuguna J, Habibi Y, Kumar R. Surface modification of inorganic nanoparticles for development of organic–inorganic nanocomposites—A review. Progress in Polymer Science. 2013;**38**(8):1232-1261

[42] Zou H, Wu S, Shen J. Polymer/ silica nanocomposites: Preparation, characterization, properties, and applications. Chemical Reviews. 2008;**108**(9):3893-3957

[43] Giovino M, Buenning E, Jimenez A, Kumar SK, Schadler L. Polymer grafted nanoparticle viscosity modifiers. Macromolecular Chemistry and Physics. 2019;**220**(7):1800543

[44] Singh SK, Singh S, Kumar A, Jain A. Thermo-mechanical behavior of TiO2 dispersed epoxy composites. Engineering Fracture Mechanics. 2017;**184**:241-248

[45] Rane AV, Kanny K, Abitha V, Thomas S. Methods for synthesis of nanoparticles and fabrication of nanocomposites. In: Synthesis of Inorganic Nanomaterials. Elsevier; 2018. pp. 121-139

[46] Naz A, Kausar A, Siddiq M. Influence of graphite filler on physicochemical characteristics of polymer/graphite composites: A review. Polymer-Plastics Technology and Engineering. 2016;**55**(6):604-625

[47] Nasir A, Kausar A, Younus A. Polymer/graphite nanocomposites: Physical features, fabrication and current relevance. Polymer-Plastics Technology and Engineering. 2015;**54**(7):750-770

[48] Duan L, D'hooge DR, Cardon L. Recent progress on flexible and stretchable piezoresistive strain sensors: From design to application. Progress in Materials Science. 2020;**114**:100617

[49] Kasaliwal GR, Pegel S, Göldel A, Pötschke P, Heinrich G. Analysis of agglomerate dispersion mechanisms of multiwalled carbon nanotubes during melt mixing in polycarbonate. Polymer. 2010;**51**(12):2708-2720

[50] Papageorgiou DG, Li Z, Liu M, Kinloch IA, Young RJ. Mechanisms of mechanical reinforcement by graphene and carbon nanotubes in polymer nanocomposites. Nanoscale. 2020;**12**(4):2228-2267

[51] Rueda MM, Auscher M-C, Fulchiron R, Perie T, Martin G, Sonntag P, et al. Rheology and applications of highly filled polymers: A review of current understanding. Progress in Polymer Science. 2017;**66**:22-53

[52] Vasileiou AA, Kontopoulou M, Docoslis A. A noncovalent compatibilization approach to improve the filler dispersion and properties of polyethylene/graphene composites. ACS applied Materials & Interfaces. 2014;**6**(3):1916-1925

[53] Le TH, Oh Y, Kim H, Yoon H. Exfoliation of 2D materials for energy and environmental applications. Chemistry–a. European Journal. 2020;**26**(29):6360-6401

[54] Tao H, Zhang Y, Gao Y, Sun Z, Yan C, Texter J. Scalable exfoliation and dispersion of two-dimensional materials–an update. Physical Chemistry Chemical Physics. 2017;**19**(2):921-960

[55] Wei Y, Sun Z. Liquid-phase exfoliation of graphite for mass production of pristine few-layer graphene. Current Opinion in Colloid & Interface Science. 2015;**20**(5-6):311-321

[56] Compton OC, Nguyen ST. Graphene oxide, highly reduced graphene oxide, and graphene: Versatile building blocks for carbon-based materials. Small. 2010;**6**(6):711-723

[57] Kuilla T, Bhadra S, Yao D, Kim NH, Bose S, Lee JH. Recent advances in graphene based polymer composites. Progress in Polymer Science. 2010;**35**(11):1350-1375

[58] Bokobza L. Natural rubber nanocomposites: A review. Nanomaterials. 2018;**9**(1):12

[59] Cataldi P, Athanassiou A, Bayer IS. Graphene nanoplatelets-based advanced materials and recent progress in sustainable applications. Applied Sciences. 2018;**8**(9):1438

[60] Shen J, Hu Y, Li C, Qin C, Shi M, Ye M. Layer-by-layer self-assembly of graphene nanoplatelets. Langmuir. 2009;**25**(11):6122-6128

[61] Yu W, Sisi L, Haiyan Y, Jie L. Progress in the functional modification of graphene/graphene oxide: A review. RSC Advances. 2020;**10**(26):15328-15345

[62] Bagherzadeh M, Farahbakhsh A. Surface functionalization of graphene. Tiwari A and Syväjärvi M, Editors. In: Graphene Materials, Scrivener Publishing LLC, Beverly, MA, United States, 2015. p. 25-66

[63] Zheng W, Shen B, Zhai W. Surface functionalization of graphene with polymers for enhanced properties. New Progress on Graphene Research. 2013;**10**:50490

[64] Ferreira FV, Cividanes LDS, Brito FS, Menezes BRC, Franceschi W, Simonetti EAN, et al. Functionalization of graphene and applications. In: Functionalizing Graphene and Carbon Nanotubes. Springer; 2016. pp. 1-29

[65] Dreyer DR, Park S, Bielawski CW, Ruoff RS. The chemistry of graphene oxide. Chemical Society Reviews. 2010;**39**(1):228-240

[66] Lee JU, Lee W, Yi JW, Yoon SS, Lee SB, Jung BM, et al. Preparation of

highly stacked graphene papers via site-selective functionalization of graphene oxide. Journal of Materials Chemistry A. 2013;**1**(41):12893-12899

[67] Ramesha G, Kumara AV, Muralidhara H, Sampath S. Graphene and graphene oxide as effective adsorbents toward anionic and cationic dyes. Journal of Colloid and Interface Science. 2011;**361**(1):270-277

[68] Parviz D, Das S, Ahmed HT, Irin F, Bhattacharia S, Green MJ. Dispersions of non-covalently functionalized graphene with minimal stabilizer. ACS Nano. 2012;**6**(10):8857-8867

[69] Bartlam C, Morsch S, Heard KW, Quayle P, Yeates SG, Vijayaraghavan A. Nanoscale infrared identification and mapping of chemical functional groups on graphene. Carbon. 2018;**139**:317-324

[70] Layek RK, Nandi AK. A review on synthesis and properties of polymer functionalized graphene. Polymer. 2013;**54**(19):5087-5103

[71] Bhuyan S, Uddin N, Bipasha FA, Islam MM, Hossain S. A review of functionalized graphene properties and its application. International Journal of Innovation Science Research. 2015;**17**(2):303-315

[72] Genorio B, Znidarsic A. Functionalization of graphene nanoribbons. Journal of Physics D: Applied Physics. 2014;**47**(9):094012

[73] Acik M, Chabal YJ. Nature of graphene edges: A review. Japanese Journal of Applied Physics. 2011;**50**(7R):070101

[74] Nanda SS, Papaefthymiou GC, Yi DK. Functionalization of graphene oxide and its biomedical applications. Critical Reviews in Solid State and Materials Sciences. 2015;**40**(5):291-315

[75] Huang T, Lu R, Su C, Wang H, Guo Z, Liu P, et al. Chemically modified graphene/polyimide composite films based on utilization of covalent bonding and oriented distribution. ACS Applied Materials & Interfaces. 2012;**4**(5):2699-2708

[76] Yousefi N, Lin X, Zheng Q, Shen X, Pothnis JR, Jia J, et al.. Simultaneous in situ reduction, self-alignment and covalent bonding in graphene oxide/epoxy composites. Carbon. 2013;**59**:406-417

[77] Vallés C, Papageorgiou DG, Lin F, Li Z, Spencer BF, Young RJ, et al. PMMA-grafted graphene nanoplatelets to reinforce the mechanical and thermal properties of PMMA composites. Carbon. 2020;**157**:750-760

[78] Potts JR, Dreyer DR, Bielawski CW, Ruoff RS. Graphene-based polymer nanocomposites. Polymer. 2011;**52**(1):5-25

[79] Zhang M, Li Y, Su Z, Wei G. Recent advances in the synthesis and applications of graphene–polymer nanocomposites. Polymer Chemistry. 2015;**6**(34):6107-6124

[80] Guan L-Z, Wan Y-J, Gong L-X, Yan D, Tang L-C, Wu L-B, et al. Toward effective and tunable interphases in graphene oxide/epoxy composites by grafting different chain lengths of polyetheramine onto graphene oxide. Journal of Materials Chemistry A. 2014;**2**(36):15058-15069

[81] Lim LP, Juan JC, Huang NM, Goh LK, Leng FP, Loh YY. Enhanced tensile strength and thermal conductivity of natural rubber graphene composite properties via rubber-graphene interaction. Materials Science and Engineering: B. 2019;**246**:112-119

[82] Kim H, Abdala AA, Macosko CW. Graphene/polymer nanocomposites. Macromolecules. 2010;**43**(16):6515-6530

[83] Punetha VD, Rana S, Yoo HJ, Chaurasia A, McLeskey JT Jr, Ramasamy MS, et al. Functionalization of carbon nanomaterials for advanced polymer nanocomposites: A comparison study between CNT and graphene. Progress in Polymer Science. 2017;**67**:1-47

[84] Strankowski M, Piszczyk Ł, Kosmela P, Korzeniewski P. Morphology and the physical and thermal properties of thermoplastic polyurethane reinforced with thermally reduced graphene oxide. Pol. J. Chem. Tech. 2015;**17**(4):88-94

[85] Utrera-Barrios S, Hernández Santana M, Verdejo R, López-Manchado MA. Design of rubber composites with autonomous self-healing capability. ACS Omega. 2020;**5**(4):1902-1910

[86] Liu X, Kuang W, Guo B. Preparation of rubber/graphene oxide composites with in-situ interfacial design. Polymer. 2015;**56**:553-562

[87] Xing W, Wu J, Huang G, Li H, Tang M, Fu X. Enhanced mechanical properties of graphene/natural rubber nanocomposites at low content. Polymer International. 2014;**63**(9):1674-1681

[88] Palermo V, Kinloch IA, Ligi S, Pugno NM. Nanoscale mechanics of graphene and graphene oxide in composites: A scientific and technological perspective. Advanced Materials. 2016;**28**(29):6232-6238

[89] Cano M, Khan U, Sainsbury T, O'Neill A, Wang Z, McGovern IT, et al. Improving the mechanical properties of graphene oxide based materials by covalent attachment of polymer chains. Carbon. 2013;**52**:363-371

[90] Dong R, Liu L. Preparation and properties of acrylic resin coating modified by functional graphene oxide. Applied Surface Science. 2016;**368**:378-387

[91] Salavagione HJ, Martínez G. Importance of covalent linkages in the preparation of effective reduced graphene oxide– poly (vinyl chloride) nanocomposites. Macromolecules. 2011;**44**(8):2685-2692

[92] Vallés C, Beckert F, Burk L, Mülhaupt R, Young RJ, Kinloch IA. Effect of the C/O ratio in graphene oxide materials on the reinforcement of epoxy-based nanocomposites. Journal of Polymer Science Part B: Polymer Physics. 2016;**54**(2):281-291

[93] Bockstal L, Berchem T, Schmetz Q, Richel A. Devulcanisation and reclaiming of tires and rubber by physical and chemical processes: A review. Journal of Cleaner Production. 2019;**236**:117574

[94] Geim AK, Novoselov KS. The rise of graphene. Nanoscience and technology: A collection of reviews from nature journals. World Scientific. 2010:11-19

[95] Zhong B, Jia Z, Dong H, Luo Y, Jia D, Liu F. One-step approach to reduce and modify graphene oxide via vulcanization accelerator and its application for elastomer reinforcement. Chemical Engineering Journal. 2017;**317**:51-59

[96] Boland CS, Khan U, Backes C, O'Neill A, McCauley J, Duane S, et al. Sensitive, high-strain, high-rate bodily motion sensors based on graphene–rubber composites. ACS Nano. 2014;**8**(9):8819-8830

[97] Zhong B, Dong H, Luo Y, Zhang D, Jia Z, Jia D, et al. Simultaneous reduction

and functionalization of graphene oxide via antioxidant for highly aging resistant and thermal conductive elastomer composites. Composites Science and Technology. 2017;**151**:156-163

[98] Dong H, Jia Z, Chen Y, Luo Y, Zhong B, Jia D. One-pot method to reduce and functionalize graphene oxide via vulcanization accelerator for robust elastomer composites with high thermal conductivity. Composites Science and Technology. 2018;**164**:267-273

[99] Hu K, Kulkarni DD, Choi I, Tsukruk VV. Graphene-polymer nanocomposites for structural and functional applications. Progress in Polymer Science. 2014;**39**(11): 1934-1972

[100] Wen Y, Yin Q, Jia H, Yin B, Zhang X, Liu P, et al. Tailoring rubber-filler interfacial interaction and multifunctional rubber nanocomposites by usage of graphene oxide with different oxidation degrees. Composites Part B: Engineering. 2017;**124**:250-259

[101] Potts JR, Shankar O, Murali S, Du L, Ruoff RS. Latex and two-roll mill processing of thermally-exfoliated graphite oxide/natural rubber nanocomposites. Composites Science and Technology. 2013;**74**:166-172

[102] Mao Y, Wen S, Chen Y, Zhang F, Panine P, Chan TW, et al. High performance graphene oxide based rubber composites. Scientific Reports. 2013;**3**(1):1-7

[103] Liu X, Sun D, Wang L, Guo B. Sodium humate functionalized graphene and its unique reinforcement effects for rubber. Industrial & Engineering Chemistry Research. 2013;**52**(41):14592-14600

[104] Tang Z, Wu X, Guo B, Zhang L, Jia D. Preparation of butadiene–styrene–vinyl pyridine rubber–graphene oxide hybrids through co-coagulation process and in situ interface tailoring. Journal of Materials Chemistry. 2012;**22**(15):7492-7501

[105] Lei Y, Tang Z, Liao R, Guo B. Hydrolysable tannin as environmentally friendly reducer and stabilizer for graphene oxide. Green chemistry. 2011;**13**(7):1655-1658

[106] Yang Z, Liu J, Liao R, Yang G, Wu X, Tang Z, et al. Rational design of covalent interfaces for graphene/elastomer nanocomposites. Composites Science and Technology. 2016;**132**:68-75

[107] Fujimoto A, Masuda T. Chemical interaction between polyphenols and a cysteinyl thiol under radical oxidation conditions. Journal of Agricultural and Food Chemistry. 2012;**60**(20):5142-5151

[108] Tang Z, Zhang C, Zhu L, Guo B. Low permeability styrene butadiene rubber/boehmite nanocomposites modified with tannic acid. Materials & Design. 2016;**103**:25-31

[109] Sirqueira AS, Soares BG. The effect of functionalized ethylene propylene diene rubber (EPDM) on the kinetics of sulfur vulcanization of normal rubber/ EPDM blends. Macromolecular Materials and Engineering. 2007;**292**(1):62-69

[110] Kim WS, Yi J, Lee DH, Kim IJ, Son WJ, Bae JW, et al. Effect of 3-aminopropyltriethoxysilane and N, N-dimethyldodecylamine as modifiers of Na+−montmorillonite on SBR/organoclay nanocomposites. Journal of Applied Polymer Science. 2010;**116**(6):3373-3387

[111] He S-J, Wang Y-Q, Wu Y-P, Wu X-H, Lu Y-L, Zhang L-Q. Preparation, structure, performance, industrialisation and application of advanced rubber/

clay nanocomposites based on latex compounding method. Plastics, Rubber and Composites. 2010;**39**(1):33-42

[112] Sadek E, El-Nashar D, Ahmed S. Effect of organoclay reinforcement on the curing characteristics and technological properties of styrene–butadiene rubber. Polymer Composites. 2015;**36**(7):1293-1302

[113] Wang Z, Liu J, Wu S, Wang W. Zhang L, Novel percolation phenomena and mechanism of strengthening elastomers by nanofillers. Physical Chemistry Chemical Physics. 2010;**12**(12):3014-3030

[114] Mensah B, Gupta KC, Kim H, Wang W, Jeong K-U, Nah C. Graphene-reinforced elastomeric nanocomposites: A review. Polymer Testing. 2018;**68**: 160-184

[115] Nawaz K, Khan U, Ul-Haq N, May P, O'Neill A, Coleman JN. Observation of mechanical percolation in functionalized graphene oxide/elastomer composites. Carbon. 2012;**50**(12):4489-4494

[116] Ren PG, Di YY, Zhang Q, Li L, Pang H, Li ZM. Composites of ultrahigh-molecular-weight polyethylene with graphene sheets and/or MWCNTs with segregated network structure: Preparation and properties. Macromolecular Materials and Engineering. 2012;**297**(5):437-443

[117] Liu K, Ronca S, Andablo-Reyes E, Forte G, Rastogi S. Unique rheological response of ultrahigh molecular weight polyethylenes in the presence of reduced graphene oxide. Macromolecules. 2015;**48**(1):131-139

[118] Paszkiewicz S, Szymczyk A, Pawlikowska D, Subocz J, Zenker M, Masztak R. Electrically and thermally conductive low density polyethylene-based nanocomposites reinforced by MWCNT or hybrid MWCNT/graphene nanoplatelets with improved thermo-oxidative stability. Nanomaterials. 2018;**8**(4):264

[119] Liu M, Kinloch IA, Young RJ, Papageorgiou DG. Modelling mechanical percolation in graphene-reinforced elastomer nanocomposites. Composites Part B: Engineering. 2019;**178**:107506

Chapter 10

Perovskite-Based Nanomaterials and Nanocomposites for Photocatalytic Decontamination of Water

Yousef Faraj and Ruzhen Xie

Abstract

The exploration of functional nanomaterials with superior catalytic activity for practical photocatalytic water decontamination is of significant importance. Perovskite-based nanomaterials, which demonstrate excellent photophysical and catalytic properties, are widely investigated as a class of adaptable materials for the photocatalytic degradation of environmental pollutants. This chapter introduces the recent progresses in using perovskite-based nanocomposites with particular emphasis on the applications for effective photocatalytic degradation of organic pollutants in wastewater. It starts by presenting the general principles and mechanisms governing photocatalytic degradation of organic pollutants in water by perovskite, along with the design criteria for perovskite-based nanocomposites. It then explains various strategies used to prepare perovskite-based nanocomposites with the aim of enhancing their photocatalytic activity. By the end of the chapter, the remaining challenges and perspectives for developing efficient perovskite-based photocatalysts with potential large-scale application are highlighted.

Keywords: perovskite, photocatalysis, water decontamination, reactive oxidation species, nanomaterials, nanocomposites

1. Introduction

The rapid growing population, urbanisation and industrial development are the major contributors of organic pollutants in water, which have a detrimental impact on the ecosystem, and cause serious problems to the living world and environment. In order to balance the ecosystem and mitigate the huge risk caused by the persistence organic substances, the removal of organic pollutants in wastewater is paramount.

Over the past two decades, solar photocatalysis has been of particular interest for the removal and degradation of organic pollutants in wastewater. In the photocatalytic water decontamination process, the production of electron-hole (e^--h^+) pairs via irradiation of the photocatalyst is the key step for the production of reactive oxidation species (ROS, i.e. hydroxyl radicals ($^{\bullet}OH$)), which is powerful oxidants and can

IntechOpen

non-selectively attack organic matters, degrading them into smaller elements and finally mineralise them to H_2O and CO_2 [1]. Under light irradiation of a photocatalyst, photons with energy equal or greater than its band gap (E_g) are absorbed by the catalyst, resulting in the formation of an electron-hole pair. Then the photogenerated conduction band electron (e_{CB-}) and valence band hole (h_{VB+}) could undergo undesired recombination or participate in a series of reactions to produce highly Reactive Oxidation Species (ROS, i.e., hydroxyl radicals ($^{\bullet}OH$)) that can mineralise any organic molecule in wastewater [2]. The necessity to find photocatalysts with unique photophysical properties that can be used efficiently in the photocatalysis process has been the driving force for the development of variety of material systems to achieve an efficient removal of organic pollutants. Different types of heterogeneous semiconductors, particularly titanium dioxide (TiO_2), ternary and other oxide systems, are the most widely studied materials for photocatalytic water decontamination. TiO_2, which is well known for its photocatalytic properties, widely used, low-cost n-type semiconductor with (E_g) of 3.2 eV, can be used for water decontamination and water splitting and building self-cleaning facades [3]. However, the major drawback of using TiO_2 in practical photocatalytic water decontamination is concerning two important aspects of its photocatalytic properties: (i) TiO_2 offers low photoconversion efficiency due to undesired recombination of electrons and holes [4] and (ii) its large band gap, which can be excited only by ultraviolet light (only 4% intensity of solar radiation) [5]. In addition, compared to other advanced oxidation processes (AOPs), such as Fenton based methods; UV/Oxidant methods and electrochemical oxidation methods, which can in-situ generate ROS during water treatment, the quantum yield of TiO_2 is low for photocatalytic ROS production, hindering its application in photocatalytic water decontamination [6]. Therefore, exploring novel photocatalysts that have unique photophysical properties, offer high photo-conversion efficiency and with superior photocatalytic activity in water decontamination application is of great importance.

Perovskite-based nanomaterial have attracted huge attentions as a promising photocatalysis nanomaterial for various environmental application due to their unique features such as high chemical and thermal stability; excellent electrical conductivity; and narrow band gap that can offer efficient use of solar energy, compared to other semiconductor photocatalysts. Perovskite-type oxides are complex metal oxides, with the general formula of ABO_3, the structure of which is shown in **Figure 1**. General structure of perovskite oxides represents a lattice that consists of larger A cations and are alkaline rare-earth metals, which are 12 fold coordinated by oxygen atoms, and small B cations that can be a divalent or trivalent transition, within oxygen octahedra. Their high stability under aggressive conditions is attributed to the existence of transition metals in their oxidation states [8, 9]. The structure of perovskites can easily be tuned by adjusting the category and proportion of their chemical compositions, which in turn inherit them diverse and unique physicochemical properties [10]. Perovskite oxides are capable of being activated by broad solar spectra to excite e^--h^+ pairs and initiate the production of ROS, which facilitate organic pollutant oxidation and mainly comprise hydroxyl-radical ($^{\bullet}OH$) and superoxide-anion radical ($O_2^{\bullet -}$) [11]. However, pure perovskites suffer from low photocatalytic efficiency, which is due to small surface area of bulk material, insufficient solar energy consumption, rapid recombination and low redox potential of e^--h^+ pairs, which are unfavorable for efficient generation of reactive species [12].

The performance of perovskites in photocatalysis process is generally influenced by their structure; composition; size and shape and synthesis process. Therefore, with the aim of enhancing their photocatalytic efficiency in the degradation of

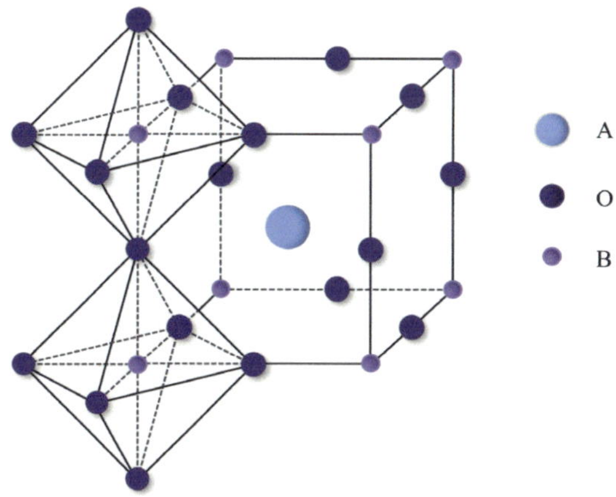

Figure 1.
ABO_3-type of perovskite structure (reprinted with permission from ref. [7]. Copyright © 2021, Elsevier).

organic pollutants, numerous studies have been carried, using various synthesis methods such as sol-gel method; hydrothermal; solvothermal; sono-chemical; microwave assisted method and co-precipitation method. In order to enhance the photocatalytic performance of perovskite, a number of strategies can be adopted, such as regulating perovskite composition through partial or full cationic substitution by certain dopant(s); rescaling its structure through downsizing or morphology alteration; hybrid modification through coating and coupling with other AOPs. It is worth pointing out that the strategy of coupling with other AOPs is beyond the scope of this chapter, therefore, no further reference will be made. The aforementioned strategies have been proven to improve perovskite's light absorption; create more active sites on the surface and inhibit e^--h^+ pairs recombination. By regulating perovskite composition through hetero-substitution of perovskite by hetero-valent or homo-valent cations in A and/or B site, the redox property of the perovskite is significantly improved and oxygen vacancies are increased, thereby promoting ROS generation [13]. Incorporating dopants into the lattice of perovskite, its inherent band gap can be reduced by shifting the top of its VB upward or CB downward, leading to an extended optical absorption improvement of its photocatalytic activity. Loading perovskite on substrates to obtain a hybrid nanostructure is an effective option for narrowing the band gap and optimisation of electronic structure to inhibit the recombination of e^--h^+ pairs. The coating strategy could address the majority of issues related to the efficient photocatalytic activity to some extent, as the coating strategy equip perovskite with an outstanding charge separation ability and strong oxidation ability. Rescaling structure and downsizing and controlling morphology of perovskites can be carried out to improve reactive sites and optimise optical absorption [14]. Smaller particle size can benefit from higher quantum efficiency due to larger accessible of reactive sites and more effective electron transfer paths. However, downsizing these particles to nanoscale increases the surface energy that prompts particles aggregation, hence elimination of the desired reactive sites and significant reduction of photocatalytic performance [15]. Controlled preparation of porous structure has been proven to equip perovskite with better optical absorption ability; increased reactive sites for photocatalytic reaction; as well as enhances the diffusion rate of organic pollutants. However, introducing pores to perovskite nanoparticles can make it physically fragile [16].

Despite intensive research studies that have been carried out on developing variety of nanoscale perovskite-based composites using different strategies, most of which with encouraging results, there is still much to be investigated. A comprehensive understanding of achieving an effective photocatalytic degradation of a wide range of organic pollutants using perovskites is highly crucial for unveiling the fundamental nature of perovskite photocatalysis for large-scale applications. In addition, to meet the requirements of designing efficient, stable and cost-effective perovskite-based composite photocatalyst with an outstanding use of solar energy for actual water remediation, a fundamental study of perovskite photocatalysis using different materials and various environmental pollutants is indispensable.

This chapter provides an overview of the state-of-the-art design and synthesis strategies for perovskite-based nanomaterials and nanocomposites for efficient water remediation. Initially the principles of photocatalysis process are described, with the emphasis on the mechanisms of photocatalytic water decontamination by perovskite and highlighting its inherent challenges. An evaluation of several strategies that have been used to develop perovskite-based nanocomposites for enhanced photocatalytic degradation of organic pollutants in water is presented. Finally, the remaining challenges and perspectives for developing novel perovskite-based photocatalysts with potential large-scale application are elucidated.

2. General principles of perovskite photocatalysis process

In photocatalytic process, perovskite uses photon as a source of energy to initiate chemical reaction. As the photocatalyst is irradiated by light with energy equal or larger than the perovskite band gap, the electrons are excited from the valance band (VB) to the conduction band (CB), as a result photoreactive species such as e^- and h^+ are created, which can be transferred to the surface of perovskite [17]. The factors affecting the photocatalytic activity of perovskite as a catalyst are namely, the excitement of the electron, separation of the electron and hole and photo-oxidation reduction reaction taking place at the surface of the catalyst [18].

2.1 Mechanisms of photocatalytic degradation of pollutants in water by perovskite

The mechanisms of photocatalytic degradation of organic pollutants consists of several steps: (1) under light irradiation perovskite absorbs photon with an appropriate energy to form photoreactive species like e^- and h^+; (2) interfacial charge transfer; (3) reduction and oxidation process to form Reactive Oxidation Species; (4) degradation of organic pollutants; and (5) desorption of pollutants/intermediates from the surface of the perovskite. The reaction mechanisms of photocatalytic degradation of pollutants in water are demonstrated by Eqs. (1)–(5) [7].

$$\text{Perovskite} + \text{Light} \rightarrow \text{Perovskite}\left(e_{cb}^- + h_{vb}^+\right) \tag{1}$$

$$e_{cb}^- + O_2 \rightarrow O_2^{--} \tag{2}$$

$$h_{vb}^{+} + H_2O/OH^{-} \rightarrow \cdot OH \quad (3)$$

$$O_2^{\cdot-}/\cdot OH + \text{Organic Pollutants} \rightarrow \text{Intermediates}, CO_2, H_2O\ldots \quad (4)$$

$$h_{vb}^{+} + \text{Organic Pollutants} \rightarrow \text{Intermediates}, CO_2, H_2O\ldots \quad (5)$$

Under light irradiation of perovskite, when the energy of photon is equal or larger than the perovskite band gap energy, the electrons are excited from the valence band (VB) of perovskite to the conduction band (CB), as a result of which the photoactive species (e^- and h^+) are formed. The photoexcited electrons would either reunite with holes or transfer to the surface of the perovskite, which can react with O_2 to form superoxide anion radical ($O_2^{\bullet-}$), while the photogenerated holes react with water to form hydroxyl radical ($^{\bullet}OH$) at the surface of the catalyst [19]. The schematic representation of the degradation mechanism is illustrated in **Figure 2**. In this process, ·OH acts as a powerful oxidising agent that attacks the organic molecules non-selectively.

Figure 3 shows the bandgap values, CB and VB positions, of several perovskite photocatalysts. It is apparent that pristine perovskites have the valance band potential energy (E_{vb}) higher than the $^{\bullet}OH/OH^-$ redox potential, which allows for the generation of ·OH during the photocatalysis process. Nonetheless, the higher position of CB compared to that of the redox potential of $O_2/O_2^{\bullet-}$, hinders the formation of $O_2^{\bullet-}$ during the photocatalytic degradation process. Therefore, during the ROS production on perovskite, in order for the electrons to react with O_2 and form $O_2^{\bullet-}$, the conduction band potential (E_{cb}) of perovskite should be more negative than the standard redox potential of $O_2/O_2^{\bullet-}$ (−0.33 eV vs. NHE). On the other hand, the valance band potential energy (E_{vb}) of perovskite should be higher than standard redox potential of $^{\bullet}OH/OH^-$ (+1.99 eV vs. NHE). In such case, the OH^- can be oxidised by the photogenerated holes and form $^{\bullet}OH$, which can attack pollutants to convert them to nontoxic forms or completely degrade them to CO_2 and H_2O [20].

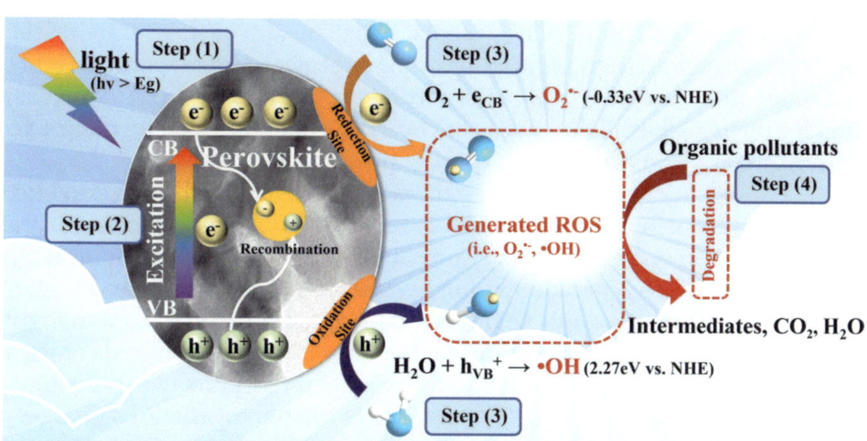

Figure 2.
Schematic representation of photocatalytic degradation of organic pollutants and ROS production by perovskite (reprinted with permission from ref. [7]. Copyright © 2021, Elsevier).

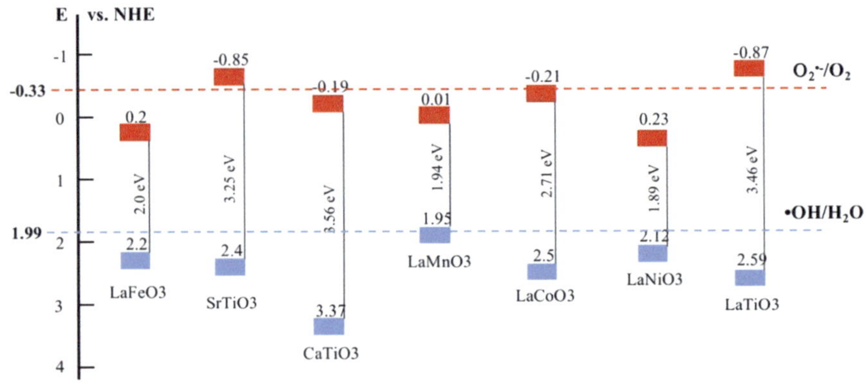

Figure 3.
Band gap values of several perovskite photocatalysts (Adapted with permission from ref. [7]. Copyright © 2021, Elsevier).

2.2 Perovskite design criteria for photocatalytic degradation of organic pollutants

The main criteria for a perovskite photocatalyst to be used in the degradation of organic pollutants in water are high capability of being activated by photons; efficiently extracting electrons for photocatalytic reaction; chemically stable; nontoxic; and cost effective. The absorption of photons the following charge generation is dependent on the physiochemical property of the perovskite and recombination.

The efficient use of solar energy still remains a great challenge. An ideal perovskite photocatalyst should have an enhanced and broaden light absorption, and capture a wide spectrum, from ultraviolet to visible light and even the near-infrared region. Therefore, it is necessary to adopt strategies that lead to optimisation of light harvesting, improving e^--h^+ separation, and generating sufficient active sites on the perovskite surface for photocatalytic reaction to take place A number of strategies have been reported in the literature, such as cationic substitution, nanostructure perovskite, coating and combined perovskite-based photocatalyst systems, in which perovskite is coupled with other AOP systems. The main aim of these state-of-the-art strategies is to enhance efficient light utilisation, improve charge separation and create richer active sites on the surface of the perovskite. Narrowing band gap is usually the option for the increased light harvesting by capturing more excited photons form a wide spectrum, and consequently enhancing photocatalytic activity [21].

Once the photogenerated charges are generated and successfully migrated to the surface of perovskite, where photocatalytic reactions take place, they can still undergo surface recombination or be trapped by undesirable reactants. In the photocatalytic process the e^--h^+ pairs are generated within several femtoseconds (fs) and undergo recombination within picoseconds (ps) to nanoseconds (ns), as depicted in **Figure 4**. However, the time span from the bulk to reactive sites is usually hundreds of ps, and the reaction time between the carriers and the adsorbed reactants requires nanoseconds (ns) to microseconds (μs) [22]. The lifetime of the photogenerated charges of some perovskites have been reported as BTO: 3.25 ns, STO: 2.06 ns, LFO: 3 ns and LMO: 2 ns, knowing that the reaction time to form $O_2^{\bullet-}$ is several nanoseconds [22]. This implies that the relatively short lifetime of the carriers on perovskite limits their application in photocatalytic degradation of organic pollutants.

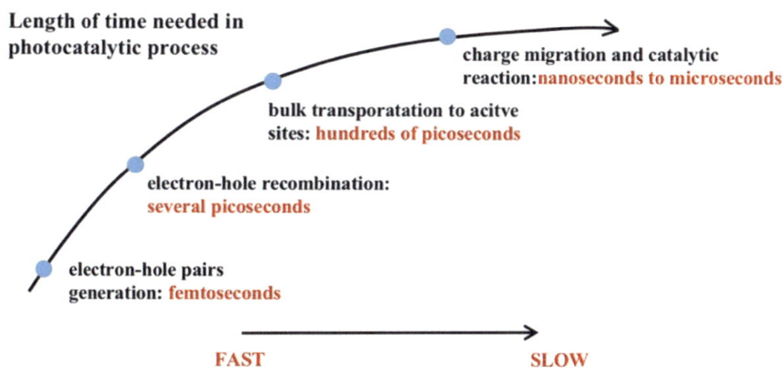

Figure 4.
Different length of time required in photocatalytic process.

In general, photocatalytic degradation takes place on the surface of the perovskite photocatalyst. Therefore, to improve photocatalytic degradation efficiency, a good adsorption of organic pollutants on the surface of perovskite is necessary. Undoubtedly, larger surface area is required to provide higher adsorption capacity towards organic pollutants and richer active sites for photocatalytic degradation reaction. A shorter diffusion pathway of charge carriers is also expected, as it reduces chance of e^{-}-h^{+} recombination.

3. Modification strategies for enhancement of perovskite photocatalytic activity

Although pure perovskites are potentially better than other oxide photocatalysts, their weak photocatalytic activity hinders their employment in industrial application. Undoubtedly, this is due to their inherent photocatalytic issues such as: (i) small surface area; (ii) insufficient solar energy utilisation; (iii) fast recombination rate of e^{-}-h^{+}. To improve carriers' utilisation, a number of modification strategies have been reported, through which nanomaterials and nanocomposite materials are developed with significantly high photocatalytic performance. Some of these strategies are described in detail in the following sections.

3.1 Partial or full cationic substitution

Poor photocatalytic degradation of pristine perovskite under visible light (>400 nm) is mainly attributed to the wide band gap [23], which hiders its potential application in the degradation of organic pollutants. The photocatalytic properties of perovskite can be modified by partial or full cationic sites substitution. The partial or full cationic substitution can narrow the band gap and inhibit the recombination of e^{-}-h^{+} , which leads to a significant enhancement of their photocatalytic activity. The substitution can be made in A-site, B-site or both sites, the detailed description of each type of substitution is provided in the following sections. It is worth mentioning that a number of factors can affect the substitution such as the types and concentration of dopant atoms, the substitution sites and so on. It is worth mentioning that the substitution of A-site, B-site or both sites in perovskite induces lattice defect and adjusts its optical and redox properties, thereby enhancing the photocatalytic degradation of organic pollutants [24].

However, choosing the right substitute is very important in maintaining the perovskite crystal structure. The stability of A-site substituted perovskite with cubic structure can be defined through Goldschmidt's tolerance factor t as shown in Eq. (6) [25]:

$$t = \frac{r_A + r_O}{\sqrt{2}\left(r_B + r_O\right)} \tag{6}$$

Where r_A, r_B and r_O are ionic radii of A-site, B-site and oxygen ion, respectively.

Since the cubic structure of perovskite is stabilised at $0.76 < t < 1.13$, then almost 90% of the natural metal elements of the Periodic Table can be incorporated into the perovskite lattice. Thus, various common metal elements, as shown in **Figure 5**, can be used to substitute A or B cationic sites in perovskite to narrow the band gap and enhance its photocatalytic activity [26, 27].

The performance of various substituted perovskites with different substitution type and dopants used in substituting A-site, B-sites and both sites in the degradation of organic pollutants is highlighted in **Table 1**.

3.1.1 A site substituted perovskite

Substitution of a metal in A-site can have a direct impact on the structure and stability of perovskite. For example, partial substitution of A′ metal on A-site, having a modified perovskite with the general formula of $A_{1-x}A_x'BO_3$, can cause the creation of vacancies in the lattice [42]. Partial substitution of A-site in perovskite is also capable of modifying the valence state of cations in B-site, leading to a better redox property and a higher photocatalytic activity. For example, substitution of La^{3+} by K^+ in A-site of $LaCoO_3$ results in modification of Co^{3+} into Co^{4+} in $La_{1-x}K_xCoO_3$ [43]. The concentration of dopants or substitutes in partial substitution of A-site is an important factor that can affect the crystal size, band gap and oxygen vacancy content. By incorporating different concentrations of La^{3+} into the lattice sites of the $SrTiO_3$(STO)

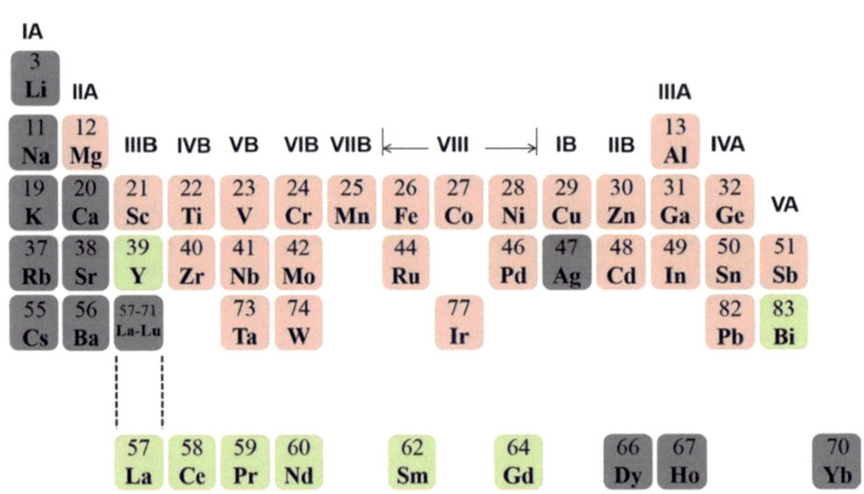

Figure 5.
Metal elements as substitutes in the ABO_3 perovskite lattice [26].

Pure perovskite	Type	Dopant	Target pollutant	Performance of substituted perovskite	Ref.
$LaFeO_3$	A-site	Eu/Gd/Dy/Nd	Safranine-O (15 mg/L)	7 times higher degradation rate than that by pure LFO	[28]
$LaFeO_3$	A-site	Ti	4-Cl-phenol (25 mg/L)	Complete removal and highest mineralization rate	[9]
$LaFeO_3$	A-site	Li	Methylene blue (78.54 mg/L)	45.7% removal compared to 35.1% by pure LFO	[29]
$LaFeO_3$	A-site	Ca	Methylene blue (10 mg/L)	77.5% removal compared to 48.9% by pure LFO	[30]
$LaFeO_3$	A-site	Bi	2,4-dichlorophenol (10 mg/L)	61% removal compared to 28% by pure PLFO	[31]
$LaTiO_3$	A-site	Ba/Sr/Ca	Congo red (100 mg/L)	75.33% degradation compared to 50.8% by pure LTO	[32]
$SrTiO_3$	A-site	Eu	RhB (5 mg/L)	95% removal, 2.6 times higher than that by pure STO	[33]
$SrFeO_3$	A-site	Pr/Sm	RhB (5 mg/L)	86% degradation efficiency compared to 43% by pure SFO	[34]
$LaFeO_3$	B-site	Mn	Methyl orange (100 mg/L)	96.4% removal higher than that by pure LFO	[35]
$LaFeO_3$	B-site	Cu	Acidpink 3B (10 mg/L)	97.4% removal compared to 62.2% by pure LFO	[36]
$SrTiO_3$	B-site	V/Mo	Methylene blue (10 mg/L)	91.5% removal compared to 59.9% by pure STO	[37]
$SrTiO_3$	B-site	Bi/Cu	Dibutyl phthalate (10 mg/L)	Higher degradation efficiency compared to pure STO	[38]
$SrTiO_3$	B-site	V	Methylene blue (10 mg/L)	Higher degradation efficiency compared to pure STO	[39]
$SrTiO_3$	A- & B-sites	La, Fe	Methyl Orange (10 mg/L)	Removal 19 times higher than that by pure STO	[40]
$SrTiO_3$	A- & B-sites	La, Cr	RhB (5 mg/L)	Removal 6 times higher than that by pure STO	[41]

Table 1.
Performance of various substituted perovskites.

host structure, different crystal defects and impurity energy levels can be formed on La-STO, resulting in different band gaps. Higher concentration of La^{3+} doping results in the decreased particle size of La-STO. However, an excessive concentration of the doping element can act as recombination center for photoinduced pairs and lead to a declined photocatalytic activity.

Various alkaline-earth metallic ions (i.e., Ba, Ca, and Sr) and rare-earth metals (i.e., La, Ce, Eu and Nd) can be used to substitute A-site of perovskite, which can result in narrowing band gap, generation of large amounts of oxygen vacancies and improved photocatalytic efficiency for the degradation of organic pollutants under visible light [32, 44]. Partially substituting La^{3+} with Li^{+} via sol-gel method yields a modified perovskite powder ($La_{0.97}Li_{0.03}FeO_3$) with improved photocatalytic degradation towards methyl blue. Since Li^{+} has lower charge than La^{3+}, the charge neutrality is maintained by forming oxygen defects on the surface of $La_{0.97}Li_{0.03}FeO_3$, which leads to improved photocatalytic activity [29]. Incorporating Ti into A-site of pure $LaFeO_3$(LFO) via solid-solid diffusion results in reduced band gap and enhanced photocatalytic activity of the modified perovskite ($La_{1-x}Ti_xFeO_3$) for the degradation of 4-chlorophenol without iron leaching. Substituting La^{3+} in pure $LaFeO_3$ (LFO) with an appropriate amount of Ca^{2+} in LFO can be a feasible strategy to improve photocatalytic degradation of methylene blue under visible light [30]. Due to smaller radius of Ca^{2+} (0.134 nm) compared with La^{3+} (0.136 nm), the substitution of Ca^{2+} affects the crystalline size and the amount of charge-compensating oxygen vacancies in $La_{1-x}Ca_xFeO_3$. The charge-compensating oxygen vacancies can act as Lewis acid sites to capture electrons, and also narrow the band gap, which increases light harvesting by capturing more excited photons [45].

3.1.2 B site substituted perovskite

The B-site element in perovskite plays a more important role than the A-site element in photocatalytic reaction, as the redox reactions generally take place at the B-site element, and it can serve as a photocatalytic active center for most of the perovskites [46]. Doping B-site with divalent or trivalent cations induces the creation of oxygen defects, leading to a relaxation structure of $AB_{1-x}B'_xO_3$ with enhanced photocatalytic degradation activity [47]. Substituting Mn into the lattice of $SrTiO_3$ (STO) via hydrothermal method yields a visible-light-responsive photocatalyst, Mn-doped STO (MSTO), having a narrower band gap and higher photocatalytic degradation efficiency towards antibiotic tetracycline [48]. Mn^{4+} species (0.067 nm) could partially substitute Ti^{4+} (0.068 nm) into STO lattice and act as impurity energy band to narrow the band gap of STO and suppress the e^{-}-h^{+} recombination, hence creating sufficient time for photogenerated holes to oxidize water and form ·OH for efficient photocatalytic degradation of tetracycline. Furthermore, substituting the B-site of perovskite by Cr yields advanced photocatalytic materials, in which the Cr^{3+} donor levels act as intermediate states for photon transition, allowing easier excitation of electrons and holes under visible light [49]. However, the presence of hexavalent Cr(VI) can hinder the photocatalytic activity of the catalyst [50]. Substitution of Fe^{3+} (0.064 nm) by a larger ionic radius Cu^{2+} (0.072 nm) in B-site of LFO results in lattice distortion, induced oxygen vacancies generation and suppressed the growth of large crystallite. This implies that a larger specific area with more accessible active sites is available for improved photocatalytic activity.

Substitution of the B-site perovskite with multiple valance cations results in coexistence of various cation states (i.e., Mn^{3+}/Mn^{4+}, Cu^{+}/Cu^{2+} and V^{3+}/V^{5+}) in perovskites [39, 51]. The presence of high valence ions can trap the photo-generated electrons in the CB, thereby the e^{-}-h^{+} pairs recombination is suppressed. While the low valence ions may supply electrons to the absorbed O_2 on the surface of the catalyst, enhancing interfacial electron transfer and increasing the photocatalytic degradation of organic pollutants in water [52]. Substitution of B-site in LFO lattice by Cu results in $LaFe_{0.85}Cu_{0.15}O_3$ with larger specific area and reduced band gap compared to LFO [53]. Under light irradiation of $LaFe_{0.85}Cu_{0.15}O_3$, the reduction of Fe^{3+} and Cu^{2+} can be accelerated and leads to the generation of Fe^{2+} and Cu^{+}. The presence of the redox couples of Fe^{2+}/Fe^{3+} and Cu^{+}/Cu^{2+} plays an important role in the creation of ·OH and other ROS, which degrade organic pollutants [54]. Doping Mn in LFO via stearic acid solution combustion method results in $LaFe_{0.5}Mn_{0.5}O_{3-\delta}$ with higher photocatalytic efficiency for methyl orange degradation under sunlight, owing to the coexistence of variable valences of Mn ions such as $Mn^{2+}/Mn^{3+}/Mn^{4+}$ [35]. The lower valence Mn^{2+} and Mn^{3+} provide electrons and reduce O_2 to generate more $O_2^{\bullet-}$, while the stable Mn^{4+} traps electrons and suppresses electron-hole recombination.

Substitution of the B-site perovskite by multiple cations can create a synergistic effect, leading to an enhanced perovskite photocatalytic activity and improved stability. The photocatalytic degradation of methylene blue can be enhanced by co-doping of Mo and V in STO [37]. The Mo and V cations incorporated into the B-site of perovskite can create impurity defects, leading to reduced band gap value and enhanced visible light utilisation. The photocatalytic activity of multiple cations doped perovskite, such as Bi and Cu doped STO, is much higher than a single cation doped perovskite [55]. The highest degradation of dibutyl phthalate can be achieved by STO co-doped with both Bi and Cu in B-stie [38].

3.1.3 A and B sites substituted perovskite

Simultaneous substitution of A- and B-sites is a feasible strategy and can increase the photocatalytic efficiency of perovskite. Since the perovskite lattice offers a great flexibility in atomic arrangement, reasonable regulation of both A- and B-sites in perovskite can produce high performance $A_{1-x}A'_xB_{1-x}B'_xO_{3-\delta}$ with improved electronic and photocatalytic properties. In general, the substitution of A-site cation leads to the generation of oxygen vacancies, whereas the substitution of B-site mainly tunes the band structure and brings about the formation of redox couples.

Simultaneous substitution of A- and B-sites of STO by La and Ni, respectively, leads to a larger surface area and new defect bands for highly efficient photocatalytic decomposition of MB compared to La or Ni mono-doped STO [56]. The substitution of both A- and B-sties of $LaCoO_3$ leads to the creation of a modified perovskite photocatalyst $La_{0.5}Ba_{0.5}Co_xMn_{1-x}O_{3-\delta}$. The partial substitution of La^{3+} by Ba^{2+} in A-site results in improved catalytic activity and structural stability. While the substitution of B-site by Mn can further enhance the photocatalytic reactivity as a result of the formation of Co−O−Mn bond, offering accelerated electron transfer between the redox couples of Co^{2+}/Co^{3+} and Mn^{3+}/Mn^{4+}. Doping perovskites with non-metal ions such as C, P, S, N, F and B is also a feasible strategy to narrow the band gap, which is necessary in enhancing photocatalytic activity of perovskite for water decontamination [57, 58].

3.2 Perovskite nanocomposites via coating strategy

Pristine perovskites show relatively weak photogenerated charge separation rate and low surface area. In addition, photo-generated e^--h^+ pairs in some narrow-band gap perovskites tend to recombine, which results in considerable energy loss. Therefore, nanoengineered perovskite particles may address the above issues and provide high surface area, full utilisation of solar energy, efficient light absorption and effective e^--h^+ separation. However, Perovskite nanoparticles can undergo agglomeration during synthesis, which endows inferior photocatalytic performance. In order to address the particle agglomeration issue and achieve efficient charge separation and good dispersion, coating perovskites on various supports has been proven to be a feasible strategy. In addition, by composing suitable cocatalysts, the photocatalytic reaction can be accelerated through lowering the activation energy. Loading perovskite on supports provides efficient charge migration and better pollutants adsorption capacity, which is crucial for improving photocatalytic activity. The loading amount can easily be tuned by simply changing the coating times and the concentration of the coating solutions. Given an appropriate support loading, abundant active sites are available for charge-transfer reactions, also the photogenerated carriers can be trapped on the supports to suppress e^{-}-h^{+} recombination. Perovskites can be coated on various supports such as carbon, silica, graphene, zeolites, semiconductor cocatalysts and so on.

3.2.1 Nanocomposite of perovskite and carbon-based materials

Carbonaceous materials have been widely studied and used in many applications, due to their outstanding characteristics such as large surface area, good electronic properties and excellent corrosion resistance. Loading perovskites on Carbon-derived materials can enhance photocatalytic degradation of organic pollutants, as they usually act as electron scavengers, owing to their large electron storage capacity. Composites of perovskite with carbon-based materials such as graphene oxide (GO) and its derivatives (graphitic carbon nitride and carbon aerogel) provide enhanced adsorption capacity towards organic pollutants, along with formed junctions, which hinders the $e^{-}h^{+}$ recombination.

Perovskite/GO composites have been proven to provide excellent photocatalytic activity in the degradation of organic pollutants. It is worth mentioning that the bandgap of the perovskite/GO composites can easily be tuned by incorporating perovskite with different proportions of GO. $LaMnO_3$/graphene composite has been reported to have a superior visible-light responsive photocatalytic activity in the degradation of diamine green B [59]. The photo-generated electrons migrate from $LaMnO_3$ (LMO) to GO across the heterojunction and temporarily stored on the surface of GO, suppressing electrons and holes recombination. The LMO particles are highly dispersed on the surface of GO, allowing them full exposure to light irradiation and high photon absorption, which improves the photocatalytic quantum efficiency. Compared to pure LMO, the decreased band gap in LMO/GO results in an obvious red shift of 30–40 nm in the light absorption edge, which enhances its photocatalytic activity. GO and STO composite can be prepared by hydrothermal method for efficient degradation of organic pollutants [60]. During heat treatment, GO is decomposed, and followed by the diffusion and dissolution of carbon species, which can penetrate into STO lattice and substitute its interlayer O^{2-} sites, thereby introducing C 2p state within the band gap. Under light irradiation, the photoexcited electrons on STO can easily be transferred to carbon and promote charge separation, resulting in an increased amount of reactive oxidation species for efficient degradation of organic pollutants.

Graphitic carbon nitride (g-C_3N_4), which is one of the most promising visible-light-driven photocatalysts, is another non-metallic material with unique layered structure and narrow band gap of 2.7–2.8 eV. However, the application of pristine g-C_3N_4 is limited by the rapid recombination of photoinduced $e^{-}h^{+}$. This issue can be addressed by adopting strategies to couple g-C_3N_4 with perovskites, thereby developing catalysts with high photocatalytic properties. $LaNiO_3$/g-C_3N_4 Z-scheme nanosheet has been prepared, in which 30 wt.% $LaNiO_3$ loading provides intimate attachment of LNO on the surface of g-C_3N_4, leading to the formation of abundant heterojunctions at the interface that are required for the spatial isolation of photogenerated charge carriers. Under light irradiation of the $LaNiO_3$/g-C_3N_4 hybrid, the accumulated electrons with stronger reducibility can reduce O_2 to yield $O_2^{\bullet-}$. As a result, the $LaNiO_3$/g-C_3N_4 composite exhibits remarkable photocatalytic activity in the degradation of tetracycline, which is 3.8 times and 3.9 times faster than those of pristine g-C_3N_4 and $LaNiO_3$, respectively. Loading p-type semiconductor LFO with n-type g-C_3N_4 nanosheets results in a hybrid p-n heterostructure photocatalyst (LFO/g-C_3N_4) that exhibits superior photocatalytic activity, compared to pristine g-C_3N_4 and LFO, in the degradation of Brilliant Blue [61].

3.2.2 Nanocomposite of perovskite and metal oxide

Metal oxide is regarded as a promising supporting material for perovskites to achieve higher photocatalytic activity in degradation of organic pollutants. By forming heterojunctions between metal oxides and perovskite, the metal oxides act as co-catalysts, serve as charge collectors to facilitate charge separation and efficiently extend the lifetime of the charge carriers. Metal oxides such as TiO_2, ZnO, and CeO_2 are abundant in nature and have been widely used as alternative catalysts to precious metals in various chemical reactions [62].

A number of metal oxides, such as ZnO, CeO_2, Al_2O_3, CuO, MnO_2 and WO_3 have been used in perovskite coating [63–66]. It can be argued that the most widely used metal oxide to couple with perovskites is TiO_2 [67–69]. The STO/TiO_2 nanofiber has been synthesised via hydrothermal method, using TiO_2 as both template and reactant [70]. Under UV light irradiation, the photogenerated electrons are transferred from STO to TiO_2 due to their close contact, thus improving the interfacial charge migration to the adsorbed substance. The electrons react with dissolved O_2 to form $O_2^{\bullet-}$ and subsequently protonated to strong oxidizing agents like H_2O_2, $HO_2^{\bullet}$ and ·OH. Reportedly, the incorporation of TiO_2 into STO can prolong the lifetime of photoinduced charge carriers. It has been demonstrated that the combination of TiO_2 and $LaNiO_3$ can significantly enhance the photocatalytic activity of the modified perovskite in degradation of methyl orange and antibiotic ciprofloxacin under visible light. The $LaNiO_3$/TiO_2 step-scheme (S-scheme) can be synthesised via a facial sol-gel method as shown in **Figure 6**, in which the S-scheme heterojunction is formed between the n-type TiO_2 and p-type $LaNiO_3$ due to the potential energy difference of VB in TiO_2 and CB in $LaNiO_3$. Exploiting the electric field and band edge bending, the electrons can spontaneously be transferred from TiO_2 CB to $LaNiO_3$ across the interfacial region until they reach similar Fermi level. O_2 is reduced to $O_2^{\bullet-}$ by the accumulative electrons in the CB of $LaNiO_3$ with more negative potential, while the remaining holes in the VB of TiO_2 oxidises H_2O to ·OH, thereby the photocatalytic activity of the coupled TiO_2 and $LaNiO_3$ is significantly promoted. In another study, STO is coated on WO_3, in which an efficient Z-scheme heterojunction is obtained [71]. Under visible light irradiation of the modified STO, the electrons in the CB of WO_3 tend to recombine with the holes in STO, thus the electrons in the CB of STO and holes in the VB of WO_3 can separate and form reactive oxidation species for efficient degradation of pollutants in water.

3.2.3 Nanocomposite of perovskite and silica-based material

Perovskites can be coated on silica-based material for efficient charge separation and enhancement of absorption capacity, thus increasing their potential application in photocatalytic degradation of pollutants in water. Coating perovskites on porous

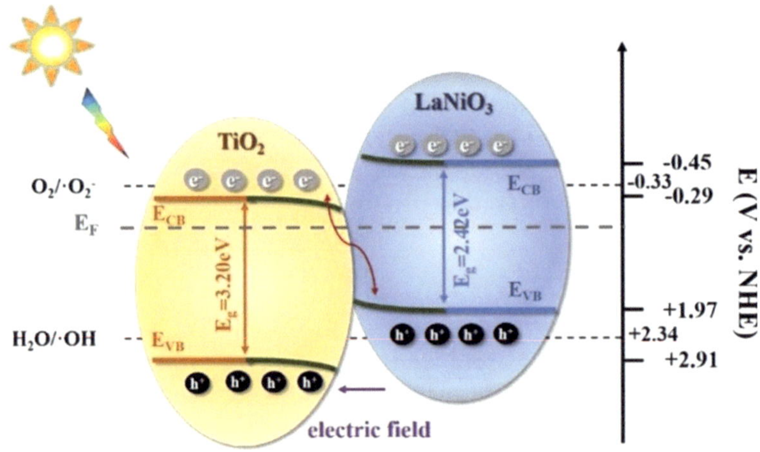

Figure 6.
Schematic representation of $LaNiO_3$ and TiO_2 nanocomposite [69].

silica can also provide easier transport of large organic molecules and the availability of more active sites, which results in efficient photo-Fenton catalytic degradation [72]. Various clay minerals such as montmorillonite, bentonite, kaolinite, illite and zeolite have been used as supporting materials for perovskites, which provide significant adsorptive capacity for the removal of toxic organic pollutants in aqueous solutions [73]. The porous silica support play two key roles in the degradation process, enhancement of the adsorption capacity via hydrogen bonds formed between the support and organic pollutants, and transportation of adsorbed substance to active sites.

Montmorillonite (MMT), which is one of the most abundant clay minerals and possesses ample –OH groups on the surface, has been used as a support for LFO to form a nanocomposite of LFO/MMT for effective removal of Rhodamine B in water. The presence of –OH groups on the surface of MMT allows LFO to be uniformly distributed on the surface of montmorillonite via Si–O–Fe bonds, which results in LFO/MMT with higher surface area and enhanced photocatalytic activity. LFO/MMT can effectively remove Rhodamine B (RhB) via synergistic effect of adsorption and photocatalytic degradation [74]. Different LFO/silica composites have been prepared and studied using several mesoporous silica materials (SBA-15, SBA-16 and siliceous mesostructured cellular foams (MCF)), along with nano-sized silica powder as supports for photocatalytic degradation of RhB [75]. Compared with other nanocomposites and pure LFO, LFO/MCF demonstrates the highest photocatalytic activity towards RhB. The superior photocatalytic activity of LFO/MCF can be attributed to the randomly distributed pores and short pore length of MCF, which allows easier and faster transportation of RhB to the active sites within the pores of LFO/MCF.

Zeolite, which is a crystalline aluminosilicate material, is another widely used supporting material for perovskites to enhance the photocatalytic degradation of organic pollutants. Using zeolite for loading perovskite can provide an effective nanocomposite with high photocatalytic activity, as its cages and pores enable easier and faster mass transfer of adsorbed substance, it provides abundant active sites for photocatalytic degradation of pollutants and can control the charge transfer process to reduce the e^--h^+ recombination [76]. To optimise physicochemical properties of zeolite, such as adsorption capacity, before loading perovskites, approaches such as heating, chemical treatment like acid-modification and metal-modification can be applied [77]. HCl can be used to modify the natural zeolite prior to the LFO loading. The acid treatment approach can remove amorphous impurities and provide zeolite with larger surface area and more available pore volume for easy incorporation of LFO. Loading 30% STO on HZSM-5 zeolite results in STO/HZSM-5 with high surface area, leading to high photocatalytic degradation rate towards Reactive Brilliant Red-X3B [78]. The HZSM-5 is capable of mediating electron migration and extending the lifetime of photogenerated charge carriers, thus leading to high photocatalytic activity towards organic pollutants.

3.3 Synthesis of various perovskite nanostructure

Constructing perovskite with various nanostructures can lead to full utilisation of solar energy, efficient light absorption and effective e^--h^+ separation. Perovskite with various structures such as nanoparticles with/without Hierarchical porous structure, core-shell structure, nanotubes, nanocubes and nanofilm have also been proven to offer significantly high photocatalytic response [79]. A well-designed nanoscale and hierarchically porous perovskite structure usually provides high surface area and excellent light absorption efficiency to take full advantage of reflection, refraction and scattering of photons [80]. A smaller size perovskite increases the availability of multiple reactive sites for enhanced photocatalytic reaction and higher degradation efficiency [81]. Downsizing perovskites to nanoscale with desired morphologies and functional properties holds a tremendous opportunity for obtaining excellent photocatalysts as decreasing the particle size leads to an increase in the quantum yields of perovskite photocatalytic reactions [82]. In smaller size perovskite particles, shorter time is required for the photoinduced charge carriers to be diffused from the bulk to the surface of perovskite, thereby suppressing the recombination of the electrons and holes. Besides, the properties and the unique architecture that are achieved by downsizing perovskites to nanoscale allow for direct use of visible light to remove pollutants without chemical addition. However, with decreasing the perovskite particle size to nanoscale the surface tension of perovskite nanoparticles significantly increases, which inevitably leads to particle aggregation.

In a bid to enhance the photocatalytic properties of perovskite with higher degradation efficiency of organic pollutants in water, a number of studies have been carried out to produce various novel nanoscale structures with unique properties [83–86]. The approaches that can be used to downsize perovskite particle and control its size are ball milling, ultrasonic treatment, micro emulsion method, addition of chelating agents and controlling the calcination temperature [87, 88]. A number of studies have used ball milling to downsize perovskite particle size. By controlling ball milling time duration, the particle size and defects of STO can be adjusted, for example, by increasing ball milling treatment time the particle size can be significantly decreased [89]. The smaller STO particles can provide shorter charge carriers transport path to the surface, where the photocatalytic reaction take place, thus decreasing the chance of photoinduced e^--h^+ pairs recombination. On the other hand, longer ball milling treatment time results in the creation of defects like oxygen vacancies, which act as a mediator and further facilitates the charge separation to accelerate the degradation of organic pollutant molecules. Nanoscale (15 nm average thickness and 70–80 nm length) floral-like LFO 3D structure has been prepared via hydrothermal method using polyvinylpyrrolidone (PVP) as a chelating agent. Apparently, the photocatalytic activity of the nanoscale engineered LFO for the degradation of organic pollutants is much higher than that of the bulk LFO, owing to the higher surface area (90.25 m^2/g as compared to bulk LFO of 8 m^2/g) that provides abundant active sites and efficient separation of e^--h^+ pairs via facile charge transport on small-sized nanosheets.

Over the last decade, several hierarchical nanostructure perovskites with different and effective morphology have been synthesised to enhance perovskite photocatalytic performance water decontamination [90, 91]. Hierarchical porous structure is one of the morphologies that provides larger surface area and more active sites, both of which enhance the contact between perovskite and organic pollutants during photocatalytic reaction [92]. Porous structure enhances the rate of mass transfer of pollutants within the perovskite, leading to an excellent photocatalytic activity with fast reaction kinetics for organic pollutant degradation. Porous nanofiber structure has also been synthesised for the enhancement of physical and chemical properties of perovskites compared to their granulate counterpart. Perovskite nanofiber structure can be synthesised using several approaches such as electrospinning method, template synthesis, hydrothermal method, self-assembly and solvothermal method, amongst which electrospinning method is widely regarded as a simple and easily controllable method for the fabrication of nanofibers [93]. The enhanced photocatalytic efficiency and higher removal of pollutants in water achieved by perovskite nanofiber structure can be attributed to the abundant and reachable active sites and the ultralong 1D nanostructure, both of which provide effective directing photogenerated electrons transportation [94, 95], as illustrated in **Figure 7**. The $LaCoO_3$ nanofiber structure has been shown to exhibit a higher photocatalytic activity than the $LaCoO_3$ nanoparticles in the degradation of RhB, owing to the favorable features of nanofiber structure such as larger surface area with more photoactive sites [23, 96]. The LFO ribbon-like porous ultrafine nanofibers have been synthesised by electrospinning, which exhibit higher specific surface area and more active sites for enhanced light absorption and photocatalytic degradation of MB.

The interfacial area of perovskite nanoparticles can be maximised by developing the Core-shell heterostructure with 3D hierarchical contact between the core and shell layers, in which a broader platform for charge carrier migration is provided. Various methods have been developed for synthesis of core-shell nanostructure

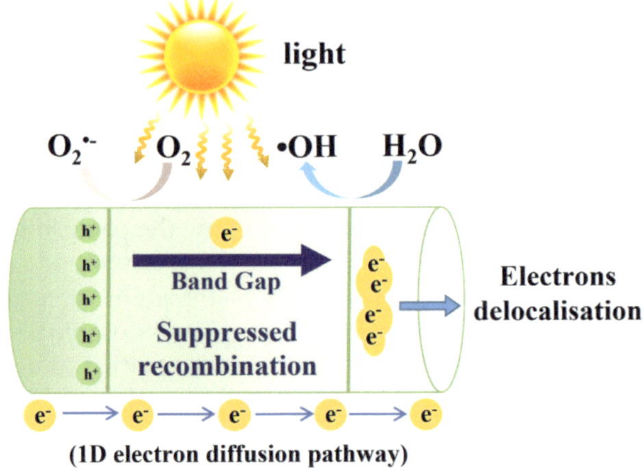

Figure 7.
Schematic representation of electron-hole pair separation in perovskite nanofiber systems.

perovskite such as surface functionalisation, template-sacrificial method and self-assembly method [97–99]. A heterostructure core-shell morphology enhances light absorption and multiple reflection of incident light, thus high light-harvesting efficiency, which makes core-shell perovskite structure ideal for photocatalytic applications, where charge separation is highly desirable [100]. A unique structure and heterojunction perovskite-based particles have been developed by coupling perovskites ($SrTiO_3$, $LaFeO_3$, $LaMnO_3$, $LaNiO_3$) with semiconductors (TiO_2, ZnO and SnO_2) as a core or shell to achieve an effective charge separation [101, 102]. LFO/TiO_2 core/shell heterostructure has been developed as photocatalysts with significantly improved photocatalytic activity for the degradation of myclobutanil pesticide under solar light and visible light irradiation [103]. The LFO shell acts as photosensitizer of TiO_2 for visible-light harvesting, and the full interphase contact between the core LFO and shell TiO_2 provides extensive charge transfer by driving electrons to the TiO_2 core and holes reversely to the LFO shell, which results in remarkably enhanced ·OH and $O_2^{\bullet -}$ generation for photodegradation of organic pollutants. A core-shell perovskite, (Ba,Sr)TiO_3 as core and TiO_2 as shell, has been reported to have an enhanced photocatalytic activity, due to charge separation efficiency across the core-shell interface.

Amongst the aforementioned nano structures, nanotube array has been widely studied for the enhancement of the photocatalytic activities of perovskites, due to their tunable size, uniformly aligned tubular structure, large internal surface area and fast electron transmission [104, 105].

4. Conclusions and future trends

Perovskites are highly crystalline and stable materials with unique and excellent features, which render them to be the best candidates amongst other semiconductor photocatalysts for photocatalytic degradation of organic pollutants. Their structure allows for tunning and adjusting their physiochemical properties through regulating their chemical composition, which offer a wide scope in developing novel nanocomposites.

Knowing that pristine perovskites suffer from a number of limitations such as low photocatalytic efficiency, insufficient solar energy consumption, rapid recombination of electron-hole and low redox potential, the great flexibility in regulating their physiochemical properties has been capitalised with the aim of developing perovskite-based materials with efficient photocatalytic activity in water remediation. A large number of perovskite-based nanocomposites have been developed and studied, using various synthesis methods and strategies such as partial and full cationic substitution using certain dopants, structure rescaling through downsizing and morphology alteration, coating and coupling with other advanced oxidation processes. Designing and preparing novel perovskite-based nanocomposites via these strategies, particularly partial and full cationic substitution, is elegant and can produce reasonably efficient photocatalysts, albeit it is still challenging.

Capitalising on the ability of computational tools, such as Density Functional Theory (DFT) based band structure, which are very effective for designing and understanding novel materials, will lead to further understanding of quantitative properties of these materials and shorten the selection process. Most of the perovskite-based nanocomposites exhibit promising photocatalytic performance, which can be considered as a viable solution to the problem of persistent organic pollutants faced by the living world and environment. However, a comprehensive study on these novel materials, particularly their ability in photocatalytic degradation of a wide range of organic pollutants is limited. The majority of the developed perovskite-based nanocomposites have been tested on synthetic wastewater under laboratory conditions, which does not quite represent natural wastewater with organic pollutants. It is worth pointing out that in natural wastewater organic pollutants can affect photocatalytic efficiency through radical scavenging and attenuation of radiation in photocatalytic process. Using actual water and studying the perovskite-based materials provide more opportunities for large applications. Rescaling the structure of perovskite nanocomposites and alteration of their morphologies, for example creation of hierarchical pore structure, may be sufficient to enhance some aspects of photocatalytic activity in the degradation of a particular dye or pollutant. Further work on the relationship between the structure and physicochemical properties of these nanocomposites will be beneficial for further enhancement of photocatalytic activities of these materials and providing a novel insight over structure-property relation. Partial and full substitution strategy using different dopants is mainly used to promote light absorption, which in turn enhances the photocatalytic activity. A comprehensive knowledge on the effects of these dopants on the photophysical properties of the modified perovskite materials will further reveal the capabilities of perovskite-based nanocomposites and pave the way for future development perovskite-based photocatalysts with superior photocatalytic activity. The main focus of almost all the studies has been on the small production of small laboratory-scale materials, incorporating 3D printing and testing the produced materials will certainly be beneficial for large-scale synthesis, leading to industrial production. Membrane filtration is usually the main separation process that is used for separating the catalysts, however, using membrane to recover perovskite-based catalysts makes the water treatment process costly and quite complex, particularly due to the occurrence of membrane fouling. Therefore, recovery and recycling the perovskite-based nanocomposites after water decontamination process merit thorough investigations.

Acknowledgements

The authors acknowledge the funding support from Chengdu Science and Technology Bureau (2019-GH02-00053-HZ).

Author details

Yousef Faraj[1]* and Ruzhen Xie[2]

1 University of Chester, Chester, UK

2 Sichuan University, Chengdu, China

*Address all correspondence to: y.faraj@chester.ac.uk

References

[1] Hejazi R, Mahjoub AR, Khavar AHC, Khazaee Z. Novel visible-light-responsive rGO-ZnO@Bi2MoO6 nanocomposite with enhanced light harvesting and Z-scheme charge transfer for photodegradation and detoxification of RhB. Solid State Sciences. 2019;**2019**:105934

[2] Robertson PKJ, Robertson JMC, Bahnemann DW. Removal of microorganisms and their chemical metabolites from water using semiconductor photocatalysis. Journal of Hazardous Materials. 2012;**211**:161-171

[3] Yao XX, Hu XL, Zhang WJ, Gong XY, Wang X, Pillai SC, et al. Mie resonance in hollow nanoshells of ternary TiO_2-Au-CdS and enhanced photocatalytic hydrogen evolution. Applied Catalysis B: Environmental. 2020;**276**:119153

[4] Devi LG, Kavitha R. A review on non metal ion doped titania for the photocatalytic degradation of organic pollutants under UV/solar light: Role of photogenerated charge carrier dynamics in enhancing the activity. Applied Catalysis B: Environmental. 2013;**140**:559-587

[5] Salas SE, Rosales BS, de Lase H. Quantum yield with platinum modified TiO_2 photocatalyst for hydrogen production. Applied Catalysis B: Environmental. 2013;**140**:523-536

[6] Loeb SK, Alvarez PJJ, Brame JA, Cates EL, Choi W, Crittenden J, et al. The technology horizon for photocatalytic water treatment: Sunrise or sunset? Environmental Science & Technology. 2019;**53**:2937-2947

[7] Wei K, Faraj Y, Yao G, Xie R, Lai B. Strategies for improving perovskite photocatalysts reactivity for organic pollutants degradation: A review on recent progress. Chemical Engineering Journal. 2021;**414**:128783

[8] Shu LN, Sunarso J, Hashim SS, Mao JK, Zhou W, Liang FL. Advanced perovskite anodes for solid oxide fuel cells: A review. International Journal of Hydrogen Energy. 2019;**44**:31275-31304

[9] Garcia-Munoz P, Lefevre C, Robert D, Keller N. Ti-substituted $LaFeO_3$ perovskite as photoassisted CWPO catalyst for water treatment. Applied Catalysis B: Environmental. 2019;**248**:120-128

[10] Wang HL, Zhang LS, Chen ZG, Hu JQ, Li SJ, Wang ZH, et al. Semiconductor heterojunction photocatalysts: Design, construction, and photocatalytic performances. Chemical Society Reviews. 2014;**43**:5234-5244

[11] Sulaeman U, Yin S, Sato T. Visible light photocatalytic activity induced by the carboxyl group chemically bonded on the surface of $SrTiO_3$. Applied Catalysis B: Environmental. 2011;**102**:286-290

[12] Luo XS, Bai LM, Xing JJ, Zhu XW, Xu DL, Xie BH, et al. Ordered mesoporous cobalt containing perovskite as a high-performance heterogeneous catalyst in activation of peroxymonosulfate. ACS Applied Materials & Interfaces. 2019;**11**:35720-35728

[13] Sun J, Wei HZ, An LY, Jin CY, Wu HL, Xiong ZA, et al. Oxygen vacancy mediated La1-xCexFeO3-delta perovskite oxides as efficient catalysts for CWAO of acrylic acid by A-site Ce doping. Applied Catalysis B: Environmental. 2019;**245**:20-28

[14] Ye TN, Dong ZH, Zhao YN, Yu JG, Wang FQ, Guo SK, et al. Controllable

fabrication of perovskite $SrZrO_3$ hollow cuboidal nanoshells. CrystEngComm. 2011;**13**:3842-3847

[15] Li XZ, Shi HY, Zhu W, Zuo SX, Lu XW, Luo SP, et al. Nanocomposite LaFe1-xNixO_3/Palygorskite catalyst for photo-assisted reduction of NOx: Effect of Ni doping. Applied Catalysis B: Environmental. 2018;**23**:92-100

[16] He Z, Sun XY, Gu X. $SrTiO_3$ nanoparticles and nanofibers: Synthesis and comparison of photocatalytic properties. Journal of Material Science. 2017;**28**:13950-13955

[17] Karthikeyan C, Arunachalam P, Ramachandran K, Al-Mayouf AM, Karuppuchamy S. Recent advances in semiconductor metal oxides with enhanced methods for solar photocatalytic applications. Journal of Alloys and Compounds. 2020;**828**:154281

[18] Herrmann JM. Heterogeneous photocatalysis: Fundamentals and applications to the removal of various types of aqueous pollutants. Catalysis Today. 1999;**53**:115-129

[19] Luo J, Li R, Chen YQ, Zhou XS, Ning XM, Zhan L, et al. Rational design of Z-scheme LaFeO3/SnS2 hybrid with boosted visible light photocatalytic activity towards tetracycline degradation. Separation and Purification Technology. 2019;**210**:417-430

[20] Jiang DL, Wang TY, Xu Q, Li D, Meng SC, Chen M. Perovskite oxide ultrathin nanosheets/g-C3N4 2D-2D heterojunction photocatalysts with significantly enhanced photocatalytic activity towards the photodegradation of tetracycline. Applied Catalysis B: Environmental. 2017;**201**:617-628

[21] Zhang Q, Huang Y, Peng SQ, Zhang YF, Shen ZX, Cao JJ, et al. Perovskite LaFeO3-$SrTiO_3$ composite for synergistically enhanced NO removal under visible light excitation. Applied Catalysis B: Environmental. 2017;**204**:346-357

[22] Schneider J, Matsuoka M, Takeuchi M, Zhang JL, Horiuchi Y, Anpo M, et al. Understanding TiO_2 photocatalysis: Mechanisms and materials. Chemical Reviews. 2014;**114**:9919-9986

[23] Li Y, Gao XP, Li GR, Pan GL, Yan L, Zhu HY. Titanate nanofiber reactivity: Fabrication of $MTiO_3$ (M = Ca, Sr, and Ba) perovskite oxides. Journal of Physical Chemistry C. 2009;**113**:4386-4394

[24] Jamil HAATS. Nano sized Fe doped strontium titanate for photocatalytic degradation of dibutyl phthalate under visible light. Advanced Materials Letters. 2016;**6**:467-471

[25] Goldschmidt VM. Die Gesetze der Krystallochemie. Naturwissenschaften. 1926:477-485

[26] Ji QQ, Bi L, Zhang JT, Cao HJ, Zhao XS. The role of oxygen vacancies of ABO(3) perovskite oxides in the oxygen reduction reaction. Energy & Environmental Science. 2020;**13**:1408-1428

[27] Haruna A, Abdulkadir I, Idris SO. Photocatalytic activity and doping effects of $BiFeO_3$ nanoparticles in model organic dyes. Heliyon. 2020;**6**:e03237

[28] Dhiman M, Singhal S. Effect of doping of different rare earth (europium, gadolinium, dysprosium and neodymium) metal ions on structural, optical and photocatalytic properties of $LaFeO_3$ perovskites. Journal of Rare Earths. 2019;**37**:1279-1287

[29] Hou L, Sun GF, Liu K, Li GFM. Preparation, characterization and investigation of catalytic activity of

Li-doped $LaFeO_3$ nanoparticles. Journal of Sol-Gel Science and Technology. 2006;**40**:9-14

[30] Li FT, Liu Y, Liu RH, Sun ZM, Zhao DS, Kou CG. Preparation of Ca-doped $LaFeO_3$ nanopowders in a reverse microemulsion and their visible light photocatalytic activity. Materials Letters. 2010;**64**:223-225

[31] Humayun M, Raziq F, Zhang XL, Yan R, Li ZJ, Qu Y, et al. Synthesis of ZnO/Bi-doped porous $LaFeO_3$ nanocomposites as highly efficient nano-photocatalysts dependent on the enhanced utilization of visible-light-excited electrons. Applied Catalysis B: Environmental. 2018;**231**:23-33

[32] Bradha M, Vijayaraghavan T, Suriyaraj SP, Selvakumar R, Ashok AM. Synthesis of photocatalytic La(1-x) A(x)TiO(3.5-delta) (A=Ba, Sr, Ca) nano perovskites and their application for photocatalytic oxidation of congo red dye in aqueous solution. Journal of Rare Earths. 2015;**33**:160-167

[33] Wang Y, Ma W, Song Y, et al. Enhanced photocatalytic performance of $SrTiO_3$ powder induced by Eu dopants. Journal of Rare Earths. 2020;**39**:541-547

[34] Ha MN, Chao ZL, Zhao Z. Doping effects on mixed-phase crystalline perovskite AxSr1−xFeO3−δ (A = Pr, Sm; $0 \leq x \leq 0.8$) nanoparticles and their application for photodegradation of rhodamine B. Research on Chemical Intermediates. 2018;**45**:1493-1508

[35] Wei ZX, Wang Y, Liu JP, Xiao CM, Zeng WW. Synthesis, magnetization and photocatalytic activity of $LaFeO_3$ and LaFe0.5Mn0.5−xO3−δ. Materials Chemistry and Physics. 2012;**136**:755-761

[36] Fu XX, Yang QH, Sang LX. Studies on photocatalytic activity of perovskite type LaFe1-xCuxO_3. Chemical Journal of Chinese Universities Chinese. 2002;**23**:283-286

[37] Hoang VQT, Phan TQP, Senthilkumar V, Doan VT, Kim YS, Le MV. Enhanced photocatalytic activities of vandium and molybdenum co-doped strontium titanate under visible light. International Journal of Applied Ceramic Technology. 2019;**16**:1651-1658

[38] Jamil TS, Abbas HA, Youssief AM, Mansor ES, Hammad FF. The synthesis of nano-sized undoped, Bi doped and Bi, Cu co-doped $SrTiO_3$ using two solegel methods to enhance the photocatalytic performance for the degradation of dibutyl phthalate under visible light. Comptes Rendus Chimie. 2017;**20**:97-106

[39] Bantawala H, Shenoy US, Bhat DK. Vanadium-doped $SrTiO_3$ nanocubes: Insight into role of vanadium in improving the photocatalytic activity. Applied Surface Science. 2020;**513**:145858

[40] Abdi M, Mahdikhah V, Sheibani S. Visible light photocatalytic performance of La-Fe co-doped $SrTiO_3$ perovskite powder. Optical Materials. 2020;**102**:109803

[41] Tonda S, Kumar S, Anjaneyulu O, Shanker V. Synthesis of Cr and La-codoped $SrTiO_3$ nanoparticles for enhanced photocatalytic performance under sunlight irradiation. Physical Chemistry Chemical Physics. 2014;**16**:23819-23828

[42] Wang M, Zhao TT, Dong XL, Li M, Wang HQ. Effects of Ce substitution at the A-site of LaNi0.5Fe0.5O_3 perovskite on the enhanced catalytic activity for dry reforming of methane. Applied Catalysis B: Environmental. 2018;**224**:214-221

[43] Xu JF, Liu J, Zhao Z, Xu CM, Zheng JX, Duan AJ, et al. Easy synthesis of three-dimensionally ordered macroporous La1-xKxCoO_3 catalysts and their high activities for the catalytic combustion of soot. Journal of Catalysis. 2011;**282**:1-12

[44] Zheng JQ, Zhu YJ, Xu JS, LuQ B, Qi C, Chen F, et al. Microwave-assisted rapid synthesis and photocatalytic activity of mesoporous Nd-doped $SrTiO_3$ nanospheres and nanoplates. Materials Letters. 2013;**100**:62-65

[45] Wang Q, Liu ZQ, Liu DM, Liu GS, Yang M, Cui FY, et al. Ultrathin two-dimensional BiOBrxI1-x solid solution with rich oxygen vacancies for enhanced visible-light-driven photoactivity in environmental remediation. Applied Catalysis B: Environmental. 2018;**236**:222-232

[46] Hou LW, Zhang H, Dong LH, Zhang L, Duprez D, Royer S. A simple non-aqueous route to nano-perovskite mixed oxides with improved catalytic properties. Catalysis Today. 2017;**287**:30-36

[47] Sulaeman U, Yin S, Sato T. Solvothermal synthesis and photocatalytic properties of chromium-doped SrTiO3 nanoparticles. Applied Catalysis B: Environmental. 2011;**105**:206-210

[48] Wu GL, Li P, Xu DB, Luo BF, Hong YZ, Shi WD, et al. Hydrothermal synthesis and visible-light-driven photocatalytic degradation for tetracycline of Mn-doped $SrTiO_3$ nanocubes. Applied Surface Science. 2015;**333**:39-47

[49] Zhu HK, Fang MH, Huang ZH, Liu YG, Chen K, Guan M, et al. Novel chromium doped perovskites A(2) ZnTiO(6) (A= Pr, Gd): Synthesis crystal structure and photocatalytic activity under simulated solar light irradiation. Applied Surface Science. 2017;**393**:348-356

[50] Reunchan P, Umezawa N, Ouyang SX, Ye JH. Mechanism of photocatalytic activities in Cr-doped SrTiO3 under visible-light irradiation: An insight from hybrid density-functional calculations. Physical Chemistry Chemical Physics. 2012;**14**:1876-1880

[51] Peng Q, Shan B, Wen YW, Chen R. Enhanced charge transport of $LaFeO_3$ via transition metal (Mn, Co, Cu) doping for visible light photoelectrochemical water oxidation. International Journal of Hydrogen Energy. 2015;**40**:15423-15431

[52] Wu SX, Ma Z, Qin YN, He F, Jia LS, Zhang YJ. XPS study of copper doping TiO_2 photocatalyst. Acta Physico-Chimica Sinica. 2003;**19**:967-969

[53] Phan TTN, Nikoloski AN, Bahri PA, Li D. Heterogeneous photo-Fenton degradation of organics using highly efficient Cu-doped $LaFeO_3$ under visible light. Journal of Industrial and Engineering Chemistry. 2018;**61**:53-64

[54] Peng JL, Lu XH, Jiang X, Zhang YH, Chen QX, Lai B, et al. Degradation of atrazine by persulfate activation with copper sulfide (CuS): Kinetics study, degradation pathways and mechanism. Chemical Engineering Journal. 2018;**354**:740-752

[55] Kato H, Kudo A. Visible-light-response and photocatalytic activities of TiO2 and $SrTiO_3$ photocatalysts codoped with antimony and chromium. The Journal of Physical Chemistry. B. 2002;**106**:5029-5034

[56] Li F, Yu K, Lou LL, Su ZQ, Liu SX. Theoretical and experimental study of La/Ni co-doped SrTiO3 photocatalyst. Material Science Engineering B Advanced. 2010;**172**:136-141

[57] Zou F, Jiang Z, Qin XQ, Zhao YX, Jiang LY, Zhi JF, et al. Template-free synthesis of mesoporous N-doped SrTiO3 perovskite with high visible-light-driven photocatalytic activity. Chemical Communications. 2012;**48**:8514-8516

[58] Khazaee Z, Mahjoub AR, Khavar AHC, Srivastava V, Sillanpaa M. Synthesis of layered perovskite Ag,F-Bi2MoO6/rGO: A surface plasmon resonance and oxygen vacancy promoted nanocomposite as a visible-light photocatalyst. Journal of Photochemistry and Photobiology A. 2019;**379**:130-143

[59] Hu J, Ma JH, Wang L, Huang H, Ma LW. Preparation, characterization and photocatalytic activity of Co-doped $LaMnO_3$/graphene composites. Powder Technology. 2014;**254**:556-562

[60] Chang CW, Hu CC. Graphene oxide-derived carbon-doped $SrTiO_3$ for highly efficient photocatalytic degradation of organic pollutants under visible light irradiation. Chemical Engineering Journal. 2020;**383**:123116

[61] Wu Y, Wang H, Tu W, Liu Y, Tan YZ, Yuan XZ, et al. Quasi-polymeric construction of stable perovskite-type LaFeO3/g-C3N4 heterostructured photocatalyst for improved Z-scheme photocatalytic activity via solid p-n heterojunction interfacial effect. Journal of Hazardous Materials. 2018;**347**:412-422

[62] Nousir S, Keav S, Barbier-Jr J, Bensitel M, Brahmi R, Duprez D. Deactivation phenomena during catalytic wet air oxidation (CWAO) of phenol over platinum catalysts supported on ceria and ceria-zirconia mixed oxides. Applied Catalysis B: Environmental. 2008;**84**:723-731

[63] Jian D, Gao PX, Cai WJ, Allimi BS, Alpay SP, Ding Y, et al. Synthesis, characterization, and photocatalytic properties of ZnO/(La,Sr)CoO3 composite nanorod arrays. Journal of Materials Chemistry. 2009;**19**:970-975

[64] Song S, Xu LJ, He ZQ, Chen JM, Xiao XZ, Yan B. Mechanism of the photocatalytic degradation of CI reactive black 5 at pH 12.0 using SrTiO3/CeO_2 as the catalyst. Environmental Science & Technology. 2007;**41**:5846-5853

[65] Zhang RJ, Wan YJ, Peng JL, Yao G, Zhang YH, Lai B. Efficient degradation of atrazine by $LaCoO_3/Al_2O_3$ catalyzed peroxymonosulfate: Performance, degradation intermediates and mechanism. Chemical Engineering Journal. 2019;**372**:796-808

[66] Yang JS, Li LM, Yang XS, Song S, Li J, Jing FL, et al. Enhanced catalytic performances of in situ-assembled $LaMnO_3$/delta-MnO_2 hetero-structures for toluene combustion. Catalysis Today. 2019;**327**:19-27

[67] Gong C, Zhang ZY, Lin S, Wu Z, Sun L, Ye CQ, et al. Electrochemical synthesis of perovskite $LaFeO_3$ nanoparticle-modified TiO_2 nanotube arrays for enhanced visible-light photocatalytic activity. New Journal of Chemistry. 2019;**43**:16506-16514

[68] Kumar RD, Thangappan R, Jayavel R. Synthesis and characterization of $LaFeO_3/TiO_2$ nanocomposites for visible light photocatalytic activity. Journal of Physics and Chemistry of Solids. 2017;**101**:25-33

[69] Chen C, Zhou JL, Geng JF, Bao RY, Wang ZH, Xia JX, et al. Perovskite $LaNiO_3/TiO_2$ step-scheme heterojunction with enhanced photocatalytic activity. Applied Surface Science. 2020;**503**:144287

[70] Cao TP, Li YJ, Wang CH, Shao CL, Liu YC. A facile in situ hydrothermal

method to $SrTiO_3/TiO_2$ nanofiber heterostructures with high photocatalytic Aactivity. Langmuir. 2011;**27**:2946-2952

[71] Liu X, Jiang JZ, Jia YS, Qiu JM, Xia TL, Zhang YH, et al. Insight into synergistically enhanced adsorption and visible light photocatalytic performance of Z-scheme heterojunction of $SrTiO_3$(La,Cr)-decorated WO_3 nanosheets. Applied Surface Science. 2017;**412**:279-289

[72] Phan TTN, Nikoloski AN, Bahri PA, Li D. Adsorption and photo-Fenton catalytic degradation of organic dyes over crystalline $LaFeO_3$-doped porous silica. RSC Advances. 2018;**8**:36181-36190

[73] Cao X, Luo SQ, Liu C, Chen JW. Synthesis of bentonite-supported Fe_2O_3-doped TiO_2 superstructures for highly promoted photocatalytic activity and recyclability. Advanced Powder Technology. 2017;**28**:993-999

[74] Phan TTN, Nikoloski AN, Bahri PA, Li D. Facile fabrication of perovskite-incorporated hierarchically mesoporous/macroporous silica for efficient photoassisted-Fenton degradation of dye. Applied Surface Science. 2019;**491**:488-496

[75] Li HL, Zhu JJ, Xiao P, Zhan YY, Lv KL, Wu LY, et al. On the mechanism of oxidative degradation of rhodamine B over $LaFeO_3$ catalysts supported on silica materials: Role of support. Microporous and Mesoporous Materials. 2016;**221**:159-166

[76] Li HL, Zhang WJ, Liu YX. HZSM-5 zeolite supported boron-doped TiO_2 for photocatalytic degradation of ofloxacin. Journal of Materials Research and Technology. 2020;**9**:2557-2567

[77] Wang XY, Ozdemir O, Hampton MA, Nguyen AV, Do DD. The effect of zeolite treatment by acids on sodium adsorption ratio of coal seam gas water. Water Research. 2012;**46**:5247-5254

[78] Zhang WJ, Du L, Bi FF, He HB. A novel SrTiO3/HZSM-5 photocatalyst prepared by sol-gel method. Materials Letters. 2015;**157**:103-105

[79] Qin CX, Li ZY, Chen GQ, Zhao Y, Lin T. Fabrication and visible-light photocatalytic behavior of perovskite praseodymium ferrite porous nanotubes. Journal of Power Sources. 2015;**285**:178-184

[80] Faisal M, Harraz FA, Ismail A, El-Toni AM, Al-Sayari SA, Al-Hajry A, et al. Polythiophene/mesoporous $SrTiO_3$ nanocomposites with enhanced photocatalytic activity under visible light. Separation and Purification Technology. 2018;**190**:33-34

[81] Huo YS, Yang H, Xian T, Jiang JL, Wei ZQ, Li S, et al. A polyacrylamide gel route to different-sized $CaTiO_3$ nanoparticles and their photocatalytic activity for dye degradation. Journal of Sol-Gel Science and Technology. 2014;**71**:254-259

[82] Souza AE, Santos GTA, Barra BC, Macedo WD, Teixeira SR, Santos CM, et al. Photoluminescence of $SrTiO_3$: Influence of particle size and morphology. Crystal Growth & Design. 2012;**12**:5671-5679

[83] Thirumalairajan S, Girija K, Mastelaro VR, Ponpandian N. Photocatalytic degradation of organic dyes under visible light irradiation by floral-like $LaFeO_3$ nanostructures comprised of nanosheet petals. New Journal of Chemistry. 2014;**38**:5480-5490

[84] Su HJ, Jing LQ, Shi KY, Yao CH, Fu HG. Synthesis of large surface area $LaFeO_3$ nanoparticles by SBA-16

template method as high active visible photocatalysts. Journal of Nanoparticle Research. 2010;**12**:967-974

[85] Li SD, He ZX, Wang XL, Gao K. Fabrication of unique ribbon-like porous LaFeO3 nanofibers photocatalyst via electrospinning. Applied Physics A Materials. 2014;**117**:1381-1386

[86] Li L, Rohrer GS, Salvador PA. Heterostructured ceramic powders for photocatalytic hydrogen production: Nanostructured TiO_2 shells surrounding microcrystalline (Ba,Sr)TiO_3 cores. Journal of the American Ceramic Society. 2012;**95**:1414-1420

[87] Buscaglia V, Randall CA. Size and scaling effects in barium titanate. An overview. Journal of European Ceramic Society. 2020;**40**:3744-3758

[88] Singh H, Rajput JK. Novel perovskite nanocatalyst (BiFeO3) for the photodegradation of rhodamine B/tartrazine and swift reduction of nitro compounds. Journal of the Iranian Chemical Society. 2019;**16**:2409-2432

[89] Kumar V, Choudhary S, Malik V, NagarajanR,KandasamiA,SubramanianA. Enhancement in photocatalytic activity of $SrTiO_3$ by tailoring particle size and defects. Physica Status Solidi A: Applications and Materials Science. 2019;**216**:1900294

[90] Humayun M, Qu Y, Raziq F, Yan R, Li Z, Zhang X, et al. Exceptional visible-light activities of TiO_2-coupled N-doped porous perovskite $LaFeO_3$ for 2,4-dichlorophenol decomposition and CO_2 conversion. Environmental Science & Technology. 2016;**50**:13600-13610

[91] Cruz del Álamo A, González C, Pariente MI, Molina R, Martínez F. Fenton-like catalyst based on a reticulated porous perovskite material: Activity and stability for the on-site removal of pharmaceutical micropollutans in a hospital wastewater. Chemical English Journal. 2020;**401**:126113

[92] Afzal S, Quan X, Zhang JL. High surface area mesoporous nanocast LaMO3 (M = Mn, Fe) perovskites for efficient catalytic ozonation and an insight into probable catalytic mechanism. Applied Catalysis B: Environmental. 2017;**206**:692-703

[93] Dong B, Li ZC, Li ZY, Xu XR, Song MX, Zheng W, et al. Highly efficient $LaCoO_3$ nanofibers catalysts for photocatalytic degradation of Rhodamine B. Journal of the American Ceramic Society. 2010;**93**:3587-3590

[94] Yun HJ, Lee H, Joo JB, Kim W, Yi J. Influence of aspect ratio of TiO_2 nanorods on the photocatalytic decomposition of formic acid. Journal of Physical Chemistry C. 2009;**113**:3050-3055

[95] Liu Y, Zhang MY, Li L, Zhang XT. One-dimensional visible-light-driven bifunctional photocatalysts based on Bi4Ti3O12 nanofiber frameworks and Bi2XO6 (X = Mo, W) nanosheets. Applied Catalysis B: Environmental. 2014;**160**:757-766

[96] Zhu K, Vinzant TB, Neale NR, Frank AJ. Removing structural disorder from oriented TiO_2 nanotube arrays: Reducing the dimensionality of transport and recombination in dye-sensitized solar cells. Nano Letters. 2007;**7**:3739-3746

[97] Pan CS, Xu J, Wang YJ, Li D, Zhu YF. Dramatic activity of C_3N_4/$BiPO_4$ photocatalyst with core/shell structure formed by self-assembly. Advanced Functional Materials. 2012;**22**:1518-1524

[98] Kumar S, Khanchandani S, Thirumal M, Ganguli AK. Achieving enhanced visible-light-driven photocatalysis using Type-II $NaNbO_3$/CdS core/shell heterostructures. ACS Applied Materials & Interfaces. 2014;**6**:13221-13233

[99] Wei YZ, Wang JY, Yu RB, Wan JW, Wang D. Constructing $SrTiO_3$-TiO_2 heterogeneous hollow multi-shelled structures for enhanced solar water splitting. Angewandte Chemie International. 2019;**58**:1422-1426

[100] Zhang N, Xu YJ. Aggregation- and leaching-resistant, reusable, and multifunctional Pd@CeO_2 as a robust nanocatalyst achieved by a hollow core-shell strategy. Chemistry of Materials. 2013;**25**:1979-1988

[101] Wang SB, Ren Z, Song WQ, Guo YB, Zhang MW, Suib SL, et al. ZnO/perovskite core-shell nanorod array based monolithic catalysts with enhanced propane oxidation and material utilization efficiency at low temperature. Catalysis Today. 2015;**258**:549-555

[102] Do JY, Im Y, Kwak BS, Park SM, Kang M. Preparation of basalt fiber@perovskite $PbTiO_3$ core-shell composites and their effects on CH_4 production from CO_2 photoreduction. Ceramics International. 2016;**42**:5942-5951

[103] Garcia-Munoz P, Fresno F, Ivanez J, Robert D, Keller N. Activity enhancement pathways in $LaFeO_3$@TiO_2 heterojunction photocatalysts for visible and solar light driven degradation of myclobutanil pesticide in water. Journal of Hazardous Materials. 2020;**400**:123099

[104] Chen X, He XF, Yang X, Wu ZS, Li YF. Construction of novel 2D/1D g-C3N4/CaTiO3 heterojunction with face-to-face contact for boosting photodegradation of triphenylmethane dyes under simulated sunlight. Journal of Taiwan Institute of Chemical Engineers. 2020;**107**:98-109

[105] Deng AJ, Chen JF, Li HL, Ren JL, Sun R, Zhao LH. Photo-degradation of methyl orange by polysaccharides/$LaFe_{0.8}Cu_{0.2}O_3$ composite films. BioResources. 2014;**9**:2717-2726

Chapter 11

Encapsulation of Metal Nanoparticles (MNPs) as Catalyst

Masoud Safari and Valiollah Nobakht

Abstract

Metal nanoparticles (MNPs) are the main agents in heterogeneous catalysis. Hence, utilizing the effective physico-chemical methods to engage them to achieve the highest catalysts performance with well-controlled size, shape, and surface properties seems to be essential. The encapsulation of metal nanoparticles is a promising approach that enhances the catalytic activity of the materials. Not only the encapsulating structures can adjust the catalytic properties of metal nanoparticles, particularly selectivity, but also prevents them from agglomeration and sintering. In this chapter, the various encapsulating structures consist of yolk/core-shell and mesoporous structures, and encapsulating materials that are divided into three parts, including inorganic materials, metal–organic frameworks, and organic materials are presented.

Keywords: encapsulation, yolk/core-shell, mesoporous structures, catalyst, metal nanoparticles

1. Introduction

Catalysts have a significant impact on contemporary chemical processes. The role of nano-catalysts in the form of metal nanoparticles (MNPs), such as nanoparticles of transition metals and particularly the group VIII metal elements in the fourth period with d-band center electronic structure, is rising significantly in the heterogeneous catalysis, due to their high active surface areas and special electronic properties [1, 2]. For instance, Fe-, Co-, Ni-, Pt-, Rh-, and Pd-based catalysts have been extensively utilized in various processes to produce varieties of fuels and chemicals, consisting of Fischer Tropsch Synthesis, steam and dry reforming, methanation, and CO oxidation [1]. MNPs expose influential activities in various catalytic reactions. One of the main difficulties in using the MNPs is their weak stability in the practical catalytic processes. In recent years, an exquisite and effective strategy has been applied to optimize the performance of MNPs by encapsulating them that provides unique advantages toward catalysis, particularly under harsh conditions by restricting the aggregation and sintering of MNPs. The concept of the encapsulated metals nanoparticles that improve a wide range of catalytic reactions, especially under rough environments has been presented as "chainmail nano-catalyst" in some of the literatures [3].

In addition, Encapsulation adjusted the energy distinctions between the highest occupied molecular orbital to the lowest unoccupied molecular orbital [4]. In fact, inside the encapsulated structure of MNPs, the electron of MNPs can penetrate through the shell to enhance the catalytic performance on the external surface. Therefore, the shell is able to restrict the medium of reactants and products from direct contact by the MNPs and protect the MNPs from damage in harsh conditions. Overall, these encapsulated MNPs represent impressive catalysts that can be applied in numerous catalytic reactions under harsh conditions and also propose a tunable structure alongside unique electronic properties [3].

In this chapter, a comprehensive and compressed presentation of the encapsulated MNPs advantages, a categorization of the encapsulating layers, and some methods of encapsulated nano-catalysts preparation are reported, respectively.

2. Advantages of encapsulated MNPs

In view of an increasing range of recent research, implementing an encapsulated structure for nano-catalysts is the efficient way to protect these catalysts' properties from any deactivation agents and prevent them from the decrease of catalytic activity. The advantages of encapsulated structure can be expressed as follows and in going into detail their effective specifications and involved parameters in this structure will be explained.

1. Provide effective metal-support interactions (MSI)

2. Prevent from the coalescence nanoparticles

3. Protect from catalyst deactivation by stabilizing the active metal species in catalysts against sintering

4. Tandem reactions can be enabled and promoted by encapsulated configuration

5. Improvement of catalytic selectivity

One of the most crucial characteristics of the encapsulated nano-catalysts is the strong interaction between the MNPs and the encapsulating materials. Interactions between MNPs and support materials, metal-support interaction (MSI), can profoundly enhance the catalytic activity and tunability to selective reactions and products [5]. MSI can modify electronic properties, geometric morphologies, or chemical compositions of MNPs to make active sites have specific properties and catalytic activities [6, 7]. The geometric result of MSI is the decoration of the metal surface, partial coverage or total encapsulation of the metal can be conducted by the support. Since morphological modifications can take place on the catalyst as a result of the MSIs and encapsulated structure paves the way to enhance them by contributing a maximal interfacial area through participating the MNPs in close contact with the encapsulating materials [6–8]. Furthermore, strong metal-support interaction (SMSI) has been mostly applied to develop bifunctionality at the interface of metal–metal oxides systems. The bifunctional outcome leads to a synergistic effect through an improvement of catalyst activity and selectivity by creating new reaction sites at the interface between the metal and the support, and the spillover phenomenon occurs at the interface by transferring reactants from metal or support to the interface, which all of these phenomena are soared at encapsulated structure [6–8]. Therefore, the optimized catalytic performance can be gained by the unique MSI among the MNPs and the encapsulating materials [5].

Metal nano-catalysts have a high surface-to-volume ratio, which leads to coalescence during catalysis, especially at high reaction temperatures where this is prone to modification, resulting in a noticeable decrease in active surface areas and, eventually, a reduction in catalytic activity and selectivity [2]. In contrast, encapsulation by encasing of MNPs in nano-shells or nanopores prohibits them from coalescence, therefore, the efficiency of the active nano-catalysts surface area is preserved remarkably.

Catalyst deactivation is a principal difficulty in the heterogenous catalytic processes since the catalyst activity and selectivity will be reduced with time on stream and the catalyst regeneration or replacement will be included at a noticeable rate of time and resources [8, 9]. Therefore, a great number of researches have been conducted to gain long-term stability in heterogeneous catalysis, particularly by applying an optimized structure that includes the maximized performance. In the midst of the principal reasons of catalyst deactivation, sintering of MNPs, which occurs by an irreversible mechanism that significantly reduces catalyst recyclability, has attracted more attention toward preventing it from happening. Catalyst sintering reduces the metal's active surface area and can occur via particle coalescence or Ostwald ripening [10, 11]. Not only may the MNPs engage in migration and coalescence particularly when they are exposed to severe conditions, but also the deactivation of catalysts via sintering can be intensified. Nevertheless, by encapsulating MNPs in a highly nanopores support, this spatial confinement can dramatically suppress their migration and coalescence and can prevent sintering without limiting the catalysis and thus stabilize MNPs under harsh reaction conditions [2, 8–11].

Chemical processes are conducted by multiple steps. To minimize energy consumption and economic cost, integrating multiple steps of reactions can be a crucial solution. A tandem catalysis reaction engages sequential reactions within one condition through the coupling of appropriate catalysts in which sequential transformation of the substrate is conducted via two or more mechanistically distinct reaction steps [12–14]. The encapsulated structure of MNPs provides an integration of various interfaces in one nanostructure in a controllable manner that converts them as effective and principal catalysts in the tandem catalysis reactions. Moreover, encapsulating MNPs in porous shells could promote tandem reactions via modifying the reaction sequences or the molecular-sieving effect [2, 15].

Due to preventing the formation of unwanted products, increasing the waste of chemicals, reducing the essential refinement stages, and optimizing the selective products, it is critical to control the chemical reactions. The heart of a chemical reaction is its catalysts so it is crucial to design an appropriate catalyst to enhance the selectivity. The activity and particularly selectivity of heterogeneous catalysts depend on their surfaces structure and their active sites. First of all, the encapsulated structures in various catalysts support by providing their unique morphology and can act as sieves for molecules that allow the transfer of molecules with selective sizes, resulting in high shape selectivity. In addition, catalytic selectivity of the encapsulated nano-catalysts can be improved by utilizing suitable functional groups in their materials that have a significant influence on the reactant adsorption and reduce the mass transfer limitations by increasing their coefficient diffusions. In addition to the controlled porosity of the encapsulated structures, owing to high active surface areas of the encapsulating materials, there are more sites to involve the suitable functional groups to enhance the selectivity of the catalysts [12–18].

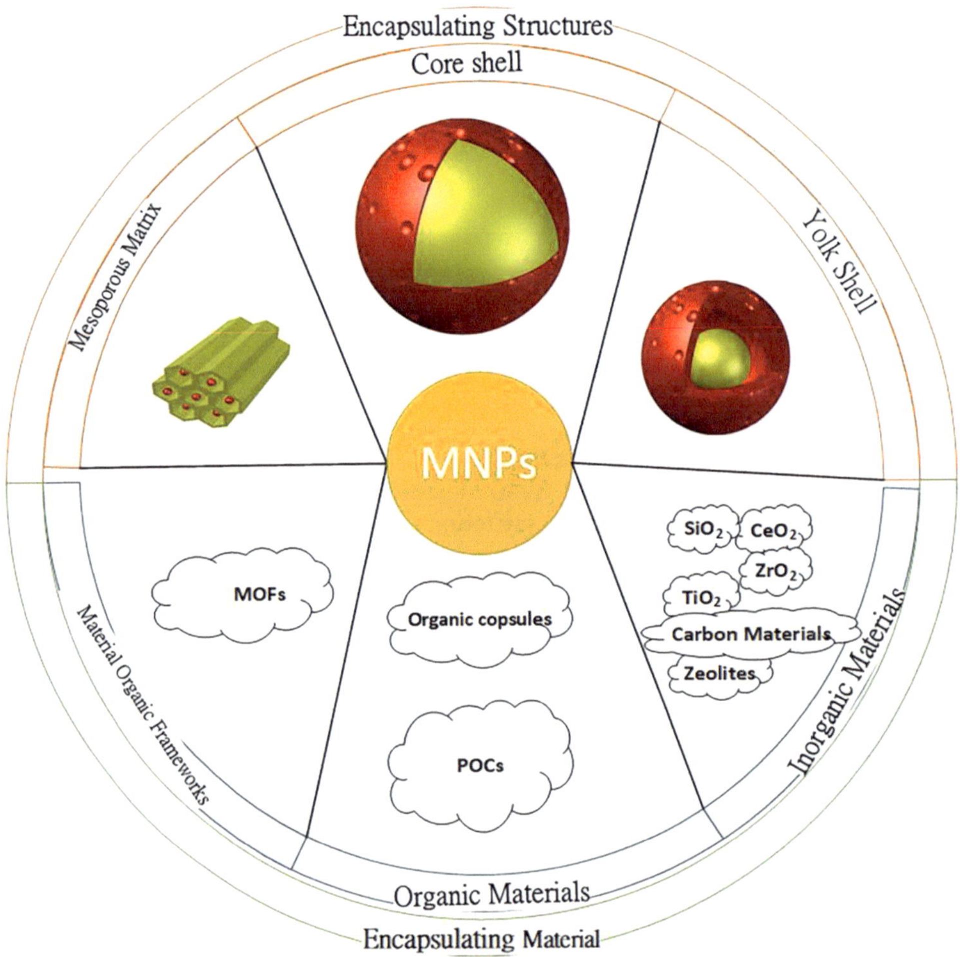

Figure 1.
Encapsulation of MNPs as catalyst.

Although each one of mentioned nano-catalysts encapsulated structure benefits are effective separately, there is a remarkable relationship among their impacts. Indeed, each of them causes the other one or implements it in parallel and intensifies each other.

In contrast, although the encapsulation structures can enhance the catalytic performances, mass transfer may be restricted by this structure to some extent, which is disadvantageous to the catalytic process. This issue should be modified by the precise sketch through the choice of materials and methods synthesis of the catalysts with encapsulation structures [1]. In this chapter, the classification of the encapsulation of catalysts is undertaken with the type of encapsulating materials because some of them can form into two distinct groups of morphologies or exhibit with an individual structure, such as organic materials. Thus, in follow firstly describe each type of morphology and then the encapsulation of catalysts is presented in three comprehensive parts from the view of encapsulating materials: (1) Inorganic Materials, (2) Metal–Organic Frameworks (MOFs), and (3) Organic Materials as it is depicted in **Figure 1**.

3. Encapsulating structures

Encapsulation provides a unique catalyst immobilization technique that by encapsulating of MNPs with porous layers creates an impressive porosity for the reactants to be able to reach the metal surface. Various strategies and materials have been utilized to make these layers. An encapsulated structure can conduct by designing and fabricating the unique chainmail catalyst via a wide range of shells [3]. As a result, a vast majority of structural designs and strategies in such encapsulated catalysts have been thrived. Although a large range of various materials, such as metallic state of metals, alloys of metals, metal carbides, metal oxides, metal phosphides, and metal nitrides, can be encapsulated, a variety range of coating layers can be utilized for the shells that contain various forms, such as shells, tubes, sheaths, matrices, and films. In this chapter, the encapsulated catalysts structures on the basis of the morphology are categorized into two main groups: (1) Yolk/Core-Shells, and (2) Mesoporous Structures [1–3], as shown in **Figure 1**.

3.1 Yolk/core-shell structures

Encapsulation compels the complex of support and MNPs to implement a non-planar geometry, which leads to a higher reactivity. The core-shell morphology is allocated to the encapsulated structure that encases nanoparticles in a confined case by an outer shell [1]. In contrast, another approach to encapsulate MNPs inside an individual architecture shell which is a state of eggs with a void space between the MNP "yolk" and the porous coating "shell," this structure has added profits due to their specific role as nanoreactors [19]. Although Yolk-shell is a "core-void-shell" structure, which is similar to the core-shell structure, a void space between the core and the shell depicts the distinction between these two structures [1]. Recently, yolk-shell materials have attracted interest due to their particular combination of high thermal stability with monodisperse and narrow particle-size distributions that convert them to an effective catalyst for heterogeneous catalysis. Not only do these specifications have a significant role in catalyst performance, but also they can facilitate kinetic and mechanistic researches [20].

In these yolk/core-shell catalysts, the vast majority of the catalytic attributes of the throughout catalyst are the same and may include a tiny alteration through the change from active particle to active particle. Thus, in these structures the resulting catalytic activity of the system can be the same all over the surface of the catalysts, in comparison, in traditional solid catalysts there are no homogenous sites and to evaluate the activity of the catalysts, an average of all possible morphologies should be considered. Furthermore, the yolk/core-shell structures provide catalytic particles with a uniform size that leads to an effective SMSI for all the particles. They contribute to the new catalytic sites at the interface of themselves and MNPs and modify their actual catalytic characteristics of them. Therefore, these features cause these kinds of encapsulation structures to be placed at the noticeable position as support of MNPs catalysts and remarkable efforts have been done in synthesizing base-metal yolk/core-shell catalysts in recent years [1, 21]. In addition, there are some significant ways toward the improvement of the design and fabrication, in particular, yolk-shells to produce effective catalysts:

3.1.1 Enhance the porosity by etching

Although the encapsulation of nanoparticles by applying an additional coating layer may decline the adsorption of the reactants on the active phase, the porosity of the shells exhibits a crucial role in the improvement of mass transfer across the shells. Therefore, porosity is an essential specification that should be considered when synthesizing yolk/core-shell catalysts. Various methods to make porose yolk/core-shell, such as layer-by-layer deposition techniques and sacrificial templating procedures, in particular for yolk-shells, have been reported that provide adequate stability even under harsh reaction conditions. Moreover, the "surface-protected etching" is a contemporary strategy that outputs catalysts with porous shells and enhanced stability simultaneously. This process protects the oxide shells, in particular the inner part of the shells, by utilizing a suitable layer of the polymeric ligand as an etching agent. Thus, the oxide shells would have their original size and the selective etching of the interior creates the porous structure [22]. On the other hand, this procedure may have a noticeable restriction in which the metal loading of the catalyst in most cases is placed at a low level. To dispel this drawback, first of all, an inert chemical linker should be utilized to fix the MNPs onto the surface of the initial support with another layer of these inert chemical linkers. Next, the surface-protected etching procedure would be utilized to modify the outer shell into a mesoporous shell that exposes the MNPs to the reactant active species. Silica is one of the most popular inert chemical linkers [22–25].

3.1.2 Surface-protected calcination

In most cases, the various yolk/core-shells that are prepared through the sol–gel deposition method are amorphous. Due to the remarkable effects of the crystalline phases on the performance of catalysts, it is essential to enhance shell crystallinity. The effective pathway to modify the amorphous shells to their crystallized counterparts is calcination at high temperatures. Although calcination would improve the crystalline phases of the shell, it might reduce the porosity of the original amorphous shell considerably. Therefore, to prevent this difficulty, first of all, it is important to conduct another inert chemical layer (such as silica layer) on the top of the shell through a sol–gel process, then the calcination treatment should be applied. Eventually, the final porous morphology that will be a yolk-shell structure can be

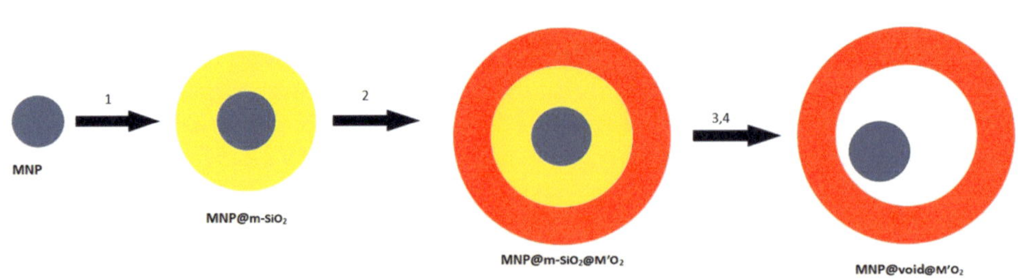

Figure 2.
Synthetic procedure of MNP@Void@M'O$_2$ (M'O$_2$: TiO$_2$, CeO$_2$, ZrO$_2$, ...) Yolk-Shell in four steps; 1: Encapsulation with sacrificed shell (m-SiO$_2$) nanoparticles, 2: Encapsulation with exterior shell (M'O$_2$), 3: Calcination, and 4: Etching.

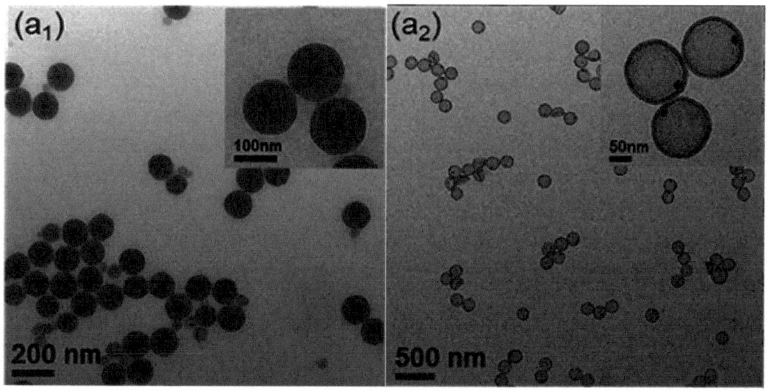

Figure 3.
TEM images of (a1) MNP@SiO_2 core-shell and (a2) MNP@Void@M'O_2 Yolk-Shell. Reprinted with permission from Ref. [26]. Copyright 2021 American Chemical Society.

achieved by scarifying the inert chemical layers through chemical etching [22–25]. **Figure 2** depicts the procedure of fabrication of a porous yolk-shell through employing sacrificed layer, calcination, and etching treatment. Two TEM images of a core-shell (a1) and yolk-shell (a2) are also illustrated in **Figure 3** [26].

3.2 Mesoporous structures

Mesoporous materials are special types of nanomaterials with ordered arrays of uniform nanochannels that are fabricated by participating self-assembly of surfactants and framework precursors. Mesoporous materials include pores with diameters in the range of 2–50 nm that by dispersion of MNPs into this porous matrix a range of heterogeneous catalysts can be formed [2]. These structures by supplying a great surface area provide a dramatic spatial dispersion of the MNPs that leads to stability enhancement in contrast to nanoparticle aggregation and coalescence in a catalytic process. In addition, the prominent porosity of these structures in catalysts paves the way to prepare an effective mass transfer that improves the catalytic performance by facilitating the contact of reactants with the MNPs as active sites of catalysts. Furthermore, the appropriate pore size distribution puts the MNPs adjacent to the mesoporous material, which leads to the enhancement of catalytic activity and stability through boosting the strong MSIs. The mesoporous materials have pores with an adequate size that cause the adsorption of pre-synthesized MNPs with small sizes directly [1, 2]. Meanwhile, the size and volume of the mesopores or in overall the size of the whole mesoporous matrix, and metal charge have a significant impact on the immobilization of metals [27]. Eventually, the specific morphology of the pores in these structures exposes the sites of active metals, which leads to facilitating the catalytic process [1, 2]. Thus, these features of mesoporous materials convert them to outstanding supports that can encapsulate the MNPs in catalysts.

Although the incipient wetness impregnation is a popular strategy for the encapsulation of MNPs in the mesoporous materials as heterogeneous catalysts, the self-assembly-based approach is the other useful strategy [2].

4. Encapsulating materials

4.1 Encapsulation of MNPs in inorganic materials

In recent decades, noticeable progress in synthesizing various inorganic nanomaterials to encapsulate the nanoparticles with unique catalytic performance, owing to their principal advantages, such as simplicity of fabrication, tunability of formation, and cost-effectiveness, converts them to popular supports. Although inorganic oxides (SiO_2, TiO_2, CeO_2, ZrO_2, etc.) are mostly applied as shell structures in the yolk/core-shell morphology, some of them, in particular silica, exist in mesoporous in some cases. In contrast, zeolite mesoporous structures are the common morphology beside the yolk/core-shell structure of them.

4.1.1 Silica

4.1.1.1 Silica nano-shells

Silica has attracted an increasing number of researcher attention owing to its inimitable physicochemical properties, such as tunable morphology, extensive surface area (≈1500 $m^2\ g^{-1}$) and pore volume, adjustable sizes (50–150 nm), shapes (hexagonal, wormhole-like, cubic, and lamellar) and morphologies (spheres, helical fibers, tubules, gyroids, crystals), ease of surface functionalization (both interior as well as exterior), unique topology, colloidal and thermal stabilities, and high dispersity [27]. Silica with each of these desirable features can be achieved through tuning the synthesis conditions, such as the temperature, pH, stirring speed, and type of silica source, and particularly the surfactant [27]. Therefore, these unique attributes of silica besides the simplicity in controlling the SiO_2 precursors convert it to a principal and useful inorganic shell for the encapsulation of MNPs. Moreover, owing to SiO_2 inert chemical features, its structure would be resistant under harsh conditions that lead to the protection of the size of MNPs under this condition. Furthermore, not only would not the chemical inertness of silica that encapsulates MNPs is influenced by the metal-oxide interactions but also intensify their catalytic properties [1, 3]. Although silica supports due to their chemical inertness have a tiny interaction with reactants that may reduce the adsorption of the reactive molecules, their porosity can restrict the effects of their weakness by intensifying their diffusion across the capsule [1]. Moreover, stability of the metal-silica nanoparticles is an outstanding challenge, particularly in catalytic applications that deal with diverse physicochemical properties especially the features of the final catalyst compounds, such as the type of MNPs, particle size, degree of silica condensation, and chemical functionalization [27].

The principal method to form the silica shells is the sol–gel. The synthesis of the silica shell is conducted through a modified Stober procedure, in which the hydrolysis and condensation of tetra ethyl orthosilicate (TEOS) take place in aqueous ethanol in the presence of a base as a catalyst, such as ammonia, to control the growth of silicate on the surface of the MNPs [1, 2]. To coat some unstable metals, such as Ag, with silica, it is essential to utilize the amines, such as dimethylamine or di-ethylamine, instead of ammonia as the base catalyst [2]. Overall, as opposed to mesoporous silica materials to encapsulate the MNP in the shell spheres of mesopores silica, firstly the metal precursors should be dispersed and the surfactant alongside the silica precursor (TEOS) should be added. This core-shell strategy contributes the best results in encapsulating some metal oxides, such as Mn, Co, and Ni [27].

Although the general configuration of silica shells achieved through the sol–gel process is microporous, the mesopore textures could also be formed that are more appropriate for catalysis applications because this shape of texture rectifies the mass transfer limitations [1, 4]. The encapsulated metal-silica nanostructure gaining through the sol–gel pathway makes a thin shell of silica which owing to the high interfacial energy a fragile interaction exists between silica shell and MNPs. Nucleation of silica prevents from occurring a suitable interaction between silica and MNPs. However, to pave the way and achieve an improved interfacial interaction, utilizing some bridging agents as surface primers is essential [4]. Generally, the surface primers are applied for encapsulation of silica over large MNPs (>10 nm) that can refer to some of them, such as amino propyltrimethoxy silane (APS), methoxy poly (ethylene glycol) thiol (MPEG-SH), and polyvinylpyrrolidone (PVP). PVP is applied for the smooth coating of silica on the MNPs [1, 2]. Despite the fact that surface primers have a critical role to create a stable and homogeneous silica coating, in some cases, they can provide a selective silica coating on the surface [4, 28]. Furthermore, the type of the silica source and the added surfactants have a crucial impact on the size of MNPs and the thickness of the SiO_2 shell in yolk/core-shell structures. Although some surfactants, such as PVP, cetyltrimethylammonium bromide, and chloride (CTAB and CTAC), could not be influential on the dispersion of the MNPs, they could have an enhancement effect on the porosity of the SiO_2 shells [1, 4].

On the other hand, ultrasmall MNPs (<10 nm) are unstable and have an aggregate tendency in alcoholic solutions. As a result, the formation of a coating of silica cannot be implemented by an appropriate outcome. To get rid of this imperfection, the silica coating should be conducted in a reverse micelles (or microemulsion) system by using polyoxyethylene nonylphenyl ether (Igepal CO-520) as a surfactant [2]. Mostly, owing to there not adequate interaction between silica and the metal surface this technique of coating may have some weaknesses that may lead to an undesirable and imperfection coating. Liu et al. [29] presented an effective procedure, ship-in-a-bottle, to fabricate a robust thin silica coating on the sub-3 nm MNPs in reverse micelles. In this method, the synthesis of MNPs and silica coating in the presence of water/cyclohexane/reverse micelle system will participate simultaneously. The combination of microemulsion system and ship-in-a bottle technique pave the way to achieve a range of various metal-silica core-shell composites [2].

4.1.1.2 Mesoporous framework of silica

Mesoporous silica nanoparticles are presented as effective support for the encapsulation of MNPs due to their outstanding specifications, such as well-ordered framework, tunable pores, high surface area, stability, and thermally toughness. In addition, they contribute a well-made 3D matrix that provides a monotonous distribution of MNPs and confinement them from aggregating there that lead to noticeably protect against the sintering of them. Moreover, in contrast to the other inorganic supports through the encapsulation of MNPs, the mesoporous silica nanoparticles can represent a wide range of applicable characteristic surfaces in both the exterior (on the surface) and interior (in the mesopore) space, where MNPs can be organized by a chemical linkage or physically immobilized by electrostatic interactions [27].

The mesoporous silica nanoparticles can be synthesized through cooperative self-assembly of surfactant and silica species. The morphology and dimensions of mesoporous silica nanoparticles are significantly influenced by the factors of reaction kinetics of sol–gel chemistry, such as assembly kinetics, silica condensation, nucleation, growth rates, and surfactant-silica interactions, in addition to pH value of the reaction medium, water content, and temperature. Mobil composition of matter (MCM)-41, and the Santa Barbara amorphous type material (SBA)-15 are the famous mesoporous silica nanoparticles that are synthesized by applying quaternary ammonium salts and Pluronic copolymer-based surfactants, respectively [27].

There are two common techniques to encapsulate the MNPs in mesoporous silica materials. First of all, in the way of an incipient wetness impregnation method dealing with the encapsulation of MNPs, a solution of the metal salt is exposed to a powder of mesoporous silica including the same pore volume as the volume of the metal salt solution. Then, the MNPs are transferred from the metal salt solution into the mesopores by utilizing the capillary force. Finally, the implementation of the calcination and reduction in H_2 subsequently will pave the way to achieve a mesoporous silica matrix that encapsulates MNPs. In some cases of the incipient wetness impregnation procedures, electrostatic interactions are the main force to conduct the diffusion of MNPs through the meso-channels that can be implemented by modifying the mesopore surface by positively charged quaternary ammonium groups [2].

The other procedure is the participation of self-assembly of the MNPs and silica precursors with a surfactant-mediated condensation technique. Firstly, a dilution of the silica precursor is mixed with the aqueous ammonia including the CTAB surfactant molecules for initial nucleation and on the contrary of the core-shell silica architecture then the desired metal precursor is subsequently added to the mixture. Eventually, to achieve the desired mesoporous silica material, the surfactant can be eliminated by either calcination procedure at high temperatures (550°C) or a variety of chemical solutions, such as acidic ethanol or ammonia in ethanol/isopropanol. Although isopropanol, as a solvent for ammonium nitrate, can be utilized to extract the surfactant, it is able to save the well-order of the mesostructures and create a dramatic impact on the surface area and pore volume of mesoporous silica nanoparticles. Hence, not only there is not any essential tuning in the pore size or volume of the mesoporous silica nanoparticles, but also there is no noticeable change in the final particle size or pore sizes in the presence of the encapsulated MNPs in the siliceous frameworks that leads to uniform encapsulating of metals in mesoporous silica nanoparticles at the basic PH [2, 27].

4.1.2 Titania nano-shells

Among the non-silica coating materials, titania is the most popular metal oxide. Despite the fact that TiO_2 is utilized as a coating layer of MNPs in a wide range of catalysts, this combination has illustrated noticeable synergy in various catalytic reactions. The synthetic routes and precursors are the main effective agents to produce a specific architecture of yolk/core-shell. The encapsulation of MNPs in titania nanoshells is the outcome of the direct coating of TiO_2 on them. Similar to the method of silica coating, sol–gel procedure is the major process of titania coating. Thus, through this method by applying titanium alkoxides like tetraisopropoxide (TTIP) in a nonaqueous solution while the existence of water, the TiO_2 coating will be implemented. Due to titanium alkoxide hydrolyze in water immediately, applying chelating agents like acetylacetone, which is a chelating agent of titanium butoxide (TBOT), controlled hydrolysis can be provided that leads to fabricate a core-shell nanostructure with a stable coating of titania [2].

To develop stable yolk/core-shells of titania, utilizing appropriate precursors have a crucial role. As evidence, a chelated complex of titanium glycolate, which is formed by reacting titanium alkoxide with ethylene glycol and is much more stable than titanium alkoxide, can provide an outstanding precursor for the titania coating of MNPs that can undertake a very controllable sol–gel process to fabricate identical metal@TiO_2 core-shell nanoparticles that are catalyzed by acetone. Moreover, this technique is applicable for a controlled coating of titania on small MNPs with a diameter from tiny sizes to 50 nm [2, 4]. Furthermore, to achieve a uniform and thin shell of titania that can be adjusted with the thickness ranging from 3 to 12 nm in the sol–gel process, an acid catalyst like citric acid to hydrolyze the TBOT at the presence of alcohol can be applied [2].

Another suitable precursor that can slow down the hydrolyze in water, titanium (IV) bis (ammonium lactate) dihydroxide (TALH), can provide the titania coating with shell thickness ranging from sub-50 nm on MNPs. Although the TALH in aqueous solutions at room temperature is stable, it can be hydrolyzed at high temperatures (approximately 65°C) that lead to control of the sol–gel process [2].

Although Zeng et al. [30] fabricated a yolk-shell nanostructure of the Au@TiO_2 through the hydrolysis of TiF_4 at high temperatures (180°C), the most popular pathway to obtain a metal/TiO_2 yolk-shell nanostructure forms a coating layer of titania on metal/SiO_2 composites, which the silica will be sacrificed [2]. The principal benefit of this procedure of MNPs encapsulation with titania at the presence of sacrificial silica is the easy formation on the contrary of the direct formation of titania nanoshell. In addition, owing to the chemical sympathy between the two oxides to achieve a suitable coating layer of TiO_2, it is not essential to implement any rectification on the silica surface to set up the interactions between titanate species and silica. Despite this coating pathway being a successful process, to improve the coating of titania in this way, utilizing a surfactant such as hydroxypropyl cellulose (HPC) is presented that enhances the colloidal dispersion of silica nanoparticles [2].

Finally, similar to encapsulation of ultrasmall MNPs with silica nano-shells, titania coating will be conducted in the same way. First of all, in a reverse micelles system nanoparticles are formed and then a coating of silica on the nanoparticles at the presence of TEOS will be done and the last layer, which is titania, will be formed through applying TBOT. Eventually, the yolk-shell nano-capsule can be achieved by thermal reduction and etching of the silica templates [2].

4.1.3 CeO_2 nano-shells

Another metal oxide that attracted significant interest is cerium oxide. The presence of oxygen vacancies at the terminating of its planes is affected considerably on the adsorption of reactant molecules in the catalytic reactions by controlling the energetics of the surface interactions [31]. Despite the fact that the encapsulation of MNPs in CeO_2 nano-shells tackle the sintering of MNPs, it provides strong metal-support interactions (SMSI) that lead to enhancing the stability and the performance of the catalysts at the series of catalytic oxidation reactions, particularly at high temperatures [2].

The core-shell nanoparticles of metal@CeO_2 could be synthesized based on the self-assembly procedure by applying a supramolecular [2]. Gorte et al. [32] implemented this strategy by applying a capping ligand of a thiolate (11-mercaptoundecanoic acid, MUA) to fabricate Pd nanoparticles that are mixed in tetrahydrofuran (THF) [2]. The carboxylic groups of MUA conducted the self-assembly of the cerium

(IV) alkoxides around the Pd nanoparticles directly by exchanging the alkoxy group on the Ce (IV) salt with the carboxylic group on the surface of the Pd nanoparticles as a result of the presence of the carboxylic group, which is a stronger ligand for Ce (IV) than the alkoxy group. Furthermore, various ranges of metal@CeO_2 core-shell nanoparticles can be fabricated by controlling and designing precisely the effective parameters of the self-assembly strategy and the sol–gel process, particularly by utilizing the appropriate chelating agent. So, some of these chelating agents and their role in the encapsulation of MNPs by CeO_2 yolk/core shells are presented as follows [2].

Ethylenediaminetetraacetic acid (EDTA) is applied as a chelating agent that chelates Ce (III) salt. Although EDTA slows down the hydrolysis of the Ce^{3+} ions, due to the negative charges of the EDTA-Ce (III) complex, the electrostatic interactions between the Ce complex precursor and MNPs have a significant role in the self-assembly strategy [2]. Moreover, triethanolamine as a chelating agent to achieve a cerium-atrane precursor has exhibited its effect on fabricating the desired metal@ CeO_2 core-shell nanoparticles by adjusting the sol–gel kinetics. Another chelating agent for the CeO_2 coating on MNPs is citric acid. It puts its effects by controlling the self-assembly way through conducting the adsorption of Ce^{3+} ions on the MNPs that were gradually oxidized to CeO_2 nanoparticles [2].

Another principal approach to synthesize metal@CeO_2 core-shell nanoparticles is the auto-redox strategy, which is based on the reduction of the high valence metal species and the oxidation of low-valence Ce (III) species. Some of metal@CeO_2 core-shell nanoparticles that are synthesized by utilizing this mechanism depict a "rice-ball" shape-like Ag@CeO_2. In addition, to encapsulate the ultrasmall MNPs and construct the uniform metal@CeO_2 core-shell nanoparticles, the auto-redox strategy could be undertaken in a reverse micelle system. Furthermore, due to any surfactants not engaged in the auto-redox methods, multicore-shell nanospheres with a diameter larger than 85 nm could be formed. Although CeO_2-encapsulated bimetallic MNPs can be formed by applying the auto-redox strategy, the self-assembly approach and the salting-out effect can be influenced the formation of these core-shell nanoparticles [2].

Another ideal encapsulation nanostructure with a hollow space between the metal core and the outer porous CeO_2 shell is the yolk-shell architecture that has an effective impact on tackling the aggregation and sintering of tiny noble MNPs in catalytic reactions [2]. To implement this kind of encapsulation, a templating method should be conducted by coating a layer of silica firstly, and then the layer of CeO_2 could participate on the MNPs through a sol–gel process. Eventually, the metal@SiO_2@ CeO_2 nanospheres can be modified into multi-yolk-shell metal@CeO_2 nanospheres by eliminating the silica template. In addition, through this process, multi-yolk-shell structured nano-catalysts can be formed, for instance, Pd@hm-CeO_2, which is presented by Zheng et al. [2, 33]. Despite silica being the most popular sacrificial template, polystyrene (PS) fibers and resorcinol-formaldehyde (RF) can be utilized as a removable template [2].

4.1.4 ZrO_2 nano-shells

Zirconium dioxide (ZrO_2) is a metal-based inorganic material that is presented as an insulator in some applications. Due to the chemical inert feature of ZrO_2, it has outstanding resistance to acids and alkalis environments that convert it to crucial catalyst support at harsh reaction conditions. In addition, it has significant heat stability that is suitable for a high thermal catalytic reaction to encapsulate and support the MNPs. Although it has low thermal conductivity and is utilized as thermal barrier coatings, it has a high refractive index, which is well-suited for various optical applications [34].

Recently, zirconia as catalyst support, particularly in the encapsulation shape, attracts numerous interests. In most cases, MNPs are encapsulated in zirconia nano-shells in the form of yolk-shell [25, 35–37]. Similar to the other yolk-shell nano-catalysts, metal@ZrO_2 yolk-shell can be synthesized which this catalyst with robust zirconia shells illustrates noticeable catalytic activity and outstanding anti-aggregation features during the time of the catalytic process and upon thermal treatment or reduction.

4.1.5 Carbon nano-shells

Apart from inorganic oxide shells, encapsulating MNPs in carbon nano-shells participated in numerous investigations. Although the core-shell nanostructures manufactured from inorganic oxide have noticeable advantages, in some conditions they illustrate some weaknesses, for instance, the dissolution of silica coatings at strong basic conditions. In contrast, carbon yolk/core-shells can tackle these difficulties and additionally demonstrate some outstanding specifications, such as high physical and chemical stability under harsh conditions, high surface area, tunable electronic structures, high electrical conductivity, good biocompatibility, and relatively low manufacturing costs. Therefore, they can be one of the best materials to encapsulate the MNPs [1, 2, 38]. With regard to the type of crystallinity of carbon shells, this section is allocated into metal @amorphous carbon and metal @graphitic carbon.

4.1.5.1 Metal @amorphous carbon

Metal @amorphous carbon can be fabricated by encapsulated MNPs through a polymer coating layer. Due to their low cost, rich chelating groups, and high compatibility with MNPs, these polymers are presented as leading carbon precursors that include resorcinol-formaldehyde (RF) resin, tannic acid, and polydopamine (PDA) [2]. To fabricate a noticeable-performance M@carbon catalysts, carbon precursors should be utilized through a suitable controlled sol–gel process to provide the target coating layer and then carbonization should be implemented which the output would be a carbon shell with appropriate thickness and tunable pore structure. For instance, the carbon nano-shell with 63 wt% would be formed during the carbonization of the RF coating shell under an inert atmosphere [1, 2]. In addition, to adopt the congruity between the inorganic cores and the RF shells, it is essential to modify the surface with CTAB or 3-aminopropyltriethxoysilane (APS) in the coating process [2].

On the other hand, although conducting the coating process after synthesizing MNPs will be yielded a controllable shell, implementing this process through a one-pot in which the formation of MNPs and the polymer coating will be done in a single step leads to achieving a more convenient pathway to synthesis M@RF core-shell nanospheres in the absence of surfactants. Through this procedure first of all, MNPs are formed from metal salts by adding formaldehyde. Next in the presence of the other precursor, ammonia, the polymerization of the RF precursors on the surface of MNPs will be undertaken. The concentration of resorcinol and formaldehyde can control the size and thickness of the RF shells. Moreover, resorcinol can reduce the surface activity of MNPs and prevent them from aggregating. Eventually, M@carbon core-shell nanospheres with a wide range of metals can be obtained after carbonization [2].

Furthermore, impregnation is another principal approach to synthesizing encapsulated MNPs in amorphous carbon. Overall, in this method firstly metal ions could be adsorbed at the sites of amino groups in pre-synthesized mesoporous aminophenol formaldehyde (APF) nanospheres. To create hollow carbon shells which, encapsulate MNPs before conducting carbonization a mesoporous silica layer should be done. Moreover, yolk-shell structures with MNPs can be achieved by coating another layer of APF on the APF@SiO_2 nanospheres. This mechanism is flexible and can be used to fabricate monometallic Au, Pt, Rh, and Ru and bimetallic Au-Pt, Au-Rh, and Pt-Rh nanoparticles [2]. What is more, some other sources of carbon such as D-glucose, saccharides including fructose and sucrose, and dopamine can be involved. Dopamine owing to its catechol and amine groups and the ability to self-polymerize on various substrates is considered as a remarkable carbon source, particularly in synthesizing yolk-shell structures containing MNPs in the yolk that encapsulated with carbon nano-shells [1, 2].

4.1.5.2 Metal @graphitic carbon

Although amorphous carbon represents significant specification in catalysis applications, graphitic carbon has further conductivity and stability, particularly in electrocatalysis. The high-temperature pyrolysis promotes the crystallization of carbon through the process of fabrication of MNPs encapsulated in graphitic carbon nano-shells (M@GC). One of the most common procedures to fabricate the M@ GC is the pyrolysis of metal–organic frameworks (MOFs) in an inert or reductive atmosphere directly. During the pyrolysis of MOFs which are the assembly of metal ions as nodes that are linked together through organic ligands as linkers, MNPs are achieved by the reduction of metal nodes, and then these MNPs can catalyze the generation and configuration of graphitic carbon from organic linkers on their surface. In addition, bimetallic alloy nanoparticles could be encapsulated in graphitic carbon through pre-encapsulating noble MNPs or metal salts in MOFs and subsequent pyrolysis [2].

Graphene is a solo layer of graphitic carbon atoms that bonded together in a honeycomb crystal lattice, which this unique structure intensifies its specifications much higher than the other carbon material and converts it to an outstanding material for encapsulating MNPs to improve their catalytic activities. As a result of diffusion restriction and chemical inertness of graphene to a various oxidizing gas, it conducts as a passivation layer to intercept some metal (Cu, Ni, etc.) from oxidation. Although the potential energy surface of graphene can transfer from 0.15 to 1 eV on various metal substrates (Ni, Co, etc.), the principal metal has an influential impact on the electronic structure of the graphene coating layer. The electronic specification of graphitic carbon can exhibit its critical role on the catalytic activity when the shell includes no more than three to four carbon layers. Hence, it is crucial to fabricate carbon encapsulated catalysts with a controllable number of graphene layers which one of the usable procedures is the chemical vapor deposition (CVD) technique [39].

Although through CVD techniques a thin film can be formed on the substrate surface which has a dramatic influence on the fabrication of carbon nano-shells, these methods have multistep synthesis processes, which lead to being complex, expensive, and difficult to implement large-scale commercialization. In addition, to achieve a graphene shell on MNPs through the CVD procedure, it is essential to provide a high temperature above 800°C, this condition may lead to melting the MNPs and gathering them together [39, 40]. Thus, it is essential to utilize a functional synthesis pathway that is presented by an arc-discharge method. Through this technique, MNPs encapsulated in graphene shells have large sizes and extensive distribution of particle sizes, in some cases empty carbon cages and CNTs were achieved simultaneously. To tackle this difficulty, applying a long pulse laser in methane or a mixture of methane and helium at room temperature are presented, the efficiency of this approach is the formation of an ordinary size of 5 nm and appropriate size distribution of 3–10 nm of core-shell structure M@GC [39–41].

On the other hand, the metal alloy can be encapsulated in highly nitrogen-doped graphene layers by one-step annealing under a nitrogen flow without adding any other carbon sources by utilizing bimetallic complexes with CN- group linkers in the form of metal–organic framework (MOF) precursors [39]. In addition, electrostatic interactions between negatively charged graphene oxide and positively charged metal oxide nanoparticles can be applied for encapsulating metal oxide in graphene shells, which this method can be followed by chemical reduction. First of all, amino propyl-trimethoxy silane (APS) paves the way for the metal oxide nanoparticles to exhibit an oxide surface positively charged. Then, electrostatic interactions put the modified metal oxide nanoparticles with negatively charged graphene oxide together. Eventually, the accumulation of them will chemically be reduced with hydrazine to gain the metal oxide nanoparticles encapsulated in graphene. Because this mechanism can provide all the applicable features, such as simplicity of operation, low cost, and optimal efficiency, it can be a functional strategy to produce a variety of graphene-encapsulated catalysts on a large scale [39, 40].

4.1.6 Zeolites

Zeolites are presented as highlight catalyst supports due to their highly crystalline, well-distributed pore structure and adjustable acidity. In general cases, the zeolite structure embodies TO_4 tetrahedra (T defines Si, Al, and P, etc.). Due to adjusting the T-O-T linkage a wide range of zeolite structures can be formed via tuning the synthesis conditions such as the composition of the gel, the nature of the structure-directing agent (SDA), or the temperature [42, 43].

Although the catalytic activity of the catalysts often depends on the host nanoparticles, it can be modified by the zeolite framework features. In particular, the modification of local geometry around active sites which is derived from steric constraints affected by the size of zeolite cavities can effectively influence the reactivity of catalysts [43]. Zeolites based on the size of their pores can be classified into small, medium, large, and extra-large pores. The catalytic activities of zeolites deeply depend on the structural and compositional features, consisting of pore sizes, channel types, and framework compositions. In comparison with the other catalyst supports, zeolites are presented as a shape selective that can selectively interact with reactants, products, and transition-states that this attribute has a significant impact on the catalytic performance of zeolites [42]. Due to zeolites being composed of tetrahedrally $[SiO_4]^{4-}$ and $[AlO_4]^{5-}$ primary units, to balance the overall electric charge of the zeolitic skeleton, some free cations are accommodated into the channels of the 3D framework, which can be substituted by other cations. Al content in the zeolite framework has the main effect on the ion exchange capacity of zeolites when the cations of zeolites are exchanged by protons, zeolites conducted as solid Brønsted acid catalysts [44]. In addition, the existence of charges in the zeolite framework, as well as extra-framework cations, can have a significant impact on the electronic and redox properties of the encapsulated complex [43]. Despite the fact that zeolites can apply as a shell through the coating of MNPs with layers of zeolite, in some cases, the encapsulation of MNPs can operate in the regular cavities and nanochannels of zeolites [42]. Hence, the encapsulation of MNPs in zeolites is considered in two parts—yolk/core-shells and mesoporous structures.

4.1.6.1 Yolk/core-shell structures of zeolites

In the approach of nanotechnology, zeolites can be utilized to make novel nanostructure synthetic materials, zeolite core-shell structured materials being the outstanding structure among them [45]. The capability to synthesize core-shell zeolite composites has depicted the principal importance of chemical adaptability and structural likeness between core and shell crystals, as well as their close crystallization conditions [46]. On one side, a core-shell structure of zeolite can be formed via crystal overgrowth in which an aluminum-free zeolite (core) was coated with aluminum-containing zeolite (shell). Fluoride ions as mineralizers can conduct the accomplished passivation of acid sites on the external surface to minimize the imperfections of the core-shell zeolite structure, so applying them is essential [47]. To increase the selectivity and catalytic activity of the core-shell zeolite with TON structure, a novel high silica zeolite, for skeletal isomerization of n-tetradecane, it is essential to break needle-like particles for the formation of new acid sites on the pore mouths of smaller broken particles since, the acid sites on the side surface of the needle-like particles, which principally catalyzed the cracking of alkanes, were passivated [47]. In addition, through utilizing the techniques of layer by layer self-assembly of polyelectrolyte the core-shell zeolite-zeolite composites consisting of single-crystal core and polycrystalline shells of various zeolite structure types can be fabricated. This approach occurs based on the coulombic forces that lead to enhancing the surface charge of the core particles by coating the layers of zeolite [45]. Moreover, core-shell structures of zeolites can perform as a multi-purpose catalyst that have several impacts in various functions simultaneously. For instance, the shape-selective attribute of zeolite shell provides this ability that a catalytically active nano/micro-sized core encapsulated with a thin selective zeolite shell can be potentially utilized as a tiny membrane reactor. To achieve this purpose, first of all, the catalytic active materials such as metal oxides will be replaced in the core. Then secondary growth will be conducted through a precursor solution to coat a layer of zeolite, subsequently, the growth of this layer can continue until a dense and well-intergrown zeolite shell is formed that can also play its role as a highly efficient zeolite membrane. Not only does this zeolite shell provide a prominent selective mass transfer between the encapsulated core and the extension, but also the participating catalytic reaction on the core can promote each of such zeolite core-shell structured particles into a tiny membrane reactor which can present appropriate potential in a wide range of reaction systems [45].

On the other hand, the other nano-shell catalyst structure of zeolite which can be pointed to it is the yolk-shell zeolite-based catalysts. This structure can be achieved by silica-zeolite core-shell materials. First of all, silica particles are partially dissolved under high pH conditions, then through controlling the recrystallization of the surface of silica spheres by the zeolite, new agglomerated zeolite nanocrystals can be formed which depict hollow capsules. MNPs prior to recrystallization can be located at the core [43]. Overall, to synthesize spherical hollow zeolitic structures should apply the sacrificial templates such as organic polymers or silica whose particles within the core can be removed in the final step respectively by calcination or etching [45].

4.1.6.2 Mesoporous framework of zeolites

Zeolites with a mesoporous matrix are a family of porous materials with an effective crystalline framework containing a finite number of well-defined and small cavities with sub-nanometer to 2 nm size. The most interesting attribute of their structures is the possibility to tune and choose the similar size of their micropores with the size of MNPs that would be encapsulated in them. In addition, it can intensify the catalytic selectivity due to enhance the efficiency of reactants and products diffusions [2]. Meanwhile, by encapsulating the MNPs inside the micropores of the zeolite framework they would be effectively enclosed through their interconnected cavities [42, 48]. Not only can mesoporous zeolites perform as an immobilize or stabilize framework to encapsulate the nanoparticle catalysts, but also they represent as a molecular sieve via molecular selecting with the proper size and shape or as a hybrid catalyst for transforming products formed firstly [43].

There are various procedures to implement the encapsulation of MNPs through the zeolite pores which are selected on the basis of nanoparticle size and their nature [43]. Although the formation of MNPs and the growth of zeolites are two parallel pathways, in most synthesis strategies both of them are simultaneously conducted in one process synthesis, due to the pore sizes of most zeolites being too small, less than 2 nm, which is not appropriate for directly encapsulating MNPs [1, 2]. Here some popular approaches to undertake the encapsulation of MNPs in zeolites are presented.

4.1.6.2.1 Hydrothermal strategy

Hydrothermal synthesis strategy is a common procedure that is done under a thoroughly alkali condition. In this technique the metal precursor is directly added to the synthetic solution, to prevent premature reduction or precipitation of the metal salt through the crystallization of the zeolite, a mercaptosilane ligand like 3-mercaptopropyl tri-methoxy silane (MPTMS) should be applied into the hydrothermal synthetic system. Despite the mercapto groups supporting the metal precursors from reducing untimely through the alkaline synthetic solution, the other group of MPTMS that is alkoxysilane prepares an appropriate condition to fabricate crystalline frameworks with silicate and aluminate. Thus, the MPTMS paves the way to conduct the encapsulation of the MNPs in the mesoporous framework of zeolite monotonously. Eventually, the crystalline framework of metal precursor-zeolite which is gained by this procedure should be calcined in air to eliminate the organic agents and reduced in H_2 to create novel MNPs through the micropores of zeolite. In addition, LTA-type zeolites including micropores with the approximate size of 0.41 nm are the appropriate option to encapsulate the MNPs through this hydrothermal synthesis strategy. To enhance this hydrothermal procedure with the highest efficiency (>90%) of zeolite-encapsulated MNPs, initially, alcohol is added to eques mixture of the mercaptosilane and metal salt, then a pre-hydrolyze at low temperature will be performed to achieve a uniform gel [2, 48].

4.1.6.2.2 Solvent-free crystallization

Solvent-free crystallization is another popular strategy to encapsulate MNPs in mesoporous zeolites. By exposing a metal/silica/alumina hybrid in the vapor of water at a high temperature the solid-phase transformation of the amorphous silica and/or alumina into a crystalline zeolite that encapsulates MNPs is initiated. Furthermore, this synthesis method can provide a pathway to fully encapsulate the pre-synthesized MNPs inside the zeolite single crystals more reliably [2]. FAU- and MFI-type zeolites are the main mesoporous zeolite types that are applied for this procedure [2, 43]. In particular, Chen et al. [49] utilized nanocrystals of MFI-type zeolite (silicate-1 or S-1) in a strong alkaline environment. The outstanding feature that promoted the crystallization involved the Kirkendall effect which led to growing the pore size of mesoporous inside the S-1 crystals to around 3 nm and may impact on the enhancement of the mass transfer in catalytic applications [2].

4.1.6.2.3 Secondary growth of zeolite on metal/zeolite seeds

The encapsulation of MNPs in zeolites can be involved in the seeded growth of zeolites on zeolite seeds that already include MNPs. This synthetic strategy is implemented in two steps—first of all, the impregnation of the zeolite seeds with a metal salt is implemented, then desiccated the mixture to achieve a dry powder, and next conducted the reduction process at a high temperature in H_2 to convert the metal salts to the MNPs, eventually the achievement products are the zeolite seeds included the MNPs. In the second step, a hydrothermal system in the presence of aluminosilicate or silicate gels and the zeolite seeds including the MNPs are involved. Consequently, the encapsulation of MNPs such as Pt, Pd, Rh, and Ag in zeolites like MFI, MOR, and BEA, particularly at the interface between the zeolite seed and sheath could be effectively conducted. Moreover, a great core-sheath interface can enhance the loading amount of MNPs which can be obtained by employing a zeolite type with a high surface area when providing the metal-containing seeds through the synthesis process [2, 48].

4.2 Metal: Organic frameworks (MOFs)

Metal–organic frameworks (MOFs) are outstanding microporous materials including two major components: bridging organic linkers and inorganic secondary building units (SBUs) of metal ions or oxo-clusters (3d transition metals, 3p metals, or lanthanides). MOFs provide an exceptional combination of inorganic and organic components with synergistic interactions among them which create great usability for a myriad of purposes. These microporous materials are fabricated by gathering metal ions with organic ligands together in appropriate solvents often during a self-assembly strategy. In addition, the organic linkers are di-topic or polytopic organic ligands like carboxylate, nitrogen-donor groups, sulfonate, or phosphonate that are able to bind with metal-containing SBUs to form crystalline framework structures with open pores. Although MOFs present crystalline structures with dramatic large and uniform internal surface areas, their porosity and chemical features can adjust respectively by tuning the pore size and modifying the organic linkers according to our requirements on various catalytic applications. Moreover, the functional groups of organic ligands such as $-NH_2$, $-NO_2$, $-SO_3H$, -Cl, and $-OCH_3$ groups can be linked on the pore walls through the one-step assembly or post-synthetic modification which can impact on the catalytic performance. As a result, more than 20,000 MOFs with various compositions and topologies have been reported in recent decades. Meanwhile, metal nodes can present Lewis acidity features which are emerged by utilizing transition metals, additionally, they can engage in redox catalysis or support the progression of coupling reactions. Furthermore, metal nodes make various coordination positions by the participation of solvent molecules which can be eliminated through a thermal approach while saving their frameworks [50–53].

Encapsulating guest particles into MOFs provides a wide range of potential for various applications, particularly in catalysis. A great number of particles can be encapsulated such as inorganic MNPs, coordination complexes, quantum dots, polyoxometalates, enzymes, and polymers through a pre- and post-synthetic strategy. In comparison with the other encapsulating materials, MOFs exhibit confinement effects and shape selectivity in a more effective route, and their synthesis conditions are more moderate. In addition, owing to the existence of a wide range of MOF structures, it is simple to adopt a suitable MOF as the encapsulating material. Hence, the encapsulation of MNPs in MOFs converts them to prominent catalysts that attract considerable attention, due to they present the unique attributes of MOFs alongside the chemical and physical properties of MNPs simultaneously. In addition, this combination of active nanoparticles and functional organic linkers of MOFs can facilitate the charge transfer interactions with active components by coordination or π···π forces, which lead to present a significant enhancement in their catalytic performances [50–53]. The composites of nanoparticles encapsulated in MOFs can be fabricated through two main strategies—(1) stabilizing pre-synthesized nanoparticles in organic or inorganic agents as core and then enclosed in the shell of MOFs which generate a core-shell structure; (2) utilizing MOFs as mesoporous templates to encapsulate nanoparticles within their cavities [50].

4.2.1 Yolk/core-shell structures of MOFs

Despite the outstanding attributes of the composites of nanoparticles encapsulated in MOFs being able to conduct the effective catalytic activity, some restricting issues still exist that should be noticed. First of all, the pores can be blocked by the encaged nanoparticles during their growth, thus restricting the diffusion of the reaction medium in the catalytic process. Secondly, the loading of guest particles is limited to them which are smaller than the pore dimensions. Furthermore, the other specifications of guest particles like shape and morphology may not have adequate adoption with the host cavities. Moreover, it is not possible to control the deposition of guests mostly and eventually the guest nanoparticles may leach in liquid phase. To tackle these difficulties, the core-shell encapsulation strategy is offered [53].

The core of a core-shell MOFs-based composite can consist of inorganic nanoparticles like metal oxides, carbon materials, and polymers or other MOFs which are encapsulated in a MOF shell. Although the shape, size, morphology, and composition of the core have a significant effect on the catalytic performance of the catalyst, this structure provides effective encapsulation due to the integration of chemical/physical properties of two distinct materials that lead to the synergetic effects. Recently, the yolk-shell or hollow structures attracted further attention because the properties of core-shell MOF-based architecture are optimized in this type of structure that presented added conductivity, hierarchical porosity, and effective diffusion. These structures can be achieved through a controllable etching of the core materials based on core-shell structures. Although carbonizing techniques at high temperature can destroy the MOFs, by employing this approach yolk-shell or hollow structures can be generated in which their porosity and active sites have been maintained [53].

4.2.1.1 Growth of MOFs on pre-synthesized MNPs

Growth of MOFs on pre-synthesized MNPs is the main approach to encapsulate MNPs in shell of MOFs which are implemented in three main strategies that are presented in follow. Although the prominent benefit of this approach is the participation of MNPs with various sizes and shapes, the control of the assembly of metal-support interface is a noticeable attribute. In addition, the principal role in fabricating an appropriate core-shell MOFs-based composite in this strategy belongs to the consistency and conformity between the MNP core and the MOF shell [2].

4.2.1.1.1 Capping agent-assisted synthesis (CAAS)

First of all, the pre-synthesized core materials and capping agents/surfactants like PVP which is an amphiphilic nonionic polymer, are simultaneously provided earlier the growth of MOF shell around them due to prevent the aggregation of active NPs and/or the self-nucleation of MOF particles [53]. In addition, capping agents intensify the compatibility between the MNPs and the MOF shell. While PVP is the most popular capping agent for encapsulation of inorganic materials in MOFs, depending on the design process of core materials can be utilized the other type of surfactants like cetyltrimethylammonium bromide (CTAB), tetradecyltrimethylammonium bromide (TTAB), cetylpyridinium bromide (CPB), and sodium dodecyl sulfate (SDS). The CTAB is the optimal surfactant in terms of shape/overgrowth control, as well as PVP, owing to the sizes and shapes of the final composite can be adjusted by changing the growth time and the quantity of CTAB in solution. Overall, despite CAAS being a popular method due to the ability to control the shape, size, and chemical nature of the encapsulated core, the multiple steps of this method restricted its utilization [53].

4.2.1.1.2 Inorganic template-assisted synthesis (ITAS)

In this method, firstly, the core material is coated by employing an inorganic material like SiO_2 as binding sites for shell MOF, or metal oxide as the metal source for the shell MOF growth, and then synthesis of MOF will be implemented. Although in contrast CAAS procedure applying metal oxide prepare some advantages due to their act as metal source or protection of reactive nanoparticles in the core, there are some restrictions to fabricate wrapped nanoparticles owing to the shell MOF chemical stability will be endangered through the etching of the sacrificing metal oxide. Despite the fact that SiO_2 can play as a useful protective layer, it can represent as a sacrificial layer to fabricate a yolk-shell structure by conducting selective etching of SiO_2. In addition, in some cases, metal oxide/metal(0) nanoparticles have also been utilized as sacrificing agents to fabricate core-shell architectures [53].

4.2.1.1.3 Epitaxial growth synthesis (EGS)

EGS is a synthetic strategy to encapsulate MOF particles as a core with the second layer of MOF. In addition, in most cases, the shell MOF topology is similar to the core MOF and active nanoparticles would be placed at the interface of the two layers prior to the epitaxial growth. Overall, this approach leads to effective integration of MOFs because synergies of various MOFs properties are simultaneously presented by an exceptional composite of MOFs @MOFs. Despite this method presenting a wide range of similarities to the CASS strategy, it can enhance the diffusion of the reaction medium toward the active sites due to the ability to fabricate a super-thin shell with a thickness size less than 10 nm [53, 54].

4.2.2 Encapsulation of the MNPs in the micropores of MOFs

Not only would MOFs provide a fully available pore space to optimize the diffusion of the reaction medium, but they also support the MNPs by enclosing them in their mesoporous matrix that increases the accessibility of active sites and enhances the catalytic activity [53]. Due to MOFs can play the role of host materials and be able to provide confined spaces for nucleation of MNPs, the encapsulation of MNPs into the cavities or channels of MOF matrix can be conducted by impregnating metal precursors in the pre-synthesized MOFs and subsequent reduction of the metal precursors in the micropores of MOFs [2]. Although the precise control of this encapsulation type of MNPs in MOFs seems to be difficult, the most effective pathway can be achieved by the double-solvent procedure prior to the reduction step. As a result of this double-solvent approach, the capillary force intensifies the mass transfer of metal precursors into the micropores of MOFs, which prevents from aggregation of MNPs and minimizes the dispersion of them on the outer surface of MOF [52].

The effective route to utilize noble metals (Pd, Ru, or Pt) in the MOFs-based catalysts is in the form of alloys with low-cost transition metals (Cu, Co, or Ni) owing to decreasing the essential amount of expensive noble metals during the catalyst's fabrication. Furthermore, the atomic and electronic specifications of their structures can be adjusted which leads to improved catalyst activity. In this approach, firstly low-cost transition MNPs are impregnated in the pre-synthesized MOFs and then are reduced by $NaBH_4$. In the next step, the salts of noble metals were exposed into a solution of MOF in which the galvanic replacement reaction with the transition MNPs with the help of an excitement process like sonication can be started. Finally, it leads to the fabrication of alloy nanoparticles like Co-Ru which is encapsulated in MOF [2, 55].

4.3 Encapsulation of MNPs in organic materials

One of the most effective approaches of ultrasmall MNPs encapsulation with meticulously composed sizes in catalysts is the encapsulation of them in organic materials, which is presented in two parts as follows [2]:

4.3.1 MNPs @organic capsules

Organic capsules, such as dendrimers, which are a family of hyperbranched polymers, have a spherical structure, which is compressed on the exterior and creates hollow space in the interior. By applying organic groups, such as tertiary amines, the interior cavity can be put into particular operation, such as interception of metal ions from the solution through encapsulating them. In addition, the dendrimers can provide a monodic encapsulation of MNPs due to form mono dispersing of them with appropriate adjustment of ultrasmall sizes. Not only can dendrimers operate as effective encapsulating materials owing to the consuming amount of them being approximately the same as the metal ions amount, but also they are able to present a stable form that is unchanged for some months. Thus, these specifications convert them as useful stabilizers in the catalytic process to prepare a monotonous dispersion of nano-catalysts particles with suitable stability [2].

Polyamidoamine (PAMAM) is the most used dendrimer for encapsulating the ultrasmall MNPs. Not only can dendrimers, particularly PAMAM, provide homogeneous catalysis, but an effectual mass transfer will also happen at their exterior surface that leads to a significant improvement in the catalytic activity of the ultrasmall MNPs. To present applicable heterogeneous catalysts of encapsulated ultrasmall MNPs in dendrimers, it seems essential to utilize mesoporous supports, such as silica. Not only do they supply high surface area, but also they prepare constructive enclosures to intensify the stabilization of ultrasmall MNPs in harsh catalytic processes. When applying the silica mesoporous matrix as support, the electrostatic interaction and hydrogen bonding between silica molecules and dendrimers enforce MNPs@ dendrimers into the mesopores of the silica matrix. Not only do these supported catalysts without removing the dendrimer present highlighted stability, selectivity, and activity in a range of catalysis processes, they illustrate adjustable catalytic specifications that are possible to alter in order to enhance the catalyst activity by modifying the active groups on dendrimers, for instance, the tertiary amines can improve the activity of the catalyst due to their electron enrichment attributes. In addition, the unique spatial morphology of dendrimers in the shape of tree may have a significant impact on the reaction substrate stability, including reactant molecules or intermediate species that can lead to decrement of the level of activation energy and improve the turnover frequency [2, 19, 56]. Furthermore, click dendrimers consist of a great ratio of triazole rings, are the other type of dendrimers that can be fabricated from the azide-alkyne cyclo addition. Their rings provide an appropriate situation to accomplish the encapsulation by adsorbing metal ions and grafting the MNPs. In addition, by adding triethylene glycol (TEG) termini to the click dendrimer convert them to an effective soluble one in aqueous solutions which are suitable to fabricate the encapsulation of ultrasmall monometallic or bimetallic like Pt-Co in water solution. Thus, the yield of catalyst performance in an aqueous solution will dramatically enhance [2].

4.3.2 Porous organic cages (POCs)

POCs with intrinsic porosity have prominent attributes, such as high surface areas, shape durability, structural adjustability, and in particular including active practical groups in their cavities that lead them to encapsulate ultrasmall MNPs in an effective way. Moreover, POCs can exhibit either in a crystalline or an amorphous structure. Furthermore, the heterogeneous catalysts that are formed by encapsulating ultrasmall nanoparticles in these POCs would be homogenized in the solutions and have a significant influence on the enhancement of the performance of catalysts. Their specifications can be modified or developed through post-synthetic modification strategies, such as polymorph selection, modular co-crystallization, and the fabrication of composite materials. Not only does the cage wrapping significantly impact porosity, during this route by employing various crystalline polymorphs distinctive physical properties can be exhibited. In addition, to fabricate a cage in the correct form, at least it is essential to provide precursors with proper and precise geometry, for instance, a mild alter in bond angles of the precursors can lead to a distinct cage that might be in size and stoichiometry [2, 57, 58].

To fabricate POGs with polyhedral organic molecules, triamine is utilized to form the top and bottom prism that include a dialdehyde with a long alkyl chain and a thioether group at the three peripheral sides. Although the long alkyl chain provides a cage profoundly soluble in an oil phase due to thriving their hydrophobic properties, the thioether group leads the cage to adsorb the metal ions and prepare suitable sites to graft the MNPs. Therefore, these specifications develop POGs capability to utilize as promising catalysts, particularly in catalyzing organic reactions [2].

Acknowledgements

A very special thanks to Mohammadreza Joharkesh, who helped to depict schemes 1, and 2.

Author details

Masoud Safari[1*] and Valiollah Nobakht[2]

1 Department of Chemical Engineering, Tarbiat Modares University, Tehran, Iran

2 Department of Chemistry, Shahid Chamran University of Ahvaz, Ahvaz, Iran

*Address all correspondence to: masoud.safari@modares.ac.ir

References

[1] Tian H, Li X, Zeng L, Gong J. Recent advances on the design of group VIII base-metal catalysts with encapsulated structures. ACS Catalysis. 2015;**5**(8): 4959-4977. DOI: 10.1021/acscatal.5b01221

[2] Gao C, Lyu F, Yin Y. Encapsulated metal nanoparticles for catalysis. Chemical Reviews. 2020;**121**(2):834-881. DOI: 10.1021/acs.chemrev.0c00237

[3] Deng J, Deng D, Bao X. Robust catalysis on 2D materials encapsulating metals: Concept, application, and perspective. Advanced Materials. 2017;**29**(43):1606967. DOI: 10.1002/adma.201606967

[4] Nemiwal M, Kumar D. TiO_2 and SiO_2 encapsulated MNPs: Synthetic strategies, properties, and photocatalytic applications. Inorganic Chemistry Communications. 2021;**128**:108602. DOI: 10.1016/j.inoche.2021.108602

[5] van Deelen TW, Mejía CH, de Jong KP. Control of metal-support interactions in heterogeneous catalysts to enhance activity and selectivity. Nature Catalysis. 2019;**2**(11):955-970. DOI: 10.1038/s41929-019-0364-x

[6] Li XG, Liu C, Sun J, Xian H, Tan YS, Jiang Z, et al. Tuning interactions between zeolite and supported metal by physical-sputtering to achieve higher catalytic performances. Scientific Reports. 2013;**3**(1):1-5. DOI: 10.1038/srep02813

[7] Cheng Q, Liu Y, Lyu S, Tian Y, Ma Q, Li X. Manipulating metal-support interactions of metal catalysts for Fischer-Tropsch synthesis. Chinese Journal of Chemical Engineering. 2021;**35**(7):220-230. DOI: 10.1016/j.cjche.2021.05.013

[8] Otor HO, Steiner JB, García-Sancho C, Alba-Rubio AC. Encapsulation methods for control of catalyst deactivation: A review. ACS Catalysis. 2020;**10**(14):7630-7656. DOI: 10.1021/acscatal.0c01569

[9] Morris DA, Calvinyl B. Heterogeneous catalyst Deactivation and Regeneration: A review. Catalysts. 2015;**5**(1):145-269. DOI: 10.3390/catal5010145

[10] Goodman ED, Schwalbe JA, Cargnello M. Mechanistic understanding and the rational design of sinter-resistant heterogeneous catalysts. ACS Catalysis. 2017;**7**(10):7156-7173. DOI: 10.1021/acscatal.7b01975

[11] Hansen TW, DeLaRiva AT, Challa SR, Datye AK. Sintering of catalytic nanoparticles: Particle migration or Ostwald ripening? Accounts of Chemical Research. 2013;**46**(8):1720-1730. DOI: 10.1021/ar3002427

[12] Lohr TL, Marks TJ. Orthogonal tandem catalysis. Nature Chemistry. 2015;**7**(6):477-482. DOI: 10.1038/nchem.2262

[13] Pei C, Gong J. Tandem catalysis at nanoscale. Science. 2021;**371**(6535):1203-1204. DOI: 10.1126/science.abh0424

[14] Wasilke JC, Obrey SJ, Baker RT, Bazan GC. Concurrent tandem catalysis. Chemical Reviews. 2005;**105**(3):1001-1020. DOI: 10.1021/cr020018n

[15] Cho HJ, Kim D, Li J, Su D, Xu B. Zeolite-encapsulated Pt nanoparticles for tandem catalysis. Journal of the American Chemical Society. 2018;**140**(41): 13514-13520. DOI: 10.1021/jacs.8b09568

[16] Climent MJ, Corma A, Iborra S, Sabater MJ. Heterogeneous catalysis

for tandem reactions. ACS Catalysis. 2014;**4**(3):870-891. DOI: 10.1021/cs401052k

[17] Chen L, Xu Q. Metal-organic framework composites for catalysis. Matter. 2019;**1**(1):57-89. DOI: 10.1016/j.matt.2019.05.018

[18] Fang Y, Powell JA, Li E, Wang Q, Perry Z, Kirchon A, et al. Catalytic reactions within the cavity of coordination cages. Chemical Society Reviews. 2019;**48**(17):4707-4730. DOI: 10.1039/C9CS00091G

[19] Lee I, Albiter MA, Zhang Q, Ge J, Yin Y, Zaera F. New nanostructured heterogeneous catalysts with increased selectivity and stability. Physical Chemistry Chemical Physics. 2011;**13**(7):2449-2456. DOI: 10.1039/C0CP01688H

[20] Galeano C, Meier JC, Soorholtz M, Bongard H, Baldizzone C, Mayrhofer KJ, et al. Nitrogen-doped hollow carbon spheres as a support for platinum-based electrocatalysts. ACS Catalysis. 2014;**4**(11):3856-3868. DOI: 10.1021/cs5003492

[21] Galeano C, Güttel R, Paul M, Arnal P, Lu AH, Schüth F. Yolk-shell gold nanoparticles as model materials for support-effect studies in heterogeneous catalysis: Au,@ C and Au,@ ZrO2 for CO oxidation as an example. Chemistry-A European Journal. 2011;**17**(30):8434-8439. DOI: 10.1002/chem.201100318

[22] Zhang Q, Lee I, Joo JB, Zaera F, Yin Y. Core-shell nanostructured catalysts. Accounts of Chemical Research. 2013;**46**(8):1816-1824. DOI: 10.1021/ar300230s

[23] Zhang Q, Lee I, Ge J, Zaera F, Yin Y. Surface-protected etching of mesoporous oxide shells for the stabilization of metal nanocatalysts. Advanced Functional Materials. 2010;**20**(14):2201-2214. DOI: 10.1002/adfm.201000428

[24] Purbia R, Paria S. Yolk/shell nanoparticles: Classifications, synthesis, properties, and applications. Nanoscale. 2015;**7**(47):19789-19873. DOI: 10.1039/c5nr04729c

[25] Arnal PM, Comotti M, Schüth F. High-temperature-stable catalysts by hollow sphere encapsulation. Angewandte Chemie International Edition. 2006;**45**(48):8224-8227. DOI: 10.1002/anie.200603507

[26] Shao Y, Song J, Li X, Ren G, Song F. Synthesis of Noble Metal M@ YSiO2 Yolk–shell nanoparticles with thin organic/inorganic hybrid outer shells via an aqueous medium phase. Langmuir. 2021;**37**(23):7237-7245. DOI: 10.1021/acs.langmuir.1c00875

[27] Kankala RK, Zhang H, Liu CG, Kanubaddi KR, Lee CH, Wang SB, et al. Metal species-encapsulated mesoporous silica nanoparticles: Current advancements and latest breakthroughs. Advanced Functional Materials. 2019;**29**(43):1902652. DOI: 10.1002/adfm.201902652

[28] Wang F, Cheng S, Bao Z, Wang J. Anisotropic overgrowth of metal heterostructures induced by a site-selective silica coating. Angewandte Chemie. 2013;**125**(39):10534-10538. DOI: 10.1002/anie.201304364

[29] Zhang T, Zhao H, He S, Liu K, Liu H, Yin Y, et al. Unconventional route to encapsulated ultrasmall gold nanoparticles for high-temperature catalysis. ACS Nano. 2014;**8**(7):7297-7304. DOI: 10.1021/nn502349k

[30] Li J, Zeng HC. Size tuning, functionalization, and reactivation of Au in TiO_2 nanoreactors. Angewandte Chemie International Edition. 2005; **44**(28):4342-4345. DOI: 10.1002/anie.200500394

[31] Fronzi M, Assadi MH, Hanaor DA. Theoretical insights into the hydrophobicity of low index CeO_2 surfaces. Applied Surface Science. 2019;**478**:68-74. DOI: 10.1016/j.apsusc.2019.01.208

[32] Cargnello M, Wieder NL, Montini T, Gorte RJ, Fornasiero P. Synthesis of dispersible Pd@ CeO2 core-shell nanostructures by self-assembly. Journal of the American Chemical Society. 2010;**132**(4):1402-1409. DOI: 10.1021/ja909131k

[33] Chen C, Fang X, Wu B, Huang L, Zheng N. A multi-yolk–shell structured nanocatalyst containing sub-10 nm Pd nanoparticles in porous CeO_2. ChemCatChem. 2012;**4**(10):1578-1586. DOI: 10.1002/cctc.201200237

[34] Finsel M, Hemme M, Döring S, Rüter JS, Dahl GT, Krekeler T, et al. Synthesis and thermal stability of ZrO_2@ SiO_2 core-shell submicron particles. RSC Advances. 2019;**9**(46):26902-26914. DOI: 10.1039/c9ra05078g

[35] Huang X, Guo C, Zuo J, Zheng N, Stucky GD. An assembly route to inorganic catalytic nanoreactors containing sub-10-nm gold nanoparticles with anti-aggregation properties. Small. 2009;**5**(3):361-365. DOI: 10.1002/smll.200800808

[36] Liu S, Han MY. Silica-coated metal nanoparticles. Chemistry: An Asian Journal. 2010;**5**(1):36-45. DOI: 10.1002/asia.200900228

[37] Galeano C, Güttel R, Paul M, Arnal P, Lu AH, Schüth F. Yolk-shell gold nanoparticles as model materials for support-effect studies in heterogeneous catalysis: Au,@ C and Au,@ ZrO_2 for CO oxidation as an example. Chemistry: A European Journal. 2011;**17**(30):8434-8439. DOI: 10.1002/chem.201100318

[38] Kang J, Kim Y, Kim HM, Hu X, Saito N, Choi JH, et al. In-situ one-step synthesis of carbon-encapsulated naked magnetic metal nanoparticles conducted without additional reductants and agents. Scientific Reports. 2016;**6**(1):1-9. DOI: 10.1038/srep38652

[39] Zhang Z, Wang S. Catalysts encapsulated in nanostructured carbon systems. In: Sadjadi S, editor. Encapsulated Catalysts. Academic Press; 2017. pp. 71-122. DOI: 10.1016/B978-0-12-803836-9.00003-1

[40] Baig N, Kammakakam I, Falath W. Nanomaterials: A review of synthesis methods, properties, recent progress, and challenges. Materials Advances. 2021;**2**(6):1821-1871. DOI: 10.1039/d0ma00807a

[41] Jing S, Lu J, Yu G, Yin S, Luo L, Zhang Z, et al. Carbon-encapsulated WOx hybrids as efficient catalysts for hydrogen evolution. Advanced Materials. 2018;**30**(28):1705979. DOI: 10.1002/adma.201705979

[42] Xu D, Lv H, Liu B. Encapsulation of metal nanoparticle catalysts within mesoporous zeolites and their enhanced catalytic performances: A review. Frontiers in Chemistry. 2018;**6**:550. DOI: 10.3389/fchem.2018.00550

[43] Farrusseng D, Tuel A. Zeolite-encapsulated catalysts: Challenges and prospects. In: Sadjadi S, editor. Encapsulated Catalysts. Academic Press;

2017. pp. 335-386. DOI: 10.1016/B978-0-12-803836-9.00011-0

[44] Miyake K, Nishiyama N. Core-shell structured zeolite catalysts with enhanced shape selectivity. In: Yamashita H, Li H, editors. Core-Shell and Yolk-Shell Nanocatalysts. Singapore: Springer; 2021. pp. 181-186. DOI: 10.1007/978-981-16-0463-8_11

[45] Khan EA, Hu E, Lai Z. Preparation of metal oxide/zeolite core-shell nanostructures. Microporous and Mesoporous Materials. 2009;**118**(1-3): 210-217. DOI: 10.1016/j.micromeso.2008.08.031

[46] Zheng J, Zhang X, Wang Y, Bai Y, Sun W, Li R. Synthesis and catalytic performance of a bi-phase core-shell zeolite composite. Journal of Porous Materials. 2009;**16**(6):731. DOI: 10.1007/s10934-008-9255-2

[47] Okamoto M. Core-shell structured zeolite catalysts with minimal defects for improvement of shape selectivity. In: Yamashita H, Li H, editors. Core-Shell and Yolk-Shell Nanocatalysts. Singapore: Springer; 2021. pp. 187-198. DOI: 10.1007/978-981-16-0463-8_12

[48] Ren N, Yang YH, Shen J, Zhang YH, Xu HL, Gao Z, et al. Novel, efficient hollow zeolitically microcapsulized noble metal catalysts. Journal of Catalysis. 2007;**251**(1):182-188. DOI: 10.1016/j.jcat.2007.07.009

[49] Cui TL, Ke WY, Zhang WB, Wang HH, Li XH, Chen JS. Encapsulating palladium nanoparticles inside mesoporous MFI zeolite nanocrystals for shape-selective catalysis. Angewandte Chemie. 2016;**128**(32):9324-9328. DOI: 10.1002/anie.201602429

[50] Lu G, Li S, Guo Z, Farha OK, Hauser BG, Qi X, et al. Imparting functionality to a metal–organic framework material by controlled nanoparticle encapsulation. Nature Chemistry. 2012;**4**(4):310-316. DOI: 10.1038/nchem.1272

[51] Hu P, Morabito JV, Tsung CK. Core-shell catalysts of MNP core and metal-organic framework shell. ACS Catalysis. 2014;**4**(12):4409-4419. DOI: 10.1021/cs5012662

[52] Li G, Zhao S, Zhang Y, Tang Z. Metal-organic frameworks encapsulating active nanoparticles as emerging composites for catalysis: Recent progress and perspectives. Advanced Materials. 2018;**30**(51):1800702. DOI: 10.1002/adma.201800702

[53] Dai S, Tissot A, Serre C. Recent progresses in metal-organic frameworks based core–shell composites. Advanced Energy Materials. 2021. DOI: 10.1002/aenm.202100061

[54] Wu MX, Wang Y, Zhou G, Liu X. Core-shell MOFs@ MOFs: Diverse designability and enhanced selectivity. ACS Applied Materials & Interfaces. 2020;**12**(49):54285-54305. DOI: 10.1021/acsami.0c16428

[55] Chen F, Shen K, Chen J, Yang X, Cui J, Li Y. General immobilization of ultrafine alloyed nanoparticles within metal–organic frameworks with high loadings for advanced synergetic catalysis. ACS Central Science. 2019;**5**(1):176-185. DOI: 10.1021/acscentsci.8b00805

[56] Li B, Yang X, Xia L, Majeed MI, Tan B. Hollow microporous organic capsules. Scientific Reports. 2013;**3**(1): 1-6. DOI: 10.1038/srep02128

[57] Tozawa T, Jones J, Swamy S, et al. Porous organic cages. Nature Materials. 2009;**8**:973-978. DOI: 10.1038/nmat2545

[58] Wang H, Jin Y, Sun N, Zhang W, Jiang J. Post-synthetic modification of porous organic cages. Chemical Society Reviews. 2021;**50**:8874-8886. DOI: 10.1039/D0CS01142H

Section 4

Polymer Nanocomposites for Diverse Engineering Applications

Chapter 12

Conducting Polymer 1-D Composites: Formation, Structure and Application

Monika Wysocka-Żołopa, Emilia Grądzka and Krzysztof Winkler

Abstract

Recent advances in the study of the synthesis, structure and applications of 1-D composites containing conducting polymers are discussed in this review. Conducting composites can form 1-D structures with metal and metal oxides, 1-D carbon nanomaterials, semiconducting materials, crystals of metalloorganic complexes. Advanced synthetic approaches allow for the formation of well-organized structures with polymeric phase deposited both on the surface of 1-D material and inside of the 1-D tubes. 1-D polymeric wires can also serve as a matrix for the formation 1-D composites with other materials. 1-D nanocomposites containing conducting polymers exhibit many exceptional properties which allow for various practical applications including energy converting and energy storage devices, electronic nanodevices, chemical, electrochemical and biochemical sensors, catalysis and electrocatalysis.

Keywords: conducting polymers, 1-D materials, nanocomposites, synthesis, application

1. Introduction

The combination of 1-D nanostructures with conducting polymers creates nanocomposites with good processability and improved physical, electrical, and mechanical properties such as conductivity, solubility, optoelectronic and magnetic properties. These systems draw considerable attention in a wide range of applications including supercapacitors, batteries, energy conversion systems, catalysts and sensors.

Many synthetic strategies, such as template-directed, template-free chemical and electrochemical method, solvothermal syntheses, electrospinning techniques, vapor-phase approaches, have been developed to prepare several classes of 1-D nanostructures including metals, metal oxides, metal complexes, and semiconductors [1–4].

Typical conducting polymers such as polypyrrole (PPY), polyaniline (PANI), polythiophene (PTH), and poly(3,4-ethylenedioxythophene) (PEDOT) are attractive polymers for composites synthesis given their low cost, easy processability, a large area of fabrication, and environmental stability. While, interest in 1-D nanostructures has increased due to their efficiency in electron transport, and their potential use in nanoelectronic devices [5, 6]. For example, structures such as nanowires, nanorods, nanotubes, or nanobelts with unique electric transporting characteristics would be more essential than irregular particles to be used as composites of solar cell devices after the introduction of conducting polymers [7, 8].

There are two main kinds of nanocomposites of conducting polymers with 1-D materials:

1. 1-D nanostructures covered with a conducting polymer. There are many techniques known for the deposition of conducting polymers onto 1-D nanomaterials. However, the encapsulation of 1-D nanostructures into the core of conducting polymers shell to obtain novel core-shell nanomaterials has become the most attractive aspect of nanocomposite synthesis. These composites are usually formed by the chemical or electrochemical polymerization of a thin layer of a conducting polymer onto different nanostructures.

2. Conducting polymers encapsulated of the 1-D nanostructures. However, there is much less work done on this type of nanocomposites.

This review aims to present general synthesis and characterization of the conducting polymers with 1-D nanostructures including the various methods used in these materials' preparation. Finally, different aspects of the practical applications of these materials are presented.

2. Synthesis and characteristic of conducting polymers with 1-D nanostructures

The synthetic methods for the fabrication of conducting polymer with 1-D nanostructures are of great importance because the structure of conducting polymers and secondary components will affect the properties of formed nanocomposites. Several template-based and template-free methods have been used to form 1-D nano-structure/polymer composites.

2.1 The conducting polymers with metal and metal oxides nanomaterials

1-D nanostructures of metal and metal oxides are of great importance for their electrical, optical, and catalytical properties as well as a wide range of applications in nanoelectronics and sensing devices [9, 10]. However, these nanostructures are very sensitive to air and moisture, which degrade the performance of the nanodevices. A polymer envelope would protect nanostructures from oxidation and corrosion, giving a good performance for a long time.

Nanostructured composites of silver nanowires with polypyrrole (Ag/PPY) have been prepared by the redox reaction in an aqueous solution at room temperature between silver nitrate and pyrrole using poly(vinyl pyrrolidone) (PVP) as assistant agent [11]. Under these conditions, a metallic nanowire coated with conducting polymer is formed. PVP is used both as a capping agent to form silver nanowires, and as a dispersant of pyrrole monomer. Silver nanowire is formed during silver nitrate reduction with pyrrole and monomer polymerizes on the surface of nanorode at the same time. A typical TEM image of these nanocables is shown in **Figure 1**.

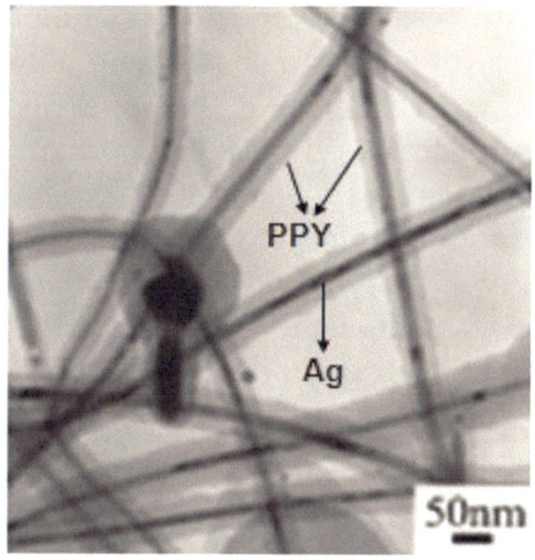

Figure 1.
TEM image of Ag/PPY nanocables (reproduced with permission from Ref. [11]).

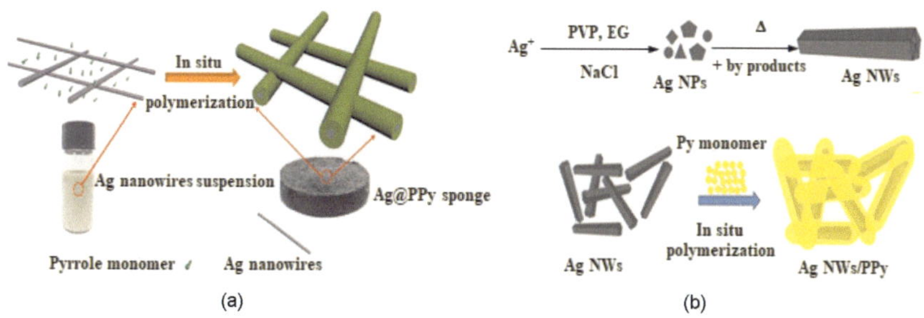

Figure 2.
Schematic illustrations of the formation of the (a) silver nanowires@polypyrrole sponge (reproduced with permission from Ref. [12]) (b) silver nanowires@polypyrrole nanocomposite (reproduced with permission from Ref. [13]).

The diameter of the layer is ~50 nm, and the diameter of the silver core is ~20 nm. The lengths of these nanocables are in the range of several to several tens' micrometers [11].

Recently, novel three-dimensional (3-D) silver nanowires@polypyrrole nanocomposites have been created [12, 13]. In the fabrication of the silver nanowires@polypyrrole sponge, the silver nanowires with the length of about 10 μm and diameter of 48.3 nm were synthesized trough solvothermal method and were used as the skeleton of the composite material. Polypyrrole was coated on the surface of silver nanowires by in situ chemical polymerization as shown in **Figure 2a** [12]. The prepared sponge exhibit good elasticity, mechanical strength and water absorption properties. The results showed that the complex permittivity and microwave absorption properties of the silver nanowires@polypyrrole sponge can be modulated by regulating the water amount. Another procedure 3-D core-shell silver nanowires@polypyyrole nanocomposite fabrication was proposed by Yuksel et al. [13]. Silver nanowires were prepared according to the polyol method and polypyrrole was synthesized using *in-situ* chemical polymerization (**Figure 2b**) [13]. These nanocomposites were applied for supercapacitors fabrication.

In addition to one-dimensional structures of silver and conducting polymer, 1-D composites of conducting polymers with copper and gold nanowires were also created. The combination of these nanowires and conducting polymer can endow new properties and exhibit synergistic effects of the nanocomposite components. For example, copper nanowires with polypyrrole [14] and polyaniline [15] have been prepared by a facile liquid-phase reduction with copper (II) chloride as precursor [14] and using a simple and reproducible approach by spontaneous chemisorption of polyaniline on the copper surface [15].

Composites containing conducting polymers with metal oxides such as ZnO, RuO_2, MnO_2, Co_3O_4, V_2O_5, MgO, Fe_2O_3, TiO_2, and NiO have been produced. For example, Fan et al. [16] adopted an electrochemical polymerization method to assemble PANI and PEDOT on the surface of different oxides nanostructures including Co_3O_4, TiO_2, and NiO nanowires, nanorods and nanoflakes. These metal oxides structures were fabricated by the hydrothermal method. The typical cyclic voltammetry (CV) curve of the TiO_2/PANI nanorods on FTO (fluorine-doped tin oxide) glass is shown in **Figure 3a**. The first redox peaks A_1 and C_1 corresponds to the change between leucoemeraldine base and emeraldine salt with anion doping upon oxidation and dedoping upon reduction. The second pair of redox peaks A_2 and C_2 is due to the conversion between emeraldine base and pernigraniline salt. The change between emeraldine salt and emeraldine base does not involve an electron transfer process, and the redox peak is not reflected in the CV curve. Also, redox peaks of TiO_2 are not observed in the studied potential range. The TiO_2/PANI nanorods display interesting electrochromic properties. Namely, these nanorods show evident electrochromism with rich reversible color changes ranging from yellow for leucoemeraldine base, green for emeraldine salt, and blue for emeraldine-base to purple for pernigraniline salt under different applied potentials. In **Figure 3b** transmittance spectra are shown for different potentials applied to the electrode covered with a thin layer of composite. Moreover, the architecture of this material is well preserved after prolonged potential cycling, and does not show evident degradation (**Figure 3c**) [16].

Multicomponent 1-D nanostructures and conducting polymers have also been made. These materials can be tailored to exhibit, besides novel electrical, magnetic and optical properties, also good processing properties. For example, the composite of Co_3O_4@PPY@MnO_2 "core-shell-shell" nanowires exhibited prominent electrochemical performance and remarkable long-term cyclic stability [17]. Co_3O_4 nanowire core backbone was grown on nickel foam by the hydrothermal and post-annealing method. Next, a conductive polypyrrole film was assembled on Co_3O_4 nanowire surface by potentiostatic deposition. The final product Co_3O_4@PPY@MnO_2 was formed by soaking Co_3O_4@PPY in aqueous $KMnO_4$. In this case, a redox reaction occurred in 3-D ordered nanowire interface. Such nanocomposites showed an effective pathway for

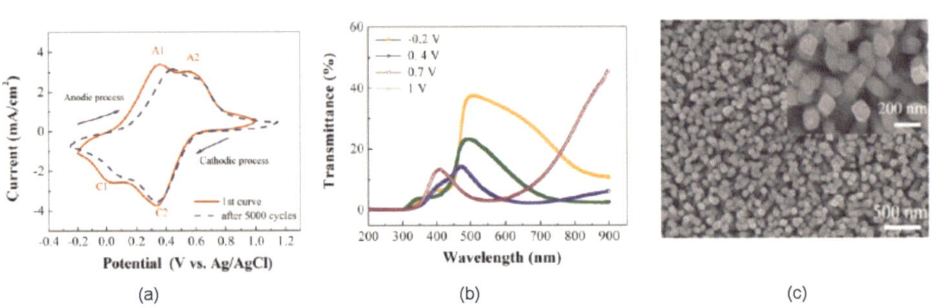

Figure 3.
Electrochromic characterization of coaxial TiO_2/PANI nanorods grown on FTO substrate: (a) CV curve in the potential range from −0.2 to 1 V at a scanning rate of 50 mV s^{-1}, (b) transmittance spectra of nanorods under different applied potentials, (c) SEM image nanorods after 5000 cycles (reproduced with permission from Ref. [16]).

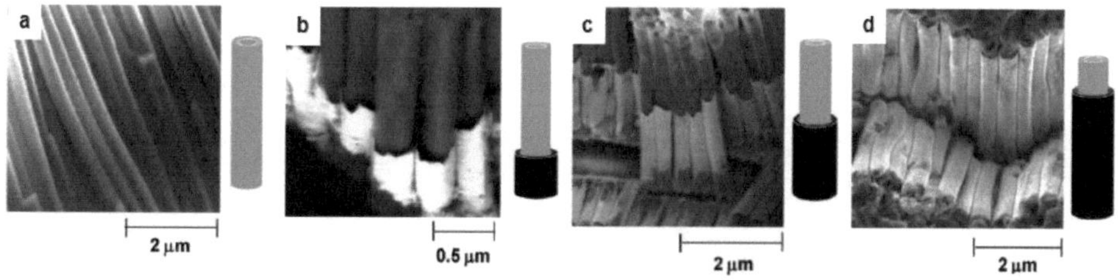

Figure 4.
SEM images of nanostructures grown in the AAO membrane (after the dissolution of AAO in 1 M NaOH for 1.5 h). (a) PANI fiber, (b–d) PANI/Au nanostructures for: (b) 1 h, (c) 1.5 h, (d) 2.5 h (reproduced with permission from Ref. [18]).

fast electron transport and accelerates the reaction kinetics between the electroactive center and current collector.

It is also possible to deposit 1-D nanostructures of metal within or around a preformed polymer nanotube. Polyaniline nanotubes prepared using AAO template were coated with gold to form PANI/Au composite [18]. The morphology of these structures in a different stage of formation is shown in **Figure 4**.

The metallic phase can be also deposited within the polymeric nanotube structure. For example, cobalt nanowires were produced within the PANI tubes [19]. Such a system exhibits unique magnetic properties. Cobalt nanowires show greatly enhanced magnetic coercivity.

2.2 The conducting polymers with carbon nanotubes or carbon fibers

For the formation of 1-D composites containing conducting polymers, carbon nanotubes and carbon fibers was used as a carbon component of composites.

Synthesis of carbon nanotubes (CNTs) and conducting polymer composite was firstly reported by Ajayan et al. [20]. Since then, a lot of attention has been paid to the fabrication of such 1-D functional composite materials with desirable electrical and mechanical properties. Composites of single-walled carbon nanotubes (SWCNTs) or multi-walled carbon nanotubes (MWCNTs) with polyaniline have been the most intensively studied. For example, aniline has been polymerized on MWCNTs electrodes to obtain PANI films with novel surface characteristics including higher current densities and more effective polymerization [21]. Nanotube electrodes were constructed with whiskers of loosely packed MWCNTs. These nanotube whiskers with typical dimensions of 0.15 cm long and 0.028 cm in diameter were used as an electrode by attaching them to the tips of copper wire covered with conductive paint. Carbon nanotubes were prepared by the electric-arc process. The PANI films formed on the nanotube electrodes were prepared by electrochemical polymerization of aniline in H_2SO_4 solution. The morphology of polyaniline film deposited on carbon nanotube electrode and corresponding cyclic voltammetric response are shown in **Figure 5** [21].

Wu et al. [22, 23] have shown that the conductivity of MWCNTs/PANI composites received by *in-situ* chemical polymerization is 50–70% higher than that of the pure PANI. Whereas, with the increase of the MWCNTs nanotubes content to 24.8 wt% the conductivity increases by two orders of magnitude. Also, the change in the sign from positive to negative of the magnetoresistance at low temperatures is observed revealing the strong coupling between the carbon nanotubes and polyaniline in these

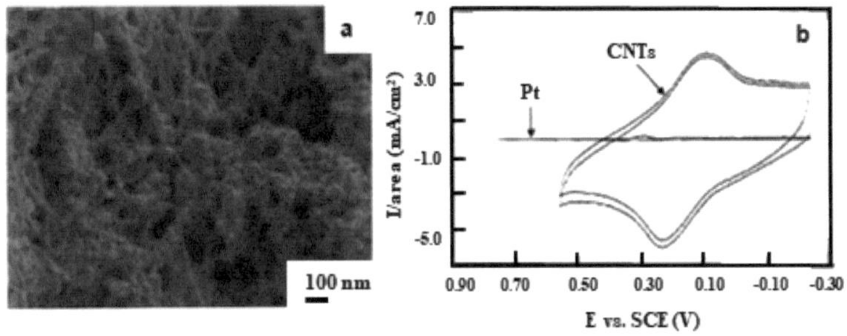

Figure 5.
PANI film deposited on carbon nanotube electrode: (a) SEM image, (b) CV curves showing much larger background currents compared to the Pt electrode. Sweep rate used 20 mV s^{-1}. The geometrical area of the CNTs is 0.016 cm^2 compared to 0.16 cm^2 for the Pt electrode (reproduced with permission from Ref. [21]).

composites. A similar effect on the conductivity of PANI was reported by Karim et al. [24]. They studied composite formed by deposition of PANI on the surface SWCNTs by using the *in-situ* chemical polymerization method [24]. The characterization of this material indicated that the conductivity and thermal stability of complex nanotubes were higher than polyaniline but lower than CNTs.

Many other conducting polymers such as PPY, PEDOT, and PTH were deposited onto the carbon nanotubes to form 1-D composites. Both *in-situ* and *ex-situ* chemical and electrochemical methods were used for polymer deposition. Similarly, such as polyaniline, these composites show considerably better electrochemical properties than that of the pristine polymer. The increase of polymeric phase conductivity (**Table 1**) and specific capacitance (**Table 2**) is observed for these materials.

There is also possible to incorporate conducting polymer into inside of carbon nanotubes. Steinmetz et al. [34] produced polyacetylene (PA) filled MWCNTs by *in-situ* polymerization using supercritical carbon dioxide ($scCO_2$) and Ziegler-Natta catalyst. The supercritical fluid method can be also used to fill nanotubes with various

Material	Method	Conductivity (S cm^{-1})	References
PANI	Chemical polymerization	~0.35	[24]
SWCNTs/PANI		3.41	
PPY	Chemical polymerization using DBSA as surfactant	22.3	[25]
MWCNTs/PPY		26	
PPY	Chemical polymerization	7.3×10^{-3}	[26]
MWCNTs (9.1 wt%)/PPY		5.6×10^{-2}	
MWCNTs (13.04 wt%)/PPY		9.6×10^{-2}	
MWCNTs (23.1 wt%)/PPY		0.23	
PTH	Chemical polymerization	$\sim 1.67 \times 10^{-6}$	[27]
SWCNTs/PTH		0.41	
PTH	γ-radiation-induced chemical polymerization	$\sim 1.23 \times 10^{-4}$	[28]
MWCNTs/PTH		3.71	

Table 1.
Average values of conductivity of pure polymers and carbon nanotubes/polymer composites.

Material	Method	Specific capacitance (F g^{-1})	References
MWCNTs/PEDOT 85/15 wt%	Chemical polymerization	95	[29]
MWCNTs/PEDOT 30/70 wt%		120	
MWCNTs/PTH	Electrochemical polymerization	110	[30]
Px-MWCNTs/PANI	Chemical polymerization	809.6	[31]
MWCNTs/PANI	Chemical polymerization	446.89	[32]
MWCNTs/PPY	Electrochemical polymerization	600	[33]

Table 2.
Average values of specific capacitance of pure polymers and carbon nanotubes/polymer composites.

organic substances to polymerize photo-conducting poly(*N*-vinyl carbazole) (PNVC) and the conducting polypyrrole inside MWCNTs and double-walled carbon nanotubes (DWCNTs) [35]. MWCNTs and DWCNTs were opened by refluxing them in concentrated HNO_3. The monomer together with an initiator was filled into these carbon nanotubes using $scCO_2$. In the case of *N*-vinyl carbazole, 2,2′-azobis-isobutyronitrile (AIBN) was used as a monomer polymerization initiator, while polymerization of pyrrole was made using $FeCl_3$ as the initiator of this process.

The structure and properties of composites of carbon 1-D nanomaterials and conducting polymers can be significantly improved by using well-organized structures of carbon nanotubes, such as aligned carbon nanotubes (ACNTs) network. In the case of ACNTs, the 1-D carbon cylinders are oriented in a parallel fashion perpendicular to the substrate. The aligned carbon nanotubes allow the polymer to be deposited on the walls of separated carbon nanotubes, limiting the thickness of the formed composite and the produced material has an open, and porous structure with a high surface area. Formation of ACNTs/conducting polymer composites was also carried out using both chemical and electrochemical polymerization. Feng et al. [36] aligned multi-walled carbon nanotubes (AMWCNTs) encapsulated by polyaniline by *in-situ* chemical polymerization. AMWCNTs grown on quartz glass sheet using catalytic pyrolysis. Next, these AMWCNTs were dipped in HCl solution containing aniline monomer for 12 h at 0°C. Aniline was adsorbed on the surface of the nanotubes. The polymerization process occurred after the addition dropwise of ammonium peroxydisulfate (APS) dissolved in HCl solution at 0°C for 4 h. **Figure 6** shows the preparation procedure for organizing AMWCNTs/PANI nanocomposite [36].

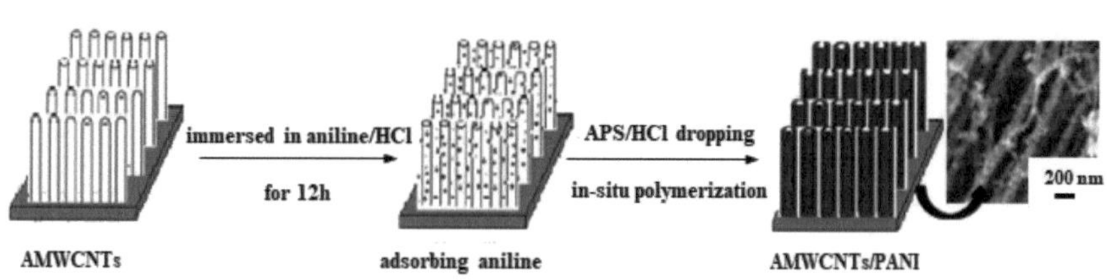

Figure 6.
The preparation procedure of organizing AMWCNTs/PANI nanotubes (reproduced with permission from Ref. [36]).

Compared to CNTs-based composites, carbon nanofibers (CNFs) have received much less attention as a component of 1-D composites, because CNTs have better mechanical properties, smaller diameter, and lower density than CNFs. However, because of their availability, relatively low price, and much easier production produce, carbon nanofibers are an excellent alternative to the more expensive carbon nanotubes [37]. Jang et al. [38] demonstrated that vapor deposition polymerization method could be effectively used for the introduction of polyaniline onto the carbon nanofibers. This process has allowed the formation of a uniform and ultrathin PANI layer of which the thickness-dependent on the amount of monomer (**Figure** 7). Besides, the increasing of PANI layer thickness results in CNFs significant increase in the specific capacitance of these composites. Good electrochemical properties are also observed for CNFs/PANI composites prepared by functionalizing carbon nanofibers with toluenediisocyanate trough amidation followed by reaction with an excess of aniline to form urea derivative and residual aniline, which was subsequently polymerized and grafted with a urea derivative [39].

Recently, graphene nanoribbons were used to make nanocomposites with polypyrrole [40]. Graphene nanoribbons were synthesized by unzipping and exfoliation of MWCNTs, while polypyrrole was prepared using a chemical polymerization process in the presence of graphene oxide nanoribbons. These nanocomposites had a

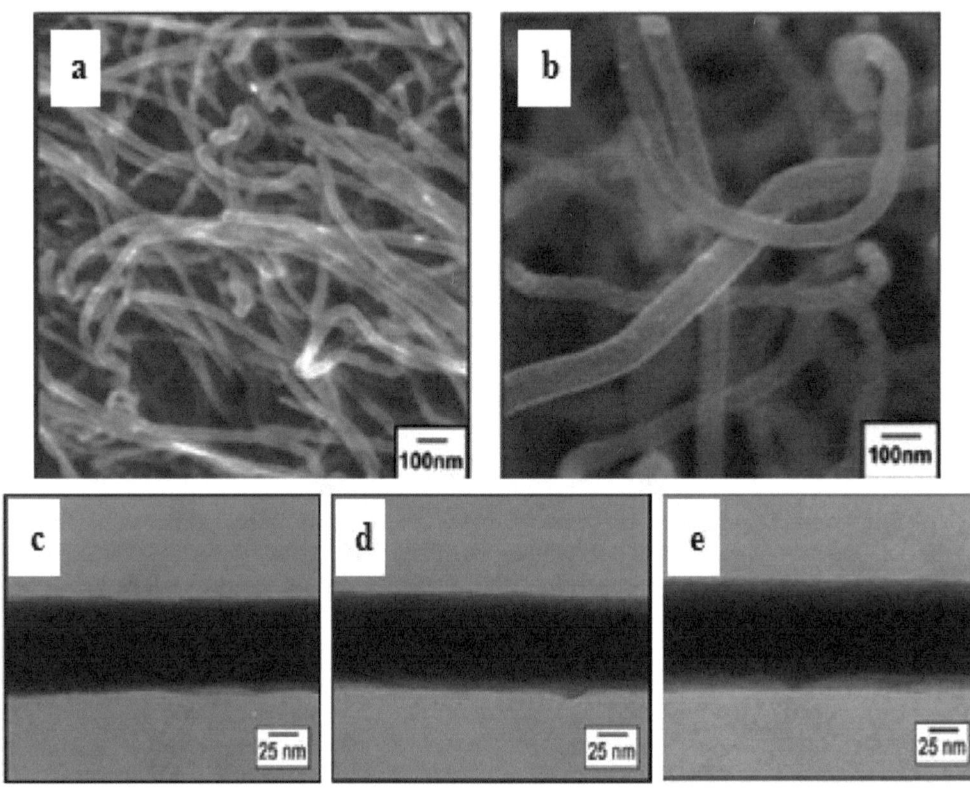

Figure 7.
SEM: (a and b), TEM: (c–e), images of pristine CNFs (a), and CNFs/PANI nanocomposites (b–e) with different the thickness of PANI layer deposited on the CNFs controlled by changing the amount of monomer: (c) 0.05 ml, (b) 0.1 ml, (c) 0.2 ml (reproduced with permission from Ref. [38]).

higher surface area than pure polypyrrole, which improved the charge storage capacity of the nanocomposites.

2.3 The other 1-D nanocomposites

To enhance processability, electrochemical, thermal, and mechanical stability, conducting polymers are often combined with other 1-D nanostructures, such as semiconducting materials, crystals of metalloorganic complexes.

Similar to the preparation of 1-D metal or metal oxide/conducting polymer nanocomposites, semiconducting selenides and sulfides can be incorporated into conducting polymers trough chemical or electrochemical methods. Alivisatos et al. [41] obtained CdSe/poly-3(hexylthiophene) (P3HT) nanorods. CdSe nanorods were dispersed with P3HT in a mixture of pyridine and chloroform and spin-cast to create a uniform film consisting of dispersed nanorods in the polymer. Such material was used to fabricate efficient hybrid solar cells with an external quantum efficiency of over 54% and monochromatic power conversion efficiency of 6.9%. Template techniques of synthesis 1-D nanomaterials, have been demonstrated for the preparation of sulfides with conducting polymers. Thus, Lin et al. [42] reported the preparation of CdS/PANI coaxial nanocables by the electrochemical synthesis in the AAO membrane. The diameter of the CdS nanowires was about 70 nm, which was the same as the pore diameter of the AAO membrane. The outer diameter of the PANI was about 90 nm. Guo et al. [43] synthesized CdS/PPY heterojunction nanowires by template technique using also porous AAO membrane. These nanowires had a smooth surface with diameters in the range of 200–400 nm. In addition to the template technique, an *in-situ* polymerization at the interfacial layer between chloroform and water has been developed for the preparation of Cu_2S nanorods coated with polypyrrole layer [44]. Smooth and uniform coaxial Cu_2S/PPY nanocables have been fabricated by controlling the reaction conditions, such as the molar ratio of pyrrole to oxidant and concentration of pyrrole in chloroform. The thickness of the PPY layer on the surface of Cu_2S nanorods depends on the polymerization time (**Figure 8**). The hydrothermal reaction was also applied to prepare Bi_2S_3/PPY nanocomposites [45]. The PPY coating on the surface Bi_2S_3/PPY nanorods was smooth and uniform in thickness.

1-D nanocrystals and conducting polymer composites were also created. Crystals of β-akaganeite (β-Fe^{3+}O(OH,Cl)) with PEDOT [46, 47] and crystals of iridium complex ($[IrCl_2(CO)_2]^-$) with PPY [48] are examples of 1-D composites containing conducting

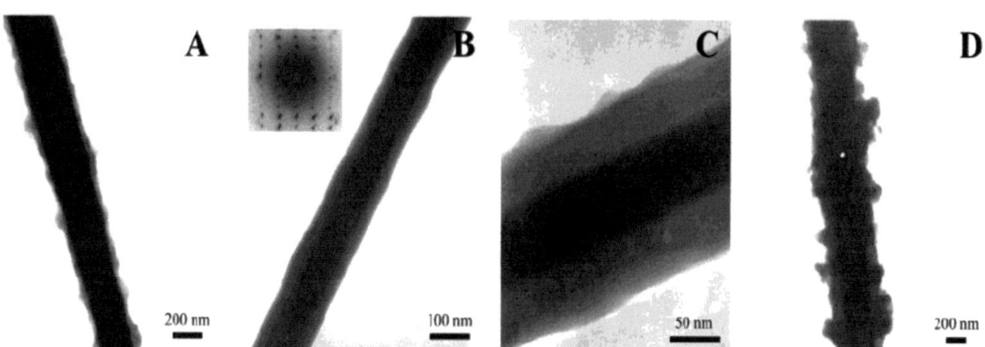

Figure 8.
TEM images of Cu_2S/PPY nanorods were obtained with a pyrrole polymerization time of: (a) 1 h, (b) 2 h, (c) 3.5 h, (d) 5 h. inset of (B): SAED of the single Cu_2S/PPY nanorod (reproduced with permission from Ref. [44]).

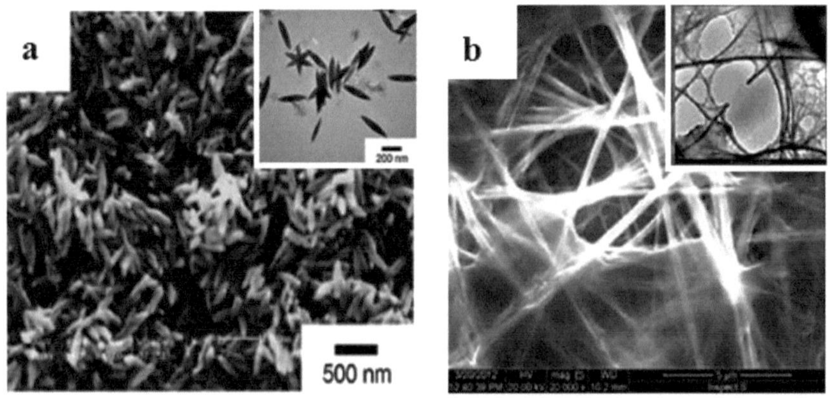

Figure 9.
SEM and inset TEM images of: (a) PEDOT/β-Fe^{3+}O(OH,Cl) nanospindles (reproduced with permission from Ref. [46]), and (b) 1-D-$IrCl_2(CO)_2$/PPY nanocomposites (reproduced with permission from Ref. [48]).

polymer and crystal metalloorganic complexes. The advantage of fabrication of these materials is the possibility of their preparation by *in-situ* one-step or two-step approach, respectively. Namely, PEDOT/β-Fe^{3+}O(OH,Cl) nanospindles (**Figure 9a**) were synthesized in an aqueous solution trough one-step chemical oxidation polymerization using monomer EDOT, and $FeCl_3{\bullet}6H_2O$ as an oxidant in the presence of CTAB and poly (acrylic acid) (PAA). Under these conditions, the polymerization of EDOT, and the hydrolyzation of $FeCl_3$ to form β-Fe^{3+}O(OH,Cl) occur at the same time, leading to the formation PEDOT/β-Fe^{3+}O(OH,Cl) nanocomposite [46, 47]. The 1-D-$IrCl_2(CO)_2$/PPY nanocomposites (**Figure 9b**) were synthesized in a dichloromethane solution by *in-situ* two-step electrodeposition. First, needles of the iridium complex were prepared by electrochemical oxidation of $(AsPh_4)[IrCl_2(CO)_2]$. Next, pyrrole was electropolymerized on the surface of the iridium needles [48]. Both the crystals size and thickness of the polymer can be easily controlled very simply using the reaction conditions such as concentration of compounds, different reaction times, and kind of solvent.

3. Applications of conducting polymers and 1-D nanostructures composites

Many of the targeted applications for 1-D nanostructures require their incorporation into conducting polymers. Therefore, such nanocomposites are expected to find applications in nanoelectronic devices, sensors, catalysis or electrocatalysis and energy.

3.1 Energy conversion and energy storage applications

Incorporating 1-D nanostructure into conducting polymers play an important role in fabricating materials in energy conversion devices such as solar cells, and fuel cells, and energy storage devices such as lithium-ion batteries and supercapacitors. These materials have improved conductivity, cycleability, mechanical stability, processability, and specific capacitance.

Solar cells are energy conversion devices that convert sun light to electric energy. In comparison to nanoparticles, 1-D nanostructures are excellent candidates for the preparation of solar cells because they provide high electron mobility along these nanostructures. Their hole-transporting properties may contribute to the improvement of the photovoltaic efficiency performance. Solar cell devices fabricated with aligned ZnO/P3TH and ZnO/didodecylquaterthiophene (QT) composites exhibit well-resolved characteristics with the efficiency of 0.036%, a circuit current density (J_{sc}) of 0.32 mA cm^{-2}, and a fill factor (FF) of 0.28 [49]. Additional enhancement of photovoltaic properties of ZnO/P3TH-based solar cells can be achieved by incorporation of TiO_2 into the composite structure. Yang et al. [8] demonstrated that solar cells based on ZnO-TiO_2/P3TH composites exhibited an efficiency of 0.29% for devices stored in air for 1 month. While, these devices without TiO_2 layer had an efficient of only 0.04%. CNTs/PEDOT-PSS (polystyrenesulfonate) nanotubes were also used as dye-sensitized solar cells (DSCs), which exhibit very high performance with the energy conversion efficiency of 6.5%, J_{sc} = 15.5 cm^{-2}, and FF = 0.63 [50].

Fuel cells, which convert the chemical energy of a fuel directly into electricity by electrochemical reactions, have attracted attention for applications in electric vehicles [51]. Also in this case, the introduction of 1-D composites containing polymers can significantly improve the performance of such systems. For example, Co/PPY/MWCNTs nanotube composites were used as the cathode electrocatalysts for the reduction of oxygen in polymer electrolyte fuel cells (PEMFCs), direct ethanol fuel cells (DEFCs), and direct methanol fuel cells (DMFCs) [52]. The stability of these composites for the reduction of oxygen was excellent without any noticeable loss in performance over long PEMFC operating time, which is shown in **Figure 10**.

Rechargeable batteries are widely used in daily life such as in cell phones, laptop computers, and electric vehicles. 1-D nanofibers composite of V_2O_5/PANI has been used as cathode materials in ion-Li batteries [53]. The composite showed enhanced capacitance properties in comparison to vanadium oxide nanotubes. The charge capacity of V_2O_5/PANI composite nanofibers was about 150 Ah kg^{-1} during the 10

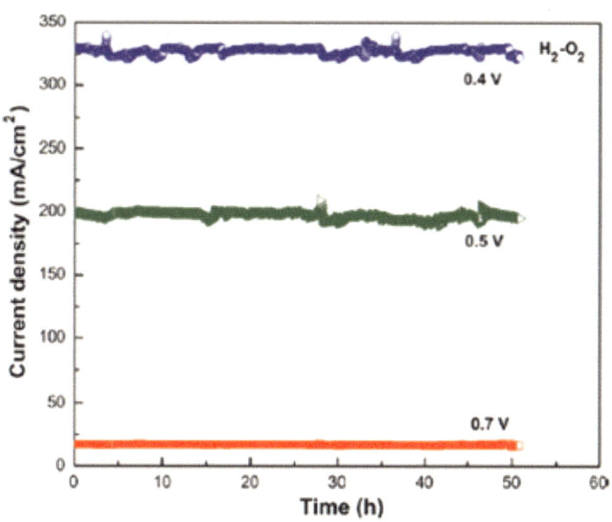

Figure 10.
Cell performance showing the stability of PEMFC at 90°C with cathode catalyst containing Co/PPY/MWCNTs and anode catalyst containing Pt/Ru/MWCNTs at ambient pressure (reproduced with permission from Ref. [52]).

initial charge/discharge cycles, while a charge capacity of 100 Ah kg^{-1} was obtained for V_2O_5 nanotubes. The composite also exhibits much better cyclability in comparison to V_2O_5 nanotubes. A significant enhancement in electrochemical performance has been also found for the silver vanadium oxides (SVO)/PANI triaxial nanowires [54]. It has been observed that the SVO/PANI triaxial nanowires exhibited a much higher current density than that of β-$AgVO_3$ nanowires, due to faster kinetics of charge transfers and higher capacity. Therefore, these composites can be used as cathode material for Li^+ ion batteries with high specific capacity and good cycle performance [54]. The composite of polypyrrole@CNTs has been used as pseudocapacitive cathodes and Fe_3O_4@carbon as anode for nonaqueous lithium-ion capacitor applications [55]. Due to the synergistic effect of the remarkable pseudocapacity of the polypyrrole and the high electrical conductivity of CNTs, the polypyrrole@CNTs composite exhibited enhanced capacitive properties and cycle life in comparison to the pristine CNTs. The rechargeable sulfate- and sodium-ion batteries based on MWCNTs-polypyrrole core-shell nanowire anode and a $Na_{0.44}MnO_2$ nanorod cathode were also studied [56]. This MWCNTs-polypyrrole core-shell nanowire//$Na_{0.44}MnO_2$ nanorod full cell delivered discharge capacities of 99.2 mAh g^{-1} and 87.2 mAh g^{-1} with a high voltage of 1.6 V at the charge-discharge current densities of 100 mA g^{-1} and 3000 mA g^{-1}, respectively, making it suitable for large scale energy storage application [56].

Supercapacitors also called electrochemical capacitors with high specific power, exceptional long cycle life compared with rechargeable batteries, and higher specific energy compared to conventional capacitors are of great interest for their potential applications in portable electronics, hybrid electronic vehicles, memory protection of computer electronics and renewable energy system [57]. Active electrode materials used in supercapacitors can be classified into three main categories: carbon, transition metal or metal oxide, and conducting polymer. Among these third groups, carbon has a relatively low specific capacitance usually under 200 F g^{-1}, metal oxides are either expensive, for example, ruthenium oxide, or poor conductors, for examples MnO_2, NiO, etc., while conducting polymers have a high specific capacitance, but their cyclic stability is poor. Therefore, nanocomposites can be useful for the construction of electrochemical capacitors with both high capacitance and good cyclic stability. PANI/CNFs composite was tested as an electrochemically active component for supercapacitors [38, 39]. The specific capacitance of PANI layer-coated CNFs showed a maximum value of 264 F g^{-1} at 20 nm thickness of PANI, whereas that of pure CNFs were as 100 F g^{-1} [38]. The specific capacitance of 557 F g^{-1} and good cycling stability were reported for CNFs/PANI by Kotal et al. [39]. The electrochemical measurements of graphene nanoribbons with PPY showed the highest specific capacitance of 2066 F g^{-1}. Therefore, these nanocomposites could be used as an electrode material for the fabrication of high-capacity supercapacitors [40]. The nanocomposite of metal wires and conducting polymers can be also used in supercapacitors [12, 13]. In the case of 3-D Ag nanowires/PPY the maximum specific capacitance, maximum power and energy density 509 F g^{-1}, 60.7 W kg^{-1}, and 4.27 Wh kg^{-1} was reported, respectively. The fabricated supercapacitors showed excellent stability of almost 90% after 10,000 charge/discharge cycles [13]. The capacitance performance of a variety of conducting polymer and 1-D carbon nanostructure composites is summarized in **Table 3**. In addition to CNTs, nanocomposites containing metal, metal oxide or metal hydroxide and conducting polymers have also been intensively investigated as building components in supercapacitors [62–65].

Composite	Formation method	Capacitance	performance	References
		Specific capacitance	Stability	
Porous PANI/CNTs	Chemical grafting and creating interpenetrating pores via templating using $CaCO_3$ nanoparticles	1266 F g^{-1} at 1 A g^{-1} in 1 M H_2SO_4	83% (10,000 cycles)	[58]
CNFs/PANI	One-step vapor deposition polymerization	264 F g^{-1} in 1 M H_2SO_4	—	[38]
PANI/ TCNF	In-situ polymerization	557 F g^{-1} in 0.5 M H_2SO_4	86% (2000) at 0.3 A g^{-1}	[39]
CNTs-GO/ PPY	Facile electrochemical synthesis	6.3 F cm^{-3} at 0.043 A cm^{-3} with PVA/H_3PO_4 gel as the solid electrolyte	87.7% (10,000 cycles)	[59]
PPY/GO/ CNTs	Electrochemi-cal co-deposition	196.7 mF cm^{-2} at 0.5 mA cm^{-2}	98.1% (5000 cycles)	[60]
PPY/GO/ MWCNTs	Facile one-step potentiostatic technique	358.69 F g^{-1} at 100 mV s^{-1} in 1 M H_2SO_4	88.69% (2000 cycles)	[61]

CNTs, carbon nanotubes.
CNFs, carbon nanofibers.
TCNF, isocyanate-functionalized CNF.

Table 3.
Capacitance performance of the composites of conducting polymers with 1-D carbon nanostructures.

3.2 Electronic nanodevices

It is generally accepted that 1-D nanostructures provide a good system to investigate the dependence of electrical transport or mechanical properties on dimensionality and size reduction. Most conducting polymers are suited for the construction of electronic devices because of their high electrical conductivity, and mechanical flexibility. Therefore, materials combined of 1-D nanostructures and conducting polymers can be potentially applicable in diodes, memory, transistors, and photovoltaic devices.

Woo et al. [66] reported the fabrication of organic light-emitting diode (OLED) using a conjugated emissive copolymer, poly(3,6-*N*-2-ethylhexyl carbazolyl cyanoterephthalidence) (PECCP) and SWCNTs dispersed in a hole conducting PEDOT in the buffer. The schematic of this device construction is shown in **Figure 11**. This composite deposited on the ITO served as an anode in LED. A cathode was a bi-layer consisting of a LiF and Al.

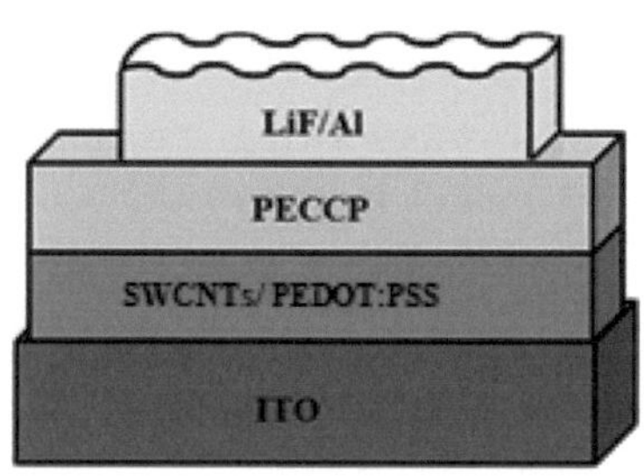

Figure 11.
Schematic of OLED construction.

By a combination of electrochemical polymerization of pyrrole and electrophoretic deposition of CNTs, new composite material has been prepared and tested for application in a triode-type field emission array (FEA) [67]. This triode-type FEA showed an emission current of 35 mA at an anode voltage of 1000 V and the gate voltage of 60 V. The emission current of the FEA was modulated by the gate voltage of 30 V. For photovoltaic applications, nanocomposite material consisting of CNTs and PANI as highly conductive and transparent has also been prepared [68]. Organic photovoltaic cells were built using this film as an anode in flexible ITO-free devices. These results indicated that novel ITO-free optoelectronic devices can be optimized with very high performance using transparent films of conjugated polymers and carbon 1-D nanomaterials.

3.3 Sensors

Conducting polymers are good candidates for chemical and biological sensors because the interactions with various analytes may influence the redox and doping states of these materials. Adding a second nanocomponent, such as carbon nanotubes, metal and metal oxide nanostructures, and biological materials, into conducting polymer is another way to increase the charge mobility of conducting polymers or to change the affinity of these composites.

Because of the large specific surface areas, these nanocomposites are good candidates for gas sensors. For example, CNTs/conducting polymer nanocomposites exhibit high sensitivity in NH_3 detection. Ammonia is one of the important industrial exhaust gases with high toxicity. Liu et al. [69] demonstrated a simple and effective method of NH_3 detection by at sensors based on MWCNTs/PANI nanocomposites. The results showed that MWCNTs/PANI had high sensitivity and quick sensor response, good reproducibility and repeatability for NH_3 detection. The mechanism of the enhanced sensitivity may be attributed to the increased surface area of PANI, providing more active sites for the adsorption of NH_3 molecules. Similar properties were also observed for MWCNTs/Au/PANI nanocomposites [70]. However, the high cost discourages its extensive application. The sensing properties of the CNFs/PPY coaxial nanocables for toxic gases, such as NH_3 and HCl detection were also studied [71]. These materials were fabricated by one-step vapor deposition polymerization. This simple process allowed the formation of ultrathin and uniform PPY layer on the CNFs surface, which thickness was dependent on the loaded amount of the monomer. The responses of the CNFs/PPY coaxial nanocables after interaction with NH_3 and HCl were dependent on the thickness of the PPY layer on CNFs and exhibited reversible and reproducible performance (**Figure 12**). The resistance change of the CNFs/PPY coaxial nanocables was negligible when the thickness of the PPY layer was smaller than 10 nm. The sensitivity of these nanocables increased significantly with increasing the PPY thickness and then stopped increasing when the PPY layer thickness was larger than the growth limit thickness point (22 nm) [71].

In recent years, conducting polymers with 1-D nanostructured composites have also been used to construct a variety of biosensors because of their large surface area and unique electronic, chemical, and mechanical properties. In biosensors, the detection of H_2O_2 is important because it is often a product in enzymatic reactions. The sensing performance of coaxial nanowires consisting of a layer of PPY uniformly coated onto aligned CNTs in H_2O_2 detection makes it attractive for the fabrication of oxidase-based glucose biosensors, because H_2O_2 is generated in the reaction between glucose and oxygen in the presence of glucose oxidase (GOX) [72]. In these materials the aligned structure of CNTs plays a significant part in

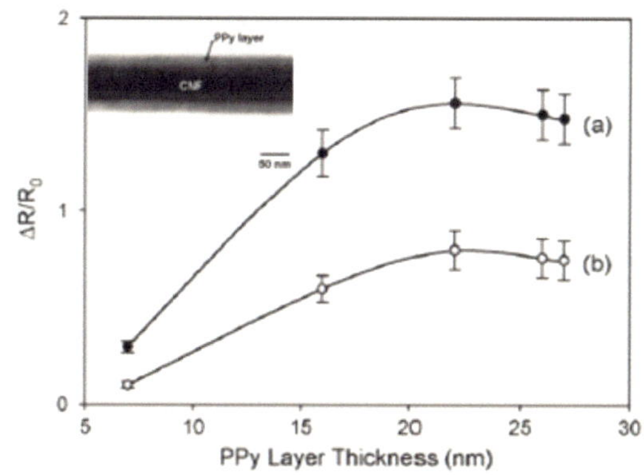

Figure 12.
Variation in normalized resistance change (absolute value) of the CNFs/PPY nanocable sensors after exposure to (a) NH_3 (20 ppm), and (b) HCl (20 ppm) vapors as a function of the PPY layer thickness. Inset TEM image of the CNFs/PPY nanocable (reproduced with permission from Ref. [71]).

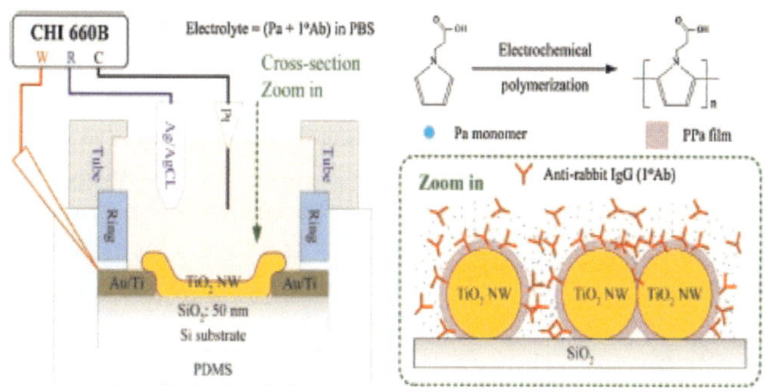

Figure 13.
The experiment setup of the electrochemical polymerization of porypyrrole propilic acid/anti-rabbit IgG immobilized TiO_2 nanowires immunosensor system (reproduced with permission from Ref. [73]).

glucose determination, shifting its oxidation potential toward less positive values and enhancing the sensitivity of glucose determination. The immunosensor was also constructed based on an antibody/conducting polymer/TiO_2 nanowires film [73]. First, TiO_2 nanowires were made by hydrothermal synthesis and spin-coated on Au/Ti microelectrodes surface patterned Si/SiO_2 substrate. Next, polypyrrole propylic acid (PPA) and antibody composite films were immobilized on the surface of TiO_2 nanowires by electrochemically polymerized using pyrrole propylic acid (PA) and anti-rabbit IgG (1°AB) mixture solution, as illustrated in **Figure 13**. The devices designed in these studies showed a linear concentration range of antigen determination between 11.2 μg/mL to 112 μg/mL. The detection sensitivity of these immunosensors was −0.64 A/(g/mL) for the 5 V of the applied voltage and the sensitivity for this voltage was better than that of 6 V and 7 V [73].

3.4 Catalysis

Catalytic materials are important for the industry and the development of various sensors. Therefore, composites of conducting polymer and 1-D nanostructures also have been studied in this area of research.

The nanowires consisting of gold-coated PANI film exhibit excellent catalytic behavior for the chemical reduction of organic dyes such as methylene blue (MB) and rhodamine B (RhB) [74]. Most of these dyes are not biodegradable and persist in the environment, but they can be disposed of by chemical reduction using a strong reducing agent as an economical route. Unfortunately, the chemical reduction of dyes is a very slow process under ambient conditions. Therefore, these Au/PANI nanowires are used as catalysts for the reduction of MB and RhB dyes in the presence of $NaBH_4$. A total catalytical reduction of MB and RhB was observed. The catalytic activity of composite was much better in comparison to the catalytic performance of individual components [75, 76].

Several 1-D semiconductor materials such as TiO_2, ZnO, MnO_2, CdS, etc., have also been used as semiconductor photocatalysists. However, their wide band gap and the low quantum yield largely limited the overall photocatalytic efficiency. The photocatalytic performance of this semiconductor can be significantly improved by incorporating them into 1-D structures with conducting polymers. For example, a novel photocatalyst, polypyrrole coated Ag/TiO_2 nanofibers, was synthesized using an electrospinning technique, followed by a surfactant *in-situ* chemical polymerization method [77]. This photocatalyst showed obvious visible-light photocatalytic activity in the decomposition of gaseous acetone. The 1.0 wt% PPY/Ag/TiO_2 sample provided the optimum photocatalytic activity. In **Figure 14**, photocurrent transient responses of PPY/Ag/TiO_2, PPY/TiO_2, Ag/TiO_2 are compared. The photocurrent followed the order: PPY/Ag/TiO_2 > PPY/TiO_2 > Ag/TiO_2 > pure TiO_2. The enhancement of PPY/Ag/TiO_2 in photocurrent indicates smaller recombination and more efficient separation of photogenerated electron-hole pairs at its interface. Besides, the recycling test revealed that the PPY/Ag/TiO_2 nanofibers were stable and effective for the removal of organic pollutants [77]. The high photoactivity of the PPY/Ag/TiO_2 nanofibers can be attributed to the synergistic effect originating from the excited-state electrons in PPY, which can be readily injected into the TiO_2 conduction band and next transported to the Fermi level of Ag (**Figure 15**) [77]. Therefore, the combination of conducting polymers with semiconducting materials is an effective strategy for improving photocatalytic activity.

The composites consisting of two or more components containing CNTs and the conducting polymers can be also used as electrocatalysts for hydrogen and alcohol fuel cells [52], as well as bio and microbial fuel cells [78]. Using these materials as a cathode or anode catalysts is an important step in reducing the use of high-cost

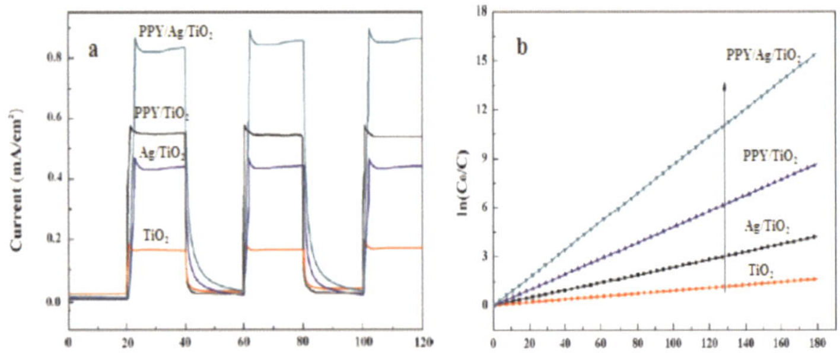

Figure 14.
(a) Photocurrent transient responses and (b) photocatalytical activity of PPY/Ag/TiO_2 and of single and two-component samples (reproduced with permission from Ref. [77]).

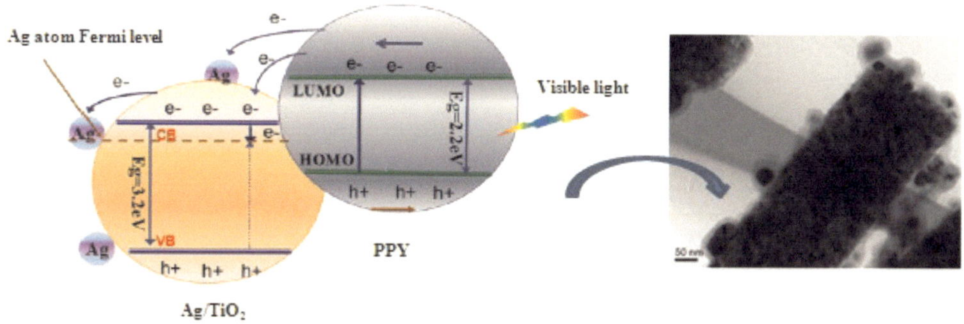

Figure 15.
Postulate mechanism of the visible-light-induced photodegradation of acetone with PPY/Ag/TiO_2 nanofibers (reproduced with permission from Ref. [77]).

platinum and platinum-based electrocatalysts to promote practical applications. Additional optimization of the catalyst structure and stability may improve catalysts performance and reduce the total cost.

4. Summary

This paper reviews study of the formation, properties and applications of conducting polymer 1-D nanocomposites. These materials have attracted much attention due to their unique physical, mechanical, chemical and electrochemical properties that provide nanostructures formed from them with multi-functionality. As described in this review, many synthetic approaches have been developed, including *in-situ* and *ex-situ* chemical and electrochemical method, template and template-free method, vapor-phase polymerization, emulsion polymerization, hydrothermal reaction, redox reaction, electrospinning, and so on, for the fabrication of conducting polymer nanocomposites with carbon nanotubes and nanofibers, 1-D metals, metal oxides, metal complexes, semiconductors, and porphyrins. Nanocomposites synthesized by these methods exhibit many exceptional properties. The size, shape and structure morphology of these composites can be controlled by the conditions of composite formation. 1-D structures exhibit high real surface area. The organization of these structures on the substrate surface provides unique conditions for reactant transport within the 1-D composite film. These composites usually exhibit good conductivity. Such properties enable wide practical applications of 1-D composites. They can be used for charge storage devices construction and as a component of electronic devices. They are applying in a chemical, biochemical and electrochemical sensors. Their catalytic and photocatalytic properties allow the use of these systems to remove environmental pollutants. So, the improvement of the syntheses is sought, which would lead to: (i) highly conducting, (ii) macroscopically uniform materials, (iii) produced at reasonable reaction time, and (iv) in high yield.

Therefore, conducting polymer 1-D nanostructures are useful materials in basic research and technology. Important of synthesis conditions and the production of new 1-D structures can have a significant impact on scientific development. Studies for the preparation of such materials are still desired. It is believed, that the combined chemical, physical, and mechanical properties of these nanocomposite materials are crucial in the development in the future more discoveries will be made in this field.

Author details

Monika Wysocka-Żołopa*, Emilia Grądzka and Krzysztof Winkler
Department of Chemistry, University of Bialystok, Bialystok, Poland

*Address all correspondence to: monia@uwb.edu.pl

References

[1] Zhou H, Wong SS. A facile and mild synthesis of 1-D ZnO, CuO, and α-Fe_2O_3 nanostructures and nanostructured arra ys. ACS Nano. 2008;**2**:944-958. DOI: 10.1021/nn700428x

[2] Cheng B, Samulski ET. Hydrothermal synthesis of one-dimensional ZnO nanostructures with different aspect ratios. Chemical Communications. 2004: 986-987. DOI: 10.1039/B316435G

[3] Wysocka M, Winkler K, Stork JR, Balch AL. Electrochemical oxidation of $(Ph_4As)[IrCl_2(CO)_2]$ in the presence of tetra(alkyl)ammonium salts. Electrocrystallization of different forms of iridium-based linear chain complexes. Chemistry of Materials. 2004;**16**: 771-780. DOI: 10.1021/cm035029r

[4] Winkler K, Wysocka-Żołopa M, Rećko K, Dobrzyński L, Vickery JC, Balch AL. Formation of a partially oxidized gold compound by electrolytic oxidation of the solvoluminescent gold (I) trimer, $Au_3(MeN{=}COMe)_3$. Inorganic Chemistry. 2009;**48**:1551-1558. DOI: 10.1021/ic801922a

[5] Hu J, Odom TW, Lieber CM. Chemistry and physics In one dimension: Synthesis and properties of nanowires and nanotubes. Accounts of Chemical Research. 1999;**32**:435-445. DOI: 10.1021/ar9700365

[6] Li Y, Qian F, Xiang J, Lieber CM. Nanowire electronic and optoelectronic devices. Materials Today. 2006;**9**:18-27. DOI: 10.1016/S1369-7021(06)71650-9

[7] Yu Z, Li L, Zhang Q, Hu W, Pei Q. Silver nanowire-polymer composite electrodes for efficient polymer solar cells. Advanced Materials. 2011;**23**: 4453-4457. DOI: 10.1002/adma.2011 01992

[8] Greene LE, Law M, Yuhas BD, Yang P. ZnO-TiO_2 core-shell nanorod/P3HT solar cells. Journal of Physical Chemistry C. 2007;**111**:18451-18456. DOI: 10.1021/ jp077593l

[9] Wiley B, Sun Y, Xia Y. Polyol synthesis of silver nanostructures: Control of product morphology with Fe (II) or Fe(III) species. Langmuir. 2005; **21**:8077-8080. DOI: 10.1021/la050887i

[10] Lu X, Yavuz MS, Tuan HT, Korgel BA, Xia Y. Ultrathin gold nanowires can be obtained by reducing polymeric strands of oleylamine-AuCl complexes fordem via aurophilic interaction. Journal of the American Chemical Society. 2008;**130**:8900-8901. DOI: 10.1021/ja803343m

[11] Chen A, Wang H, Li X. One-step process to fabricate Ag-polypyrrole coaxial nanocables. Chemical Communications. 2005:1863-1864. DOI: 10.1039/B417744D

[12] Yu L, Yang Q, Liao J, Zhu Y, Li X, Yang W, et al. A novel 3D silver nanowires@polypyrrole sponge loaded with water giving excellent microwave absorption properties. The Chemical Engineering Journal. 2018;**352**:490-500. DOI: 10.1016/j.cej.2018.07.047

[13] Yuksel R, Alpugan E, Unalan HE. Coaxial silver nanowire/polypyrrole nanocomposite supercapacitors. Organic Electronics. 2018;**52**:272-280. DOI: 10.1016/j.orgel.2017.10.012

[14] Liu K, Li Y, Zhang H, Liu Y. Synthesis of the polypyrrole encapsulated copper nanowires with excellent oxidation resistance and temporal stability. Applied Surface Science. 2018;**439**:226-231. DOI: 10.1016/j.apsusc.2018.01.020

[15] Sarvi A, Gelves GA, Sundararaj U. Facile one step-synthesis and characterisation of high aspect ratio core-shell copper-polyaniline nanowires. Canadian Journal of Chemical Engineering. 2014;**92**:1207-1212. DOI: 10.1002/cjce.21973

[16] Xia X, Chao D, Qi X, Xiong Q, Zhang Y, Tu J, et al. Controllable growth of conducting polymers shell for constructing high-quality organic/inorganic core/shell nanostructures and their optical-electrochemical properties. Nano Letters. 2013;**13**:4562-4568. DOI: 10.1021/nl402741j

[17] Han L, Tang P, Zhang L. Hierarchical Co_3O_4@ PPy@MnO2 core–shell–shell nanowire arrays for enhanced electrochemical energy storage. Nano Energy. 2014;**7**:42-51. DOI: 10.1016/j.nanoen.2014.04.014

[18] Lahav M, Weiss EA, Xu Q, Whitesides GM. Core-shell and segmented polimer-metal composite nanostructures. Nano Letters. 2006;**6**:2166-2171. DOI: 10.1021/nl061786n

[19] Cao H, Xu Z, Sang H, Sheng D, Tie C. Template synthesis and magnetic behavior of an array of cobalt nanowires encapsulated in polyaniline nanotubules. Advanced Materials. 2001;**13**:121-123. DOI: 10.1002/1521-4095(200101)13:2<121::AID-ADMA121>3.0.CO;2-L

[20] Ajayan PM, Stephan O, Colliex C, Trauth D. Aligned carbon nanotube arrays formed by cutting a polymer resin-nanotube composite. Science. 1994;**265**:1212-1214. DOI: 10.1126/science.265.5176.1212

[21] Downs C, Nugent J, Ajayan PM, Duquette DJ, Santhanam KSV. Efficient polymerization of aniline at carbon nanotube electrodes. Advanced Materials. 1999;**11**:1028-1031. DOI: 10.1002/(SICI)15214095(199908)11:12<1028::AID-ADMA1028>3.0.CO;2-N

[22] Wu TM, Lin YW. Doped polyaniline/multi-walled carbon nanotube composites: Preparation, characterization and properties. Polymer. 2006;**47**:3576-3582. DOI: 10.1016/j.polymer.2006.03.060

[23] Wu TM, Lin YW, Liao CS. Preparation and characterization of polyaniline/multi-walled carbon nanotube composite. Carbon. 2005;**43**:734-740. DOI: 10.1016/j.carbon.2004.10.043

[24] Karim MR, Lee CJ, Park YT, Lee MS. SWNTs coated by conducting polyaniline: Synthesis and modified properties. Synthetic Metals. 2005;**151**:131-135. DOI: 10.1016/j.synthmet.2005.03.012

[25] Han G, Yuan J, Shi G, Wei F. Electrodeposition of polypyrrole/multiwalled carbon nanotube composite films. Thin Solid Films. 2005;**474**:64-69. DOI: 10.1016/j.tsf.2004.08.011

[26] Long Y, Chen Z, Zhang X, Zhang J, Liu Z. Electrical properties of multi-walled carbon nanotube/polypyrrole nanocables: Percolation-dominated conductivity. Journal of Physics D: Applied Physics. 2004;**37**:1965-1969. DOI: 10.1088/0022-3727/37/14/011

[27] Karim MR, Lee CJ, Lee MS. Synthesis and characterization of conducting polythiophene/carbon nanotubes composites. Journal of Polymer Science, Part A: Polymer Chemistry. 2006;**44**:5283-5290. DOI: 10.1002/pola.21640

[28] Karim MR, Yeum JH, Lee MS, Lim KT. Synthesis of conducting polythiophene composites with multi-walled carbon nanotube by the γ-radiolysis polymerization method.

Materials Chemistry and Physics. 2008; **112**:779-782. DOI: 10.1016/j.matchemphys.2008.06.042

[29] Lota K, Khomenko V, Frackowiak E. Capacitance properties of poly(3,4-ethylenedioxythiophene)/carbon nanotubes composites. Journal of Physics and Chemistry of Solids. 2004;**64**: 295-301. DOI: 10.1016/j.jpcs.2003.10.051

[30] Fu C, Zhou H, Liu R, Huang Z, Chen J, Kuang Y. Supercapacitor based on electropolymerized polythiophene and multi-walled carbon nanotubes composites. Materials Chemistry and Physics. 2012;**132**:596-600. DOI: 10.1016/j.matchemphys.2011.11.074

[31] Potphode DD, Sinhaa L, Shirage PM. Redox additive enhanced capacitance: Multi-walled carbon nanotubes/ polyaniline nanocomposite based symmetric supercapacitors for rapid charge storage. Applied Surface Science. 2019;**469**:162-172. DOI: 10.1016/ j.apsusc.2018.10.277

[32] Pal R, Goyal SR, Rawal I. High-performance solid state supercapacitors based on intrinsically conducting polyaniline/MWCNTs composite electrodes. Journal of Polymer Research. 2020;**27**:179. DOI: 10.1007/s10965-020-02144-y

[33] Grądzka E, Dłużewski P, Wigda I, Wysocka-Żołopa M, Winkler K. Formation and electrochemical properties of multiwalled carbon nanotubes and polypyrrole composite with (n-Oc4N)Br binder. Synthetic Metals. 2021;**272**:116661. DOI: 10.1016/ j.synthmet.2020.116661

[34] Steinmetz J, Lee HJ, Kwon S, Lee DS, Goze-Bac C, Abou-Hamad E, et al. Routes to the synthesis of carbon nanotube-polyacetylene composites by Ziegler-Natta polymerization of acetylene inside carbon nanotubes. Current Applied Physics. 2007;**7**:39-41. DOI: 10.1016/j.synthmet.2020.116661

[35] Steinmetz J, Kwon S, Lee HJ, Abou-Hamad E, Almairac R, Goze-Bac C, et al. Polymerization of conducting polymers inside carbon nanotubes. Chemical Physics Letters. 2006;**431**:139-144. DOI: 10.1016/j.cplett.2006.09.070

[36] Feng W, Bai XD, Lian YQ, Liang J, Wang XG, Yoshino K. Well-aligned polyaniline/carbon-nanotube composite films grown by in-situ aniline polymerization. Carbon. 2003;**41**: 1551-1557. DOI: 10.1016/S0008-6223 (03)00078-2

[37] Al-Saleh MH, Sundararaj U. Review of the mechanical properties of carbon nanofiber/polymer composites. Composites Part A: Applied Science and Manufacturing. 2011;**42**:2126-2142. DOI: 10.1016/j.compositesa.2011.08.005

[38] Jang J, Bae J, Choi M, Yoon SH. Fabrication and characterization of polyaniline coated carbon nanofiber for supercapacitor. Carbon. 2005;**43**: 2730-2736. DOI: 10.1016/j.carbon.2005. 05.039

[39] Kotal M, Thakur AK, Bhowmick AK. Polyaniline-carbon nanofiber composite by a chemical grafting approach and its supercapacitor application. ACS Applied Materials & Interfaces. 2013;**5**: 8374-8386. DOI: 10.1021/am4014049

[40] Dream JA, Zeguine C, Siam K, Kahol PK, Mishra SR, Gupta RK. Electrochemical properties of graphene oxide nanoribbons/polypyrrole nanocomposites. Journal of Carbon Research C. 2019;**5**:18. DOI: 10.3390/ c5020018

[41] Huynh WU, Dittmer JJ, Libby WC, Whiting GL, Alivisatos AP. Controlling

the morphology of nanocrystal±polymer composites for solar cells. Advanced Functional Materials. 2003;**13**:73-79. DOI: 10.1002/adfm.200390009

[42] Xi Y, Zhou J, Guo H, Cai C, Lin Z. Enhanced photoluminescence in core-sheath CdS-PANI coaxial nanocables: A charge transfer mechanism. Chemical Physics Letters. 2005;**412**:60-64. DOI: 10.1016/j.cplett.2005.06.087

[43] Guo Y, Tang Q, Liu H, Zhang Y, Li Y, Hu W, et al. Light-controlled organic/inorganic p-n junction nanowires. Journal of the American Chemical Society. 2008;**130**:9198-9199. DOI: 10.1021/ja8021494

[44] Zhang W, Wen X, Yang S. Synthesis and characterization of uniform arrays of copper sulfide nanorods coated with nanolayers of polypyrrole. Langmuir. 2003;**19**:4420-4426. DOI: 10.1021/la020894w

[45] Ota J, Srivastava SK. Polypyrrole coating of tartaric acid-assisted synthesized Bi_2S_3 nanorods. Journal of Physical Chemistry C. 2007;**111**: 12260-12264. DOI: 10.1021/jp072906y

[46] Mao H, Lu X, Chao D, Cui L, Li Y, Zhang W. Preparation and characterization of PEDOT/β-Fe^{3+}O(OH,Cl) nanospindles with controllable sizes in aqueous solution. Journal of Physical Chemistry C. 2008;**112**: 20469-20480. DOI: 10.1021/jp807988f

[47] Mao H, Lu X, Wang C, Zhang W. Investigation on PEDOT/β-Fe^{3+}O(OH, Cl) nanospindles as a new steady electrode material for detecting iodic compounds. Electrochemistry Communications. 2009;**11**:603-607. DOI: 10.1016/j.elecom.2008.12.057

[48] Wysocka-Żołopa M, Winkler K. Structure, electrochemical properties and capacitance performance of polypyrrole electrodeposited onto 1-D crystals of iridium complex. Journal of Power Sources. 2015;**300**:472-482. DOI: 10.1016/j.jpowsour.2015.09.099

[49] Briseno AL, Holcombe TW, Boukai AI, Garnett EC, Shelton SW, Fréchet JJM, et al. Oligo- and polythiophene/ZnO hybrid nanowire solar cells. Nano Letters. 2010;**10**:334-340. DOI: 10.1021/nl9036752

[50] Fan B, Mei X, Sun K, Ouyang J. Conducting polymer/carbon nanotube composite as counter electrode of dye-sensitized solar cells. Applied Physics Letters. 2008;**93**:143103. DOI: 10.1063/1.2996270

[51] Carrette Friedrich KA, Stimming U. Fuel cells-fundamentals and applications. Fuel Cells. 2001;**1**:1. DOI: 10.1002/1615-6854(200105)1:1<5::AID-FUCE5>3.0.CO;2-G

[52] Reddy ALM, Rajalakshmi N, Ramaprabhu S. Cobalt-polypyrrole-multiwalled carbon nanotube catalysts for hydrogen and alcohol fuel cells. Carbon. 2008;**46**:2-11. DOI: 10.1016/j.carbon.2007.10.021

[53] Malta M, Louarn G, Errien N, Torresi RM. Nanofibers composite vanadium oxide/polyaniline: Synthesis and characterization of an electroactive anisotropic structure. Electrochemistry Communications. 2003;**5**:1011-1015. DOI: 10.1016/j.elecom.2003.09.016

[54] Mai L, Xu X, Han C, Luo Y, Xu L, Wu YA, et al. Rational synthesis of silver vanadium oxides/polyaniline triaxial nanowires with enhanced electrochemical property. Nano Letter. 2011;**11**:4992-4996

[55] Han C, Shi R, Zhou D, Li H, Xu L, Zhang T, et al. High-energy and

high-power nanaqueous lithium-ion capacitors based on polypyrrole/carbon nanotube composites as pseudocapacitive cathodes. ACS Applied Materials & Interfaces. 2019;**11**: 15646-15655. DOI: 10.1021/acsami. 9b02781

[56] Lim H, Jung JH, Park YM, Lee HN, Kim HJ. High-performance aqueous rechargeable sulfate- and sodium-ion battery based on polypyrrole-MWCNT core-shell nanowires and $Na_{0.44}MnO_2$ nanorods. Applied Surface Science. 2018; **446**:131-138. DOI: 10.1016/j.apsusc. 2018.02.021

[57] Simon P, Gogotsi Y. Materials for electrochemical capacitors. Nature Materials. 2008;7:845-854. DOI: 10.1142/ 9789814287005_0033

[58] Bo Y, Zhao Y, Cai Z, Bahi A, Liu C, Ko F. Facile synthesis of flexible electrode based on cotton/ polypyrrole/multi-walled carbon nanotube composite for supercapacitors. Cellulose. 2018;**25**:4079-4091. DOI: 10.1007/s10570-018-1845-9

[59] Liu R, Lee SB. MnO_2/poly(3,4-ethylenedioxythiophene) coaxial nanowires by one-step coelectrodeposition for electrochemical energy storage. Journal of the American Chemical Society. 2008;**130**:2942-2943. DOI: 10.1021/ja7112382

[60] Moon H, Lee H, Kwon J, Suh YD, Kim DK, Ha I, et al. Ag/Au/polypyrrole core-shell nanowire network for transparent, stretchable and flexible supercapacitor in wearable energy devices. Scientific Reports. 2017;7:41981. DOI: 10.1038/srep41981

[61] Huang H, Gan M, Ma L, Yu L, Hu H, Yang F, et al. Fabrication of polyaniline/ graphene/titania nanotube arrays nanocomposite and their application in supercapacitors. Journal of Alloys and Compounds. 2015;**630**:214-221. DOI: 10.1016/j.jallcom.2015.01.059

[62] Che B, Li H, Zhou D, Zhang Y, Zeng Z, Zhao C, et al. Porous polyaniline/carbon nanotube composite electrode for supercapacitors with outstanding rate capability and cyclic stability. Composites Part B Engineering. 2019;**165**:671-678. DOI: 10.1016/j.compositesb.2019. 02.026

[63] Zhou H, Zhai HJ. A highly flexible solid state supercapacitor based on the carbon nanotube doped graphene oxide/ polypyrrole composites with superior electrochemical performances. Organic Electronics. 2016;**37**:197-206. DOI: 10.1016/j.orgel.2016.06.036

[64] Zhou H, Zhai HJ, Zhi X. Enhanced electrochemical performances of polypyrrole/carboxyl graphene/carbon nanotubes ternary composite for supercapacitors. Electrochimica Acta. 2018;**290**:1-11. DOI: 10.1016/j.electacta. 2018.09.039

[65] Abdah MAAM, Razali NSM, Lim PT, Kulandaivalu S, Sulaiman Y. One-step potentiostatic electrodeposition of polypyrrole/graphene oxide/multi-walled carbon nanotubes ternary nanocomposite for supercapacitor. Materials Chemistry and Physics. 2018; **219**:120-128. DOI: 10.1016/j.match emphys.2018.08.018

[66] Woo HS, Czerw R, Webster S, Carroll DL, Park JW, Lee JH. Organic light emitting diodes fabricated with single wall carbon nanotubes dispersed in a hole conducting buffer: The role of carbon nanotubes in a hole conducting polymer. Synthetic Metals. 2001;**116**:369-372. DOI: 10.1016/S0379-6779(00)00439-2

[67] Jin YW, Jung JE, Park YJ, Choi JH, Jung DS, Lee HW, et al. Triode-type field

emission array using carbon nanotubes and a conducting polymer composite prepared by electrochemical polymerization. Journal of Applied Physics. 2002;**92**:1065-1068. DOI: 10.1063/1.1489067

[68] Salvatierra RV, Cava CE, Roman LS, Zarbin AJG. ITO-free and fexible organic photovoltaic device based on high transparent and conductive polyaniline/carbon nanotube thin films. Advanced Functional Materials. 2013;**23**:1490-1499. DOI: 10.1002/adfm.201201878

[69] He L, Jia Y, Meng F, Li M, Liu J. Gas sensors for ammonia detection based on polyaniline-coated multi-wall carbon nanotubes. Materials Science and Engineering B. 2009;**163**:76-81. DOI: 10.1016/j.mseb.2009.05.009

[70] Chang Q, Zhao K, Chen X, Li M, Liu J. Preparation of gold/polyaniline/multiwall carbon nanotube nanocomposites and application in ammonia gas detection. Journal of Materials Science. 2008;**43**:5861-5866. DOI: 10.1007/s10853-008-2827-3

[71] Jang J, Bae J. Carbon nanofiber/polypyrrole nanocable as toxic gas sensor. Sensors and Actuators B: Chemical. 2007;**122**:7-13. DOI: 10.1016/j.snb.2006.05.002

[72] Gao M, Dal L, Wallace GG. Glucose sensors based on glucose-oxidase-containing polypyrrole/aligned carbon nanotube coaxial nanowire electrodes. Synthetic Metals. 2003;**137**:1393-1394. DOI: 10.1016/S0379-6779(02)01156-6

[73] Lin CC, Chu YM, Chang HC. In situ encapsulation of antibody on TiO_2 nanowire immunosensor via electro-polymerization of polypyrrole propylic acid. Sensors and Actuators B: Chemical. 2013;**187**:533-539. DOI: 10.1016/j.snb.2013.03.045

[74] Dutt S, Siril PF, Sharma V, Periasamy S. Gold$_{core}$-polyaniline$_{shell}$ composite nanowires as a substrate for surface enhanced Raman scattering and catalyst for dye reduction. New Journal of Chemistry. 2015;**39**:902-908. DOI: 10.1039/C4NJ01521E

[75] Xuan S, Wang YXJ, Yu JC, Leung KCF. Preparation, characterization, and catalytic activity of core/shell Fe_3O_4@polyaniline@Au nanocomposites. Langmuir. 2009;**25**:11835-11843. DOI: 10.1021/la901462t

[76] Zhang B, Zhao B, Huang S, Zhang R, Xu P, Wang HL. One-pot interfacial synthesis of Au nanoparticles and Au-polyaniline nanocomposites for catalytic applications. CrystEngComm. 2012;**14**:1542-1544. DOI: 10.1039/C2CE06396D

[77] Yang Y, Wen J, Wei J, Xiong R, Shi J, Pan C. Polypyrrole-decorated Ag-TiO_2 nanofibers exhibiting enhanced photocatalytic activity under visible-light illumination. ACS Applied Materials & Interfaces. 2013;**5**:6201-6207. DOI: 10.1021/am401167y

[78] Jiang Y, Xu Y, Yang Q, Chen Y, Zhu S, Shen S. Power generation using polyaniline/multi-walled carbon nanotubes as alternative cathode catalyst in microbial fuel cells. International Journal of Energy Research. 2014;**38**:1416-1423. DOI: 10.1002/er.3155

Chapter 13

Synthesis, Experimental and Theoretical Investigations on the Optical and Electronic Properties of New Organic Active Layer for a New Generation of Organic Light-Emitting Diode

Mourad Chemek, Ali Mabrouk, Mourad Ben Braieck, Jany Wérry Ventirini and Alimi Kamel

Abstract

In this chapter, we present new attempts for the development of a new generation of high-performance organic light-emitting diodes (OLEDs). First of all, we present two strategies for obtaining a luminescent active layer. The first one is the chemical synthesis of a block copolymer based on the cross-linked Poly (N-vinyl carbazole) (PVK) and the conjugated poly(3-methylthiophene) (PMeT) system. Secondly, newly small luminescent organic molecules are chemically synthesized and studied. Photo-physical and electronic properties of the synthesized organic materials are fully investigated through experimental analysis and theoretical computations using essentially DFT and TDDFT methodologies. Optical measurements revealed the formation of a new highly luminescent organic material. Furthermore, the newly synthesized small molecules showed a high emission in the blue part. Based on the synthesized active layers, newly multi-structure OLED architectures are theoretically designed by the insertion of a single-walled carbon nanotube (SWCNTs) as a single layer. The theoretical computations show that the insertion of single-walled carbon nanotubes (SWCNTs) single layer improves the injection of electron charge carriers from the chosen cathode (Ca, Mg) to the synthesized active layers, which enhances the performance of the electronic focused devices based on the organic synthesized active layer.

Keywords: organic light-emitting diode (OLEDs), poly(N-vinyl carbazole) (PVK), single-walled carbon nanotube (SWCNTs), DFT methodologies, small molecules

IntechOpen

1. Introduction

The progress of the research in the field of Organic electronic devices based on organic materials has been actually a considerable challenge for scientific chemistry and physics [1–3]. The use of conjugated polymers such as Poly(p-phenylene vinylene) (PPV) derivatives was permitted to obtain flexible and transparent green OLEDs [4, 5]. Else, the extension of the effective conjugated length in poly(3-alkyl thiophene) (PAT) family, enhances the π-π stacking effects in the condensed state [6, 7], consequently, a poor red emission in the condensed state, is obtained with a weak photoluminescence quantum efficiency (PLQY), which is at around 1–2%, limiting also the use of PAT conjugated polymers for OLED applications [6, 7]. In order to enhance the photophysical properties of the investigated organic active layer of the organic light-emitting diode (OLEDs) and to improve the performance of the OLEDs displays, many efforts have recently been undertaken on the photoactive layer [8, 9]. Furthermore, experimental and theoretical efforts on the multilayer structure were employed in order to enhance the performance of the focused displays [8–10]. Recently, focused researchers show that Carbazole or bicarbazole based conjugated polymers are among superior host materials, showing promising performances in newly organic light-emitting devices [11–14]. Moreover, the use of small molecules is one of the new solutions to minimize the geometric distortion and the possible defect on the main chain and to enhance the investigated photophysical properties [8–10]. In this area of research, we have firstly, incorporated Poly(9-vinyl carbazole) (PVK) under the chemical polymerization of 3-methyl thiophene monomers, using the anhydrous Ferric chloride ($FeCl_3$) in chloroform solution [15]. The new polythiophene derivative is named PVK-MeT [15]. Photophysical investigations based essentially on spectroscopic analysis show a great enhancement in the luminescence properties, which encourage the investigation of the new material for a new generation of OLED devices [15, 16]. On the other hand, we present the chemical synthesis of a new blue-emitted organic small molecule [17]. Two newly synthesized organic materials are obtained and named respectively 1,4-Bis(a-cyano-4-methoxystyryl) benzene (DOMCN) and 1,4-Bis(a-cyano-4-bromostyryl)benzene (DBrCN) [17]. The chemical synthesis of the DOMCN and DBrCN molecules is done for possible investigation as an active layer for a new generation of a new performant organic blue light-emitting diode. The chemical synthesis and the photophysical properties were briefly described herein, and theoretical computations are undertaken to determine the electronic parameters of the synthesized chemical structure. Moreover, the synthesized organic materials presented in this chapter were theoretically used as an active layer in a newly designed multilayer structure for organic light-emitting display using essentially Density functional theory (DFT) computations on the geometric and electronic structure of the optimized chemical structure. Theoretical investigations show that The insertion of single-walled carbon nanotubes (SWNTs) on the prepared multilayer structure enhances the injection of the charge carriers (electron) and improves the physical and electronic performance of the electronic devices based on the investigated organic materials.

2. Experimental and computational details

Optical absorption measurements were carried out at room temperature (RT) using a Varian Cary 5G spectrophotometer. Photoluminescence excitation (PLE) measurements were recorded with Continuous-wave photoluminescence measurements (PL), which were collected on a Jobin-Yvon Fluorolog 3 spectrometer using a Xenon lamp (450 W). Quantum chemical computations were performed using density functional theory (DFT) and TDDFT methodologies of calculation [18], and using B3LYP functional coupling with the 6-31G (d) basis set [19, 20]. DFT and TDDFT methodologies of calculations are frequently used for a better description of the structural, photophysical and electronic properties of organic synthesized materials [21, 22]. The geometry of the first excited state was obtained by the re-optimization of the ground state optimized geometry with the configuration interaction singles restricted (RCIS) ab initio method (RCIS/6-31G(d))[23]. Used methodologies of calculation are implemented in Gaussian 09 program [24].

3. Chemical synthesis and experimental properties of the news organics active layers

3.1 Brief description of the chemical synthesis and Photophysical properties of the PVK-MeT

The chemical synthesis of the PVK-MeT starts with the cross-linking of the Poly(9-vinylcarbazole) (PVK) in presence of 3-(methyl thiophene) monomers in chloroform solution using the anhydrous ferric chloride ($FeCl_3$) as an oxidant. The material was obtained in the doped state and the neutral state was fully obtained by chemical treatment described elsewhere in our previous paper [15]. The neutral and doped states were identified essentially by analyzing the obtained samples by infrared and Raman analysis as well as optical absorption tools to prove the successful de-doping of the sample [15].

3.2 Photo-physical properties of the PVK-MeT

Optical absorption spectra of the synthesized material PVK-MeT was first investigated by optical absorption and PLE spectroscopies in chloroform diluted solution. Obtained spectra are presented in **Figure 1**. The synthesized material absorbs essentially in the UV-visible part, with an intense narrow absorption centered at around 250 nm and a shoulder absorption band centered at 410 nm. The PLE spectrum registered for the investigated material in the solution state of 550 nm maximum emission, shows a maximum at around 410 nm, coinciding with the maximum absorption assigned to π-π^* optical transition. As mentioned above, poly (3-alkyl thiophene) and in particular Poly(3-methyl thiophene) synthesized by Fecl3 routes absorbs essentially in the visible part with an intense absorption located at around 500 nm in the condensed state [25].

In order to study the excited state of the synthesized organic material, the sample was excited under various excited lines going from UV to visible part using first steady-state photoluminescence spectroscopy. No optical signal is obtained under UV excitation and a high emission is obtained under visible excitation, confirming PLE results. On the second steep, PVK-MeT and PMeT powder were then excited under 400 nm laser pulsed excitation and the PL obtained spectra are presented in **Figure 2**.

From **Figure 2**, it is clearly seen that the PL intensity is improved upon using the Poly(N_vinylcarbazole) (PVK) on the chemical synthesis of the Poly (3-methyl thiophene) (PMeT) by FeCl3 oxidative polymerization routes. In contrast with the PL

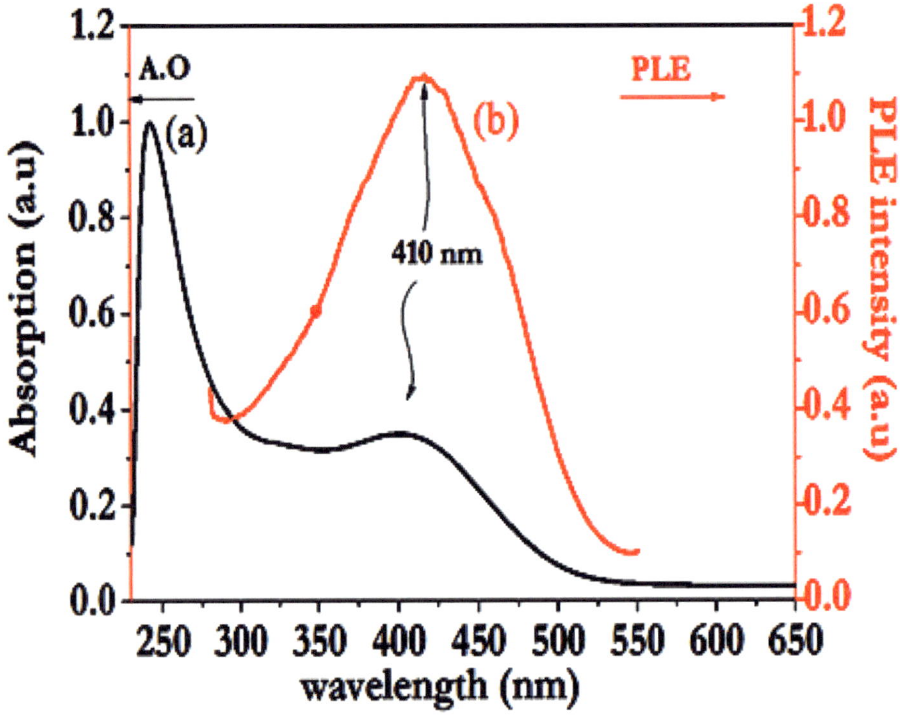

Figure 1.
Optical absorption (a) and PLE spectra (b) of the synthesized PVK-MeT.

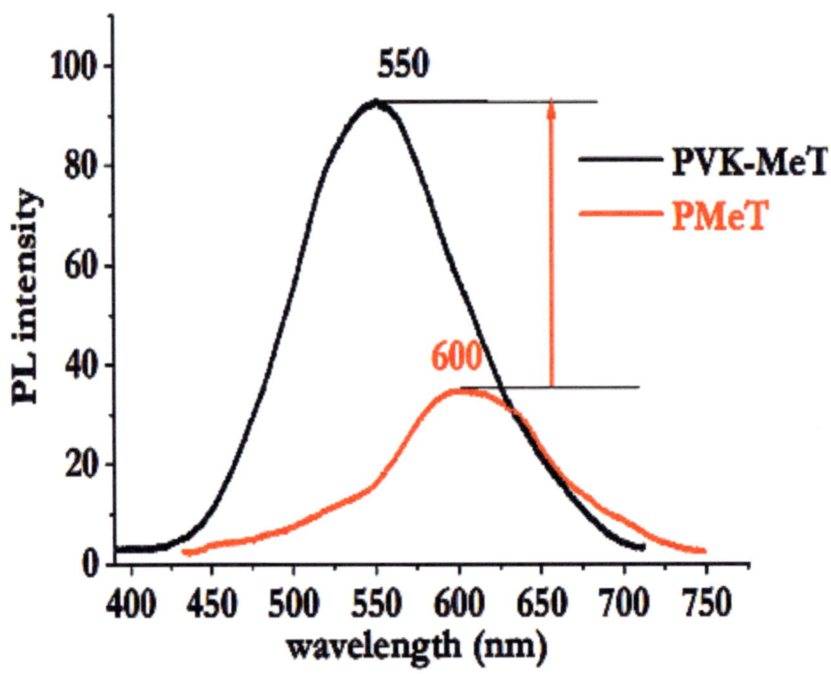

Figure 2.
Photoluminescence spectra of PVK-MeT and PMeT powder (excitation 400 nm).

spectrum of PMeT, PVK-MeT exhibits an intense emission (see **Figure 2**). Indeed, we calculate the ratio I1/I2, where, I1 and I2 present respectively the maximum intensity of the PL spectra of the PVK-MeT and that of the PMeT homopolymer. The ratio is at around 4. An emissive active layer based on thiophene derivative was successfully obtained. The improvement of the luminescence properties is accompanied by a blue shift of the maximum emission from 600 nm to 550 nm, showing a change in the spectral emission, which is shown by the CIE cartographic (See **Figure 6**). The enhancement of the luminescence properties is confirmed by the analysis of the PL decays analysis using transient photoluminescence analysis on the synthesized PVK-MeT and PMeT homopolymer powder, described in detail in our previous paper [23]. Transient photoluminescence spectroscopy confirms the change of the nature of the photogenerated species going from PMeT to PVK-MeT, which induce an increase of the PLQY from 1 to 2% for poly(3-alkylthiophene) [6, 16] to 13% for PVK-MeT in the condensed state [16].

Theoretical computations based essentially on DFT-B3LYP-6-31G(d) level of theory, was permitting to deduces first the chemical structure of the synthesized material and their electronic parameters such as HOMO, LUMO energies, and the energy gap

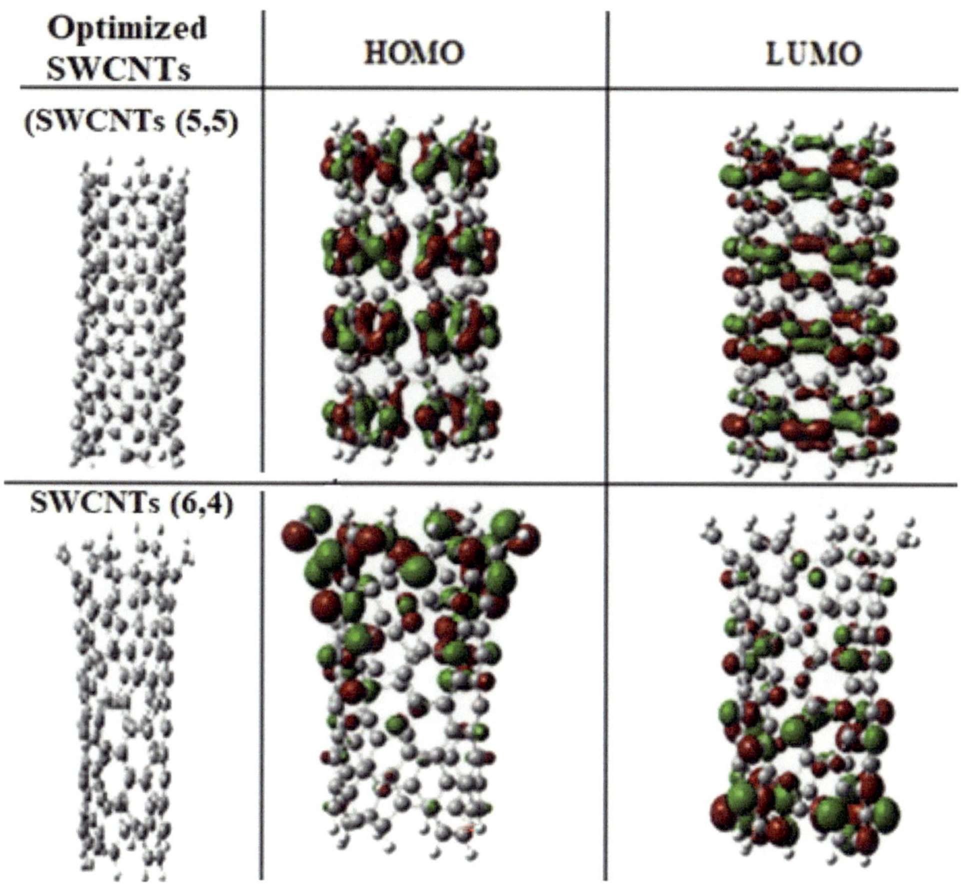

Figure 3.
Optimized chemical structure of (SWCNTs) (5,5) and (SWCNTs) (6,4) and their contour plot HOMO and LUMO.

Photoactive layer	EHOMO (eV)	ELUMO (eV)	Eg (eV)	IP (eV)	EA (eV)
DOMCN	−5.41	−2.27	3.14	5.41	2.27
DBrCN	−5.98	−2.80	3.18	5.98	2.8
Pv *PVK-MeT*	−4.8	−1.8	3	4.8	1.8

Table 1.
Electronic parameters of the investigated organic active layer.

HOMO-LUMO. The ionization potential (I.P) and electron affinity (E.A) are theoretically deduced using Koopman's theorem [26]. Obtained results are summarized in **Table 1**. Polythiophene derivatives and in particular Poly(3-methylthiophene) (PMeT) present a good absorption in the visible part and a red light emission is produced, when the PMeT homopolymer is excited by the visible line [27]. Else, in a condensed state (thin film) a low emission is produced with a low photoluminescence quantum yield (PLQY), which is at around 1–2% [6, 28]. The planarity of the chemical structure on the excited state of pol(3-alkylthioophenes) derivative, favors the formation of the non-emissive excimer species under the photo-excitation (excited state) [6, 28]. Further, the high π-conjugated system in polythiophenes favors the formation of the π-π stacking on the condensed state, which enhances the fast migration mechanism of the photo-generation species [6, 28, 29]. Consequently, non-radiative pathways are favored and a poor emission signal is obtained with fast multi-exponential decays are produced [29, 30].

In order to study the structure-properties relationship of the investigated materials, first, a simple optimization on the ground state of an oligomer of 6-thiophenes coupled at the 2–5 positions and of the deduced chemical structure of the PVK-MeT. Secondly, the fully optimized chemical structures on the first excited state are

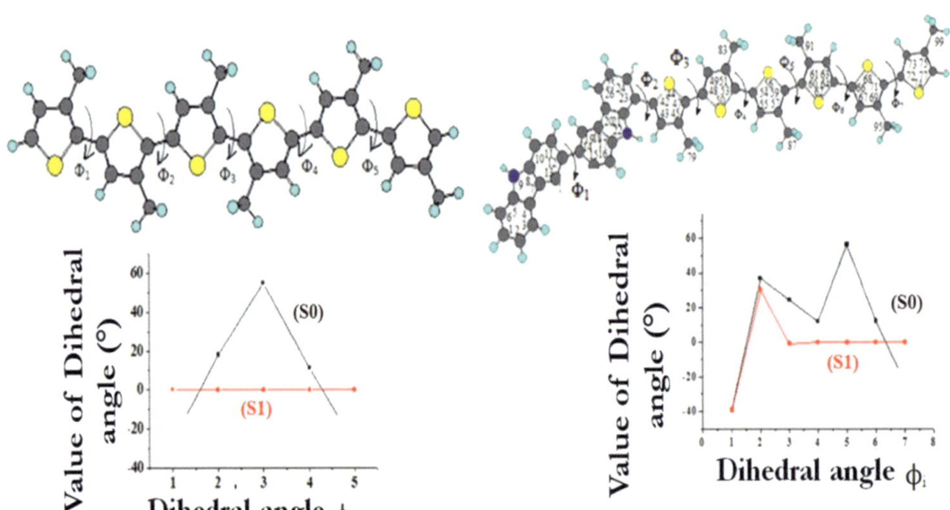

Figure 4.
Geometric parameters (dihedral angle ϕi) of the optimized chemical structure of the six-coupled non-regioregular 3-(methyl thiophene) and the PVK-MeT chemical structure on the ground (black line) (So) and the first calculated excited state (S1) (red line).

DOI: http://dx.doi.org/10.5772/intechopen.103807

successfully obtained by re-optimization of the optimized ground-state chemical structure with the RSCI/6-31G(d) level of theory. The optimized chemical structures on the ground and the first excited state are drawn with Gauss view software and presented in **Figure 4**.

Geometric parameters such as dihedral angle ϕi of 6-Oligothiophene optimized in the ground and the first excited-state state are presented in **Figure 4**. It is clearly seen from **Figure 4**, that the conformation of the non-regioregular coupled 3-methyl thiophene presents a non-planar conformation, and the distortion is higher on the TT coupling (ϕ3) in the ground state. Else, the first excited state of the non-regioregular 6-oligothiophene, obtained by the re-optimization of the optimized chemical structure by CIS/STO-3G present no distortion and a fully coplanar structure. Focused studies reveal that the luminescence in the conjugated system is only obtained when the main chain is deformed especially in the excited state [26]. Indeed, Bai et al. [6], suggest that upon excitation of the conjugated system in poly (3-alkyl thiophene), the planarity increases, which favors the formation of the nonemissive excimer species in the condensed state [6]. Consequently, a poor luminescence with low PLQY is obtained for polythiophenes derivatives in the condensed state [6]. Herein, the insertion of the cross-linked carbazole system (bicarbazole) grafted to the 6-oligothiophene system induces the increase of the distortion and the high deformation of the main chain in both grounds and excited state (**Figure 3**). These behaviors induced the decrease of the non-radiative photo-generated species and the improvement of the luminescence properties going from PMeT to PVK-MeT.

3.3 Chemical elaboration of new small molecules for OLEDs application

3.3.1 Chemical synthesis of news small molecule: DOMCN and DBrCN

To an equimolar stirred mixture of 1,4- phenylenediacetonitrile (1) (1 mmol) and aldehyde (2) (2 mmol) in 10 mL of tert-butyl alcohol (10 mL) were added a solution of tetrabutylammonium hydroxide (TBAH) (1 M solution in methanol). After 24 h of stirring at 50°C, the reaction mixture was cooled to room temperature and quenched with distilled water. Then, the obtained solution was extracted with chloroform. The organic layer was washed several times with distilled water and dried over anhydrous magnesium sulfate. The purification was carried out by precipitation in methanol. A yellow powder was filtered and dried in a vacuum for 24 h (**Figure 5**).

^{1}H NMR (CDCl 3) δ [ppm]: 7.89 (d, 4H, Ar-H), 7.71 (s, 4H, Ar-H), 7.49 (s, 2H, Vinyl-H), 6.96 (d, 4H, Ar-H), 4.04 (t, 6H, -OCH3). ^{13}C NMR (CDCl3) δ [ppm]: 160.5, 142.4, 135.1, 131.5, 126.2, 126.2, 118.5, 115.1, 107.5, 68.1.

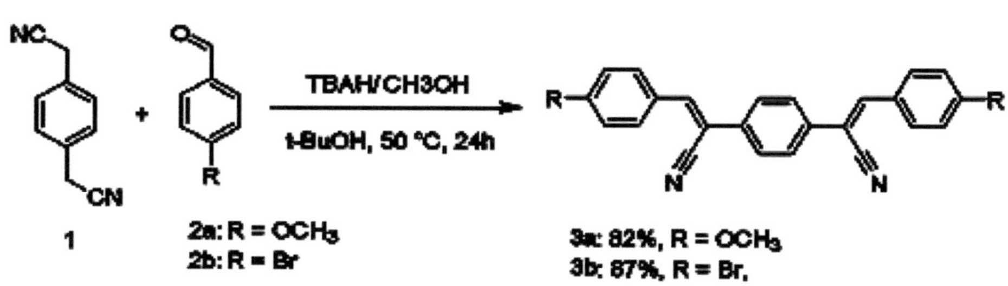

Figure 5.
The synthesis adopted procedure of the DOMCN and DBrCN.

3.3.2 Photo-physical properties of the synthesized DBrcN and DOMCN

Optical properties of the synthesized oligomer were investigated essentially with steady-state optical absorption and photoluminescence spectroscopies in a dilute solution state. Optical absorption and photoluminescence spectra are presented in **Figure 6**. Both synthesized molecules absorb in the UV and in the near-visible part. A wide absorption is detected at 250 nm and a large band absorption centered at around 360 nm and 380 nm respectively for DBrCN and DOMCN synthesized molecules, and are assigned to π-π* optical transition. From optical absorption spectra, optical band gap Eg is estimated from the onset of absorption, which is estimated to be 3 eV and 2.8 eV respectively for DBrCN and DOMCN molecules.

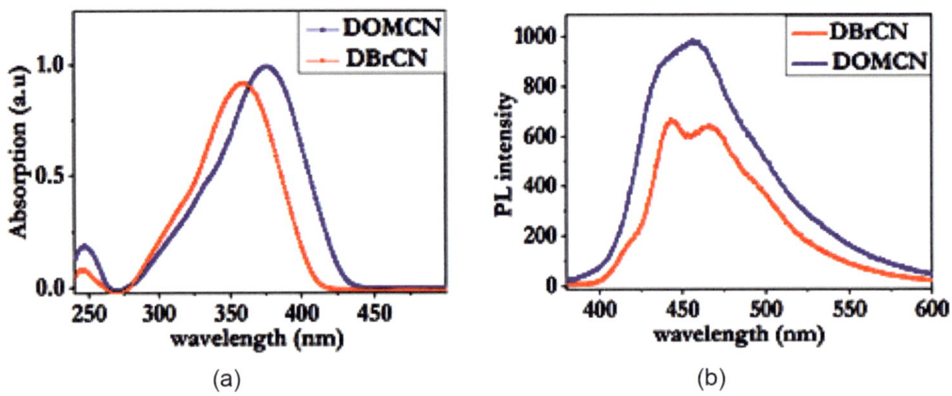

Figure 6.
(a) Optical absorption and (b) photoluminescence spectra (excitation 370 nm) of the DOMCN and DBrCN synthesized molecules recorded in chloroform solution.

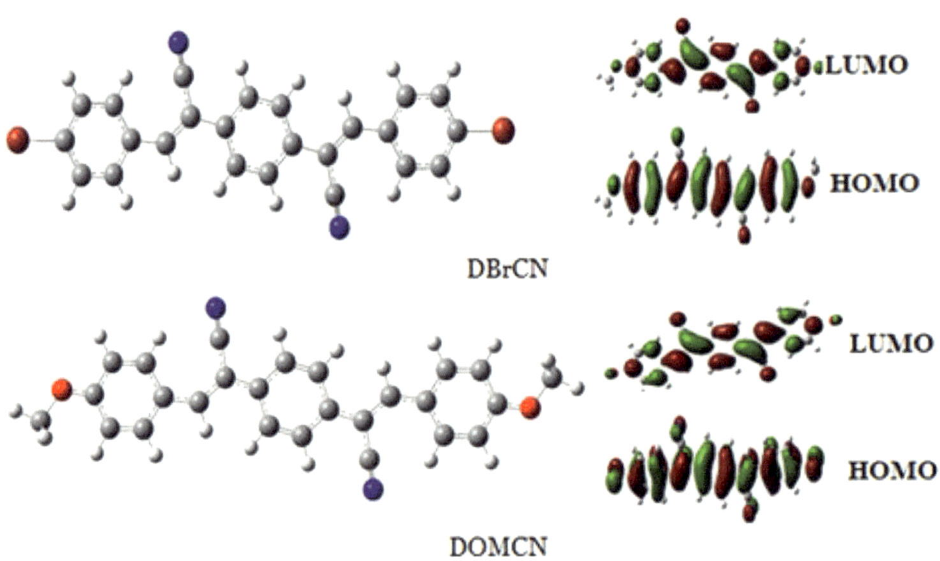

Figure 7.
Optimized chemical structure and their contour plots orbitals HOMO and LUMO of DOMCN and DBrCN.

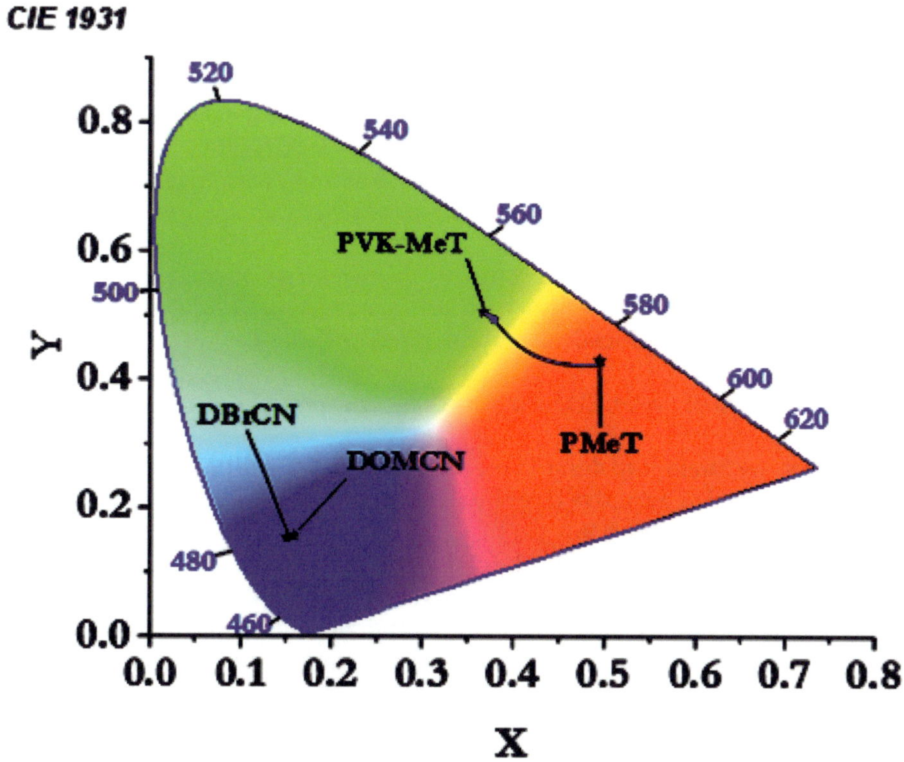

Figure 8.
C.I.E cartographic of the investigated organic material.

Under 370 nm line excitation a narrow blue emission is observed for the two materials. A change in the shape of the PL spectrum of the molecule is observed going from DBrCN and DOMCN. A vibronic spectrum with two maxima at 443 and 460 nm for DBrCN is obtained. Theoretical computations using the DFT-B3Lyp-6-31G(d) level of theory were undertaken in order to estimate the electronic parameter. **Figure 7** presents the optimized chemical structure of the investigated molecule at the ground state and the contour plot HOMO and LUMO obtained with 0.02 isovalue using GAUSSVIEW software [27].

Electronic parameters deduced from the ground state optimized chemical structure of the two compounds and the synthesized PVK-MeT are summarized in **Table 1**.

From CIE cartography (**Figure 8**), it is clearly seen the effect of the use of Poly(N_vinylcarbazole) (PVK) on the chemical polymerization of 3-methylthiophene monomers. From PMeT to PVK-MeT a decrease of x value from 0.5 for single PMET to 0.38 for PVK-MET resulted in a green-yellow color emission. Meanwhile, the synthesized small molecule DOMCN and DBrCN present a blue emission, which permits the use as an emissive layer for the blue light-emitting diode.

4. Theoretical design of new multilayer architecture based on the synthesized materials

Recently, Organic light-emitting diodes (OLEDs) have been considerably developed and considerable progress in the efficiency and durability is obtained [28, 29]. In order to improve the injection and transport of the charge carriers from the anode or cathode, special multi-layer architectures are needed [30]. In this way, we are going in this part to the theoretical design of a new multilayer architecture using the investigated materials as an active layer. The first step of the calculations consists of the optimization of the geometries of the two single-walled carbon nanotube (SWCNTs) (5,5) and (6,4) having a diameter ϕ = 0.71 nm and length L = 100 nm (150 atoms), with the DFT-B3LYP-6-31G(d) level of theory. The two optimized chemical structures are presented in **Figure 3**.

From the optimized chemical structure of the two forms of single-walled carbon nanotubes, the electronic parameters such as E_{HOMO}, E_{LUMO}, and the energy gap Eg = E_{HOMO}-E_{LUMO} were deduced and summarized in **Table 2**.

4.1 Energy diagram of an OLED based on these compounds

In this part, we attempt to design new organic architecture using the luminescent investigated material as an active layer by the insertion of single-walled carbon nanotubes as shown in **Figure 9**.

First, the anode and cathode must be chosen to facilitate the injection of the free charge carrier from anode or cathode to the active layer (luminescent organic material). **Table 3** presents the traditionally used material and their electronic characteristics parameters (anode output work (Φ_A) and the output work of the cathode (Φ_C).

By comparing electronic parameters of the investigated material and that of The traditional anode and cathode materials, cathode (Ca, Mg) can better helpfully inject the charge carriers (electron) from the cathode to the synthesized materials. Using

	E_{HOMO} (eV)	E_{LUMO} (eV)	I.P (eV)	E.A (eV)	E_g = E_{HOMO}-E_{LUMO} (eV)
SWCNTs (5,5)	–4.30	–2.93	4.3	2.93	1.37
SWCNTs (6,4)	–3.94	–3.54	4.3	3.54	0.40

Table 2.
Electronic parameters of the optimized chemical structure of SWCNTs (5,5) and SWCNTs (6,4).

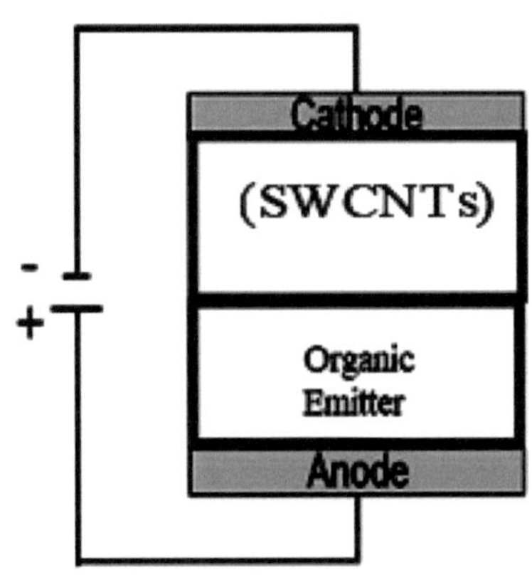

Figure 9.
Proposed multilayer structure using the investigated organic material as an emitting layer.

Anode	Cathode
ITO; Φ_A = 4.7 eV	Ca, Mg: Φ_C = 2.90 eV
ITO:PEDOT-PSS, Φ_A = 5.20 eV	Al, Ag; Φ_C = 4.30 eV
	Au; Φ_C = 5.00 eV

Table 3.
Value of the work function of the possible used cathode and anode.

Anode		Cathode	Active layer	ΔEe (eV)	Active layer- SWCNTs	ΔEe (eV)
ITO Φ_A = 4.7 eV	ITO: PEDOT-PSS Φ_A = 5.20 eV	(Ca, Mg) Φ_C = 2.90 eV	PVK-MeT	1.1	PVK-MeT/ SWCNTs (5,5)	0.03
					PVK-MeT/ SWCNTs (6,4)	0.64
			DMOCN	0.63	DOMCN / SWCNTs (5,5)	0.03
					DOMCN/ SWCNTs (6,4)	0.64
			DBrCN	0.1	DBrCN/ SWCNTs (5,5)	0.03
					DBrCN/ SWCNTs (6,4)	0.64

Table 4.
Electronic parameters of the possible used anode and cathode and the active layer with and without SWCNTs.

(Ca, Mg) as a chosen cathode in the designed multi-structure layer, and the synthesized materials (PVK-MeT, DMOCN and DBrCN) as an active layer, we calculate the electron injection barrier (ΔE_e), which is defined as the energy difference between the energy value of the LUMO of the used photoactive layer and the output work of the cathode (Φ_C). The obtained results are summarized in **Table 4**. The insertion of the single-walled carbon nanotube (SWCNTs) between the cathode and the photoactive layer

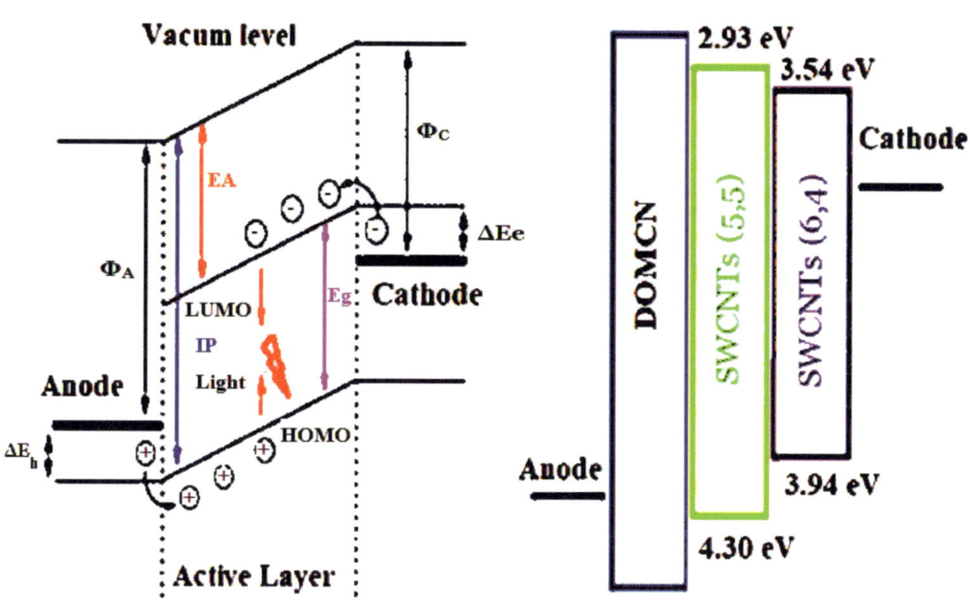

Figure 10.
Energetic diagram of the designed OLEDs based on the investigated organic material.

greatly reduces the value of ΔE_e. The reduction of ΔE_e is more important with SWCNTs (5,5) than when using the SWCNTs (6,4) reduction (see **Figure 10**).

Consequently, SWCNTs (5,5) as a layer between the (Ca, Mg) and the investigated photoactive multilayer can helpfully improve the electron injection from the cathode to the active material which improves the performance of The OLEDs based on the designed multi-structure.

5. Conclusions

In summary, we have attempted in the present chapter to enhance the optical emission and electronic properties of the chosen organic emitting layer for a new generation of the organic light-emitting diode (OLED). The first step is to use the non-conjugated Poly(Vinylcarbazole) (PVK) in the chemical synthesis of the conjugated polymers (case Poly(3-methyl thiophene)). The use of PVK under the chemical synthesis of the poor luminescent Poly(3-alkylthiophene) family permits first the partially crosslinking of the carbazole units, which enhances the PL intensity and improves their luminescence properties. Elsewhere, the use of a new family of small molecules permits also to obtain a blue luminescent organic material with reduced non-radiative recombination of hole and electron. The chemical structures of the synthesized materials were used as an active layer in newly simulated light-emitting devices. DFT simulations on the investigated materials and in the simulated devices show that the introduction of the optimized layer of single-walled carbon nanotubes (SWCNTs) single layer permits to the enhancement of the electron injection and improves the performance of the focused emitting devices.

Acknowledgements

This study was funded by the Ministry of Higher Education and Scientific Research-Tunisia: Synthèse asymétrique et ingénierie moléculaire de matériaux organiques pour l'électronique organique (LR18ES19). Authors are grateful to Dr. Florian Massuyeau and to Eric Faulques for their help on the optical measurements done at IMN Nantes (France).

Author details

Mourad Chemek[1*], Ali Mabrouk[1], Mourad Ben Braieck[1], Jany Wérry Ventirini[2] and Alimi Kamel[1]

1 Faculté des Sciences de Monastir, Synthèse asymétrique et ingénierie moléculaire de matériaux organiques pour l'électronique organique (LR18ES19), Monastir, Tunisie

2 Institut des Matériaux Jean Rouxel, Nantes cedex 3, France

*Address all correspondence to: chmek_mourad@yahoo.fr

References

[1] AlSalhi MS, Alam J, Dass LA, Raja M. Recent advances in conjugated polymers for light emitting devices. International Journal of Molecular Sciences. 2011; **12**(3):2036-2054. DOI: 10.3390/ijms12032036

[2] Grimsdale AC, Leok Chan K, Martin RE, Jokisz PG, Holmes AB. Synthesis of light-emitting conjugated polymers for applications in electroluminescent devices. Chemical Reviews. 2009;**109**(3):897-1091. DOI: 10.1021/cr000013v

[3] Zhang C, Braun D, Heeger AJ. Light-emitting diodes from partially conjugated poly(*p*-phenylene vinylene). Journal of Applied Physics. 1993;**73**(10):5177. DOI: 10.1063/1.353794

[4] Burn PL, Homes AB, Kraft A, Bradley DDC, Brown AR, Friend RH. Synthesis of a segmented conjugated polymer chain giving a blue-shifted electroluminescence and improved efficiency. Journal of Chemical Society, Chemical Communications. 1991:32-34. DOI: 10.1039/C39920000032

[5] Grem G, Leditzky G, Ullrich B, Leising G. Realization of a blue-light-emitting device using poly(p-phenylene) Advanced Materials. 1992;**4**:36. DOI: 10.1002/adma.19920040107

[6] Xu B, Holdcroft S. Molecular control of luminescence from poly (3-hexylthiophenes). Macromolecules. 1993;**26**:4457-4460. DOI: 10.1021/ma00069a009

[7] Ng SC, Xu JM, Chan HSO. Electrically conductive and fluorescent poly[1,4-bis(3-alkyl-2- thienyl) phenylenes]: Syntheses and preliminary characterization aspects. Synthetic Metals. 1998;**92**:33-37. DOI: 10.1016/S0379-6779(98)80019-2

[8] Chemek M, Chebil S, Ben Amor S, Khlaifa D, Faulques E, Haj Said A, et al. Chemical insertion of anthracene moiety into the backbone of a newly synthesized oligophenylene (OMPA): effect on the photo-physical properties. Research on Chemical Intermediates. 2021;**47**:3437-3451. DOI: 10.1007/s11164-021-04481-z

[9] Bourass M, Kohima N, Keltoum Kabbaj O, Wazzan N, Chemek M, Bouachrine M. The photophysical properties and electronic structures of bis[1]benzothieno[6,7-*d*:6′,7′-*d*′]benzo[1,2-*b*:4,5-*b*′]dithiophene (BBTBDT) derivatives as hole-transporting materials for organic light-emitting diodes (OLEDs). New Journal of Chemistry. 2019;**43**: 15899-15909. DOI: 10.1039/C9NJ02756D

[10] Wazzan N, Irfan A. Exploring the optoelectronic and charge transport properties of Pechmann dyes as efficient OLED materials. Journal of Optik. 2019;**197**:163200. DOI: 10.1016/j.ijleo.2019.163200

[11] Bezvikonnyi O, Grybauskaite-Kaminskiene G, Volyniuk D, Simokaitiene J, Danyliv Y, Durgaryan R, et al. 3, 3′-Bicarbazole-based compounds as bipolar hosts for green and red phosphorescent organic light-emitting devices. Materials Science and Engineering: B. 2020;**261**:114662. DOI: 10.1016/j.mseb.2020.114662

[12] Liu X-Y, Zhang Y-L, Fei X, Liao L-S, Fan J. 9, 9′-Bicarbazole: New molecular skeleton for organic light-emitting diodes. Chemistry A European Journal. 2019:25/4501-25/4508. DOI: 10.1002/chem.201806314

[13] Rui B, Tan Y, Liu X, Wei C, Liu Y, Li J, et al. Synthesis, characterization of bicarbazole compounds and their application in codepositing luminescent copper iodide complexes. ChemistrySelect. 2017;**2**:6556-6562. DOI: 10.1002/slct.201700742

[14] Zhou R, Jin M, Malval J, Pan H, Wan D. Bicarbazole-based oxalates as photoinitiating systems for photopolymerization under UV–Vis LEDs. Journal of Polymer Science. 2020;**60**:1-13. DOI: 10.1002/pol.20199298

[15] Chemek M, Wéry J, Bouachrine M, Paris M, Lefrant S, Alimi K. Synthesis and characterization of novel graft copolymers of Poly(N-vinylcarbazole) and Poly(3-methylthiophene) for optoelectronic applications. Synthetic Metal. 2010;**160**:2306-2314. DOI: 10.1016/j.synthmet.2010.09.001

[16] Chemek M, Massuyeau F, Wéry J, Hlel A, Ayachi S, Faulques E, et al. Photoluminescence properties of new poly(N-vinylcarbazole)-3-methylthiophene (PVK-3MeT) graft copolymer. Journal of Applied Polymer Science. 2012;**125**:126-132. DOI: 10.1002/app.35074

[17] Bouazizi H, Mabrouk A, Braiek MB, Mestiri T, Alimi K. New conjugated organic matrix-carbon nanotube functionalization: DFT modeling and spectroscopic analysis. Journal of Physics and Chemistry of Solids. 2020;**136**:109131. DOI: 10.1016/j.jpcs.2019.109131

[18] Armelin E, Bertran O, Estrany F, Salvatella R, Alemán C. Characterization and properties of a polythiophene with a malonic acid dimethyl ester side group. European Polymer Journal. 2009;**45**:2211. DOI: 10.1016/j.eurpolymj.2009.04.010

[19] Lee C, Yang W, Parr RG. Journal of Physical Review B. 1998;**37**:785

[20] Becke AD. The Journal of Chemical Physics. 1996;**104**:1040

[21] HamrounI K, Hafedh N, Chmeck M, Aloui F. Photooxidation pathway providing 15-bromo-7-cyano [6] helicene. Chiroptical and photophysical properties and theoretical investigation. Journal of Molecular Structure. 2020;**1217**:128399. DOI: 10.1016/j.molstruc.2020.128399

[22] Samaher C, Mourad C, Alimi K. Theoretical investigations about the effect of electron-withdrawing groups on proprieties of A-π-D-π-a type small molecules donor for organic solar cells. Journal of Molecular Modeling. 2021;**27**:54. DOI: 10.1007/s00894-020-04654-4

[23] Belletête M, Bedard M, Leclerc M, Durocher G. Journal of Molecular Structure (THEOCHEM). 2004;**9**:679

[24] Frisch MJ, Trucks GW, Schlegel HB, Scuseria GE, Robb MA, Cheeseman JR, et al. Gaussian09 Revision D.01. Wallingford: Gaussian Inc; 2009

[25] Zhao L, Chen Q, Li C, Shi G. Electrochemical fabrication of p-poly (3-methylthiophene)/n-silicon solar cells. Solar Energy Materials & Solar Cells. 2007;**91**:1811-1815

[26] Chemek M, Ayachi S, Hlel A, Wéry J, Lefrant S, Alimi K. A theoretical investigation on the properties of the new poly(N-vinylcarbazole)-3-methylthiophene (PVK-3MeT) synthesized graft copolymer. Journal of Applied Polymer Science. 2011;**122**:2391-2402

[27] Wang L, Xian W, Wang X, Feng Q, Pei M, Zhang G. Synthesis and characterization of three novel conjugated polythiophene derivatives. Designed Monomers and Polymers. 2013;**16**:339-348. DOI: 10.1080/15685551.2012.747151

[28] Massuyeau F, Faulques E, Lefrant S, Majdoub M, Ghedira M, Alimi K, et al. Photoluminescence properties of new PPV derivatives. Journal of Luminescence. 2011;**131**: 1541-1544

[29] Massuyeau F, Duvail JL, Athalin H, Lorcy JM, Lefrant S, Wéry J, et al. Elaboration of conjugated polymer nanowires and nanotubes for tunable photoluminescence properties. Nanotechnology. 2009;**20**(15):155701. DOI: 10.1088/0957-4484/20/15/155701

[30] Monkman AP, Bloor D, Stevens GC. Electronic structure and charge transport mechanisms in polyaniline. Synthetic Metals. 1987;**21**:71. DOI: 10.1016/0379-6779(89)90307-X

Chapter 14

Design and Fabrication of Microencapsulated Phase Change Materials for Energy/Thermal Energy Storage and Other Versatile Applications

Tahira Mahmood, Rahmat Ali, Abdul Naeem and Murtaza Syed

Abstract

Microencapsulated phase change materials have been considered as potential candidates to overcome the global energy shortage, as these materials can provide a viable method for storing thermal energy and offering consistent energy management by controllable heat release in desirable environments. Microencapsulation technology offers a method for overcoming the trouble associated with the handling of solid–liquid phase change materials (PCMs) via encapsulating PCMs with thin or tiny shells which are known as 'microcapsules'. Microcapsule shells not only keep PCMs isolated from the surrounding materials but also provide a stable structure and sufficient surface for PCMs to enhance heat transfer. Thus microencapsulation technology received remarkable attention from fundamental studies to industrial growth in recent years. In order to provide a reliable source of information on recent progress and development in microencapsulated PCMs, this chapter emphases on methods and techniques for the encapsulation of PCMs with a diversity of shell materials from traditional organic polymers to novel inorganic materials to pursue high encapsulation efficiency, excellent thermal energy-storage performance and long-term operation durability. The chapter also highlights the design of bi- and multi-functional PCM-based microcapsules by fabricating various functional shells in a multilayered structure to meet the growing demand for versatile applications.

Keywords: microencapsulation, phase change materials, designing, multifunctional microcapsules, thermal energy storage, versatile applications

1. Introduction

The rapid global economic growth and population explosion resulted in increased consumption of nonrenewable energy resources such as coal, petroleum and natural gas which not only reduces these fossil fuels sources but also leads to major global environmental issues like CO_2 emission, and global warming and air pollution. If the world energy requirements totally depend on fossil fuel which is continuously exhausting will results in energy crisis in the near future. To minimize the reliance on fossil fuels for energy production, the development of renewable energy resources and the enrichment of energy efficiency have been deliberated as the alternative strategy that could be adopted [1, 2]. The scientific development of thermal energy storage by utilizing phase change materials (PCMs) to store latent heat has been considered as a worthy solution for reducing the worldwide energy scarcity as these materials provide viable ways of keeping thermal energy and offering reliable energy management by controllable heat release in suitable environments [3]. PCMs are a class of heat storage materials, able to absorb and release sufficient amounts of latent-heat energy at a constant temperature when a state change occurs from a solid form to the liquid one and vice-versa. In addition to higher thermal energy-storage density compared to conventional heat-storage materials, PCMs can bridge the gap between energy availability and energy use to reduce energy waste [4].

The application of PCMs as a means of thermal-energy storage has been practiced since 1970s, and PCMs have been developed and designed to fulfill the desired requirements. Nowadays, PCMs have not been only applied in renewable energy effective utilization such as solar thermal energy and low-temperature waste heat utilization but also used for thermal regulation and thermal management in the fields of photovoltaic-thermoelectric systems, temperature-sensitive electronic parts or devices requiring cool or thermal protection, biological products or pharmaceutical needing cool storage, smart fibers and textiles with a thermoregulatory function, telecom shelters in tropical regions, thermal buffering of Li-ion batteries, energy-saving buildings, thermal comfort in vehicles, etc. [5].

Though PCMs due to their desirable properties is widely used in both domestic and industrial areas in recent years, their phase transition brings some difficulties during their application for thermal energy storage and management. After fusion PCMs are converted to low viscous liquids which can then easily diffuse or flow over other materials and thus cause difficulty in handling the process in the liquid state [6]. Other problems associated with the commonly used PCMs include the need of using special latent heat devices, the hysteresis of thermal response due to low thermal conductivity and supercooling, the poor heat transfer during the charging and recovery processes, absorb moisture from the atmosphere or lose water through evaporation, the leakage and loss of PCMs, etc. [7]. Due to these problems, pristine PCMs are generally not recommended for thermal energy storage applications. To avoid the problem, microencapsulation technology was introduced which involves the packing/encapsulation of PCMs into tiny closed ampules that not only protect the liquid PCMs from the interference and interaction of the surrounding materials but also give them a stable form in the liquid state. The product obtained as a result of this packing technology was named microcapsule. The microcapsules which pack the PCM core individually with a firm shell can, therefore, handle even liquids as a solid material [8]. Additionally, the development of a microcapsule shell provides a large heat transfer surface to the encapsulated PCMs and hence considerably increases the heat transfer and thermal response [9]. Thus, microencapsulation of PCMs has been accepted as a more consistent technology for liquid PCMs compare to form stable composite PCMs. Microencapsulation technology of solid–liquid PCMs has received great attention for over 20 years, and several studies can be found in the literature on this topic [10]. Usually, the microencapsulated PCMs could be prepared by making a polymeric shell via coacervation, in-situ polymerization, interfacial polymerization and suspension polymerization techniques, for which the commonly used shell materials include polyureas, poly (methyl methacrylate), melamine-formaldehyde resins, polystyrene, urea-formaldehyde resins, and bio-based polymers such as Arabic gum, agar and gelatin. Moreover, many inorganic materials such as titanium dioxide (TiO_2), silica (SiO_2), calcium carbonate and aluminum oxide have been reported in the recent literature that could also be used as shell materials for encapsulating PCMs [11–13]. These inorganic shells have shown much higher mechanical strength and rigidity than the polymeric ones and can form a much more secure barrier around PCMs to protect them from damaging interaction with the environment.

Currently, the researchers are interested in the design and development of multifunctional microencapsulated PCMs. One of the potential approaches to achieve the additional functionality involve the use of inorganic functional shell assembly on the microencapsulated PCM core. In this way, not only the additional functions for microencapsulated PCMs along with the wall materials is achieved but also allows the establishment of signal or multilayered shells with various designed functions. Pointing at the high-tech designs and versatile applications of microencapsulated PCMs for thermal energy storage and thermal management, this chapter provides a reliable source of information on recent progress and development in microencapsulation technology for solid–liquid PCMs and especially introduces the diverse designs for PCMs-based microcapsules with various special functions. Moreover, a thorough analysis of the trend in the development and applications of microencapsulated PCMs is also presented. The chapter also highlights the design of bi- and multi-functional PCM-based microcapsules by fabricating various functional shells in a multilayered structure to offer a great potential to meet the growing demand for versatile applications.

2. Phase change materials (PCMs)

PCMs due to their higher latent heat values can store and release a large amount of heat energy during melting and solidifying processes [14]. These materials have been thought to act as a storage medium with numerous applications such as cooling of food products, buildings, textiles, solar systems, spacecraft thermal systems and waste heat recovery systems [15]. On the basis of phase conversion PCMs are categorized into solid–liquid, solid–solid, solid–gas and liquid–gas [1]. Among these categories, solid–liquid PCMs due to their high density, favorable phase equilibrium, minor volume changes and low vapor pressure at the operating temperature during phase transition are more suitable for thermal energy storage systems. Furthermore, solid–liquid PCMs show little or no subcooling during freezing, melting/freezing at the same temperature and phase separation, and sufficient crystallization rate.

PCMs possess high chemical stability, nontoxic, nonexplosive and noncorrosive nature, do not undergo degradation after long-term thermal cycles, and have good chemical properties capable of completing reversible freezing/melting cycle [6]. Solid–liquid PCMs can be divided into three major types: (a) organic PCMs (b) inorganic PCMs and (c) eutectic PCMs [16]. Organic PCMs include paraffin and nonparaffin (alcohols, fatty acids and glycols) materials [2]. Inorganic PCMs generally include salt hydrates, metallic compounds and metal alloys with the advantages of a broader range of transition temperature, high thermal conductivity and high latent heat storage capacity, low cost and nonflammable nature. In contrast, lack of thermal stability, phase segregation, supercooling, corrosion and decomposition, are problems that dominate their benefits [17]. Eutectic PCMs are the combination of two or more low melting components, each of which freezes and melts congruently to make a mixture of the components' crystals upon crystallization [6]. Eutectic PCMs can be prepared for a specific application by mixing organic–organic, inorganic–inorganic, or a combination of the two PCMs at a given ratio. These PCMs have high thermal conductivity and density without segregation and supercooling, while their specific heat capacity and latent heat are much lower than those of paraffin/salt hydrates [18].

3. Microencapsulation shell materials

In recent years, microencapsulation of PCMs has been widely used to avoid the leakage and reaction of the PCMs with the surrounding environment during the solid–liquid phase transition.

Microencapsulation of PCMs can also be responsible for relatively constant volume, high thermal cycling stability and large heat transfer area for PCM-based thermal storage [19]. Shell/wall materials play a vital role in controlling various physical properties like morphology, mechanical and thermal properties of the produced microcapsules [7]. On the basis chemical nature shell material can be divided into three groups: (a) organic, (b) inorganic and (c) organic–inorganic hybrid materials [20].

3.1 Organic shells

Organic shell materials include synthetic and natural polymeric materials, which have excellent structural flexibility, good sealing properties and high resistance to the volume change associated with repeated phase transformations of PCMs [21]. The organic shell materials most frequently used consist of urea-formaldehyde (UF) resin [22], melamine-formaldehyde resin [23], and acrylic resin [24]. Many workers around the world used MF resin as the wall material due to its good chemical compatibility, low cost and thermal stability [25]. Mohaddes et al. effectively utilized MF as the shell material for encapsulation of *n*-eicosane for application to textiles [26]. Fabrics doped with this type of microcapsules have higher thermoregulation capacity and low thermal delay efficiency. Among the group of acrylic resins, the copolymers of methacrylate have significant thermal stability, chemical resistance, nontoxic nature and easy preparation. Alkan et al. has shown that *n*-eicosane microencapsulation with polymethylmethacrylate (PMMA) shell had good thermal stability [27]. Ma et al. successfully encapsulated binary core materials, butyl stearate and paraffin using poly(methylmethacrylate-*co*-divinylbenzene) (P(MMA-*co*-DVB)) copolymer as the shell material [28]. The microcapsules so obtained possess a uniform size of 5–10 μm with a uniform spherical shape and dense surface. Moreover, the phase transition temperature of these microcapsules can be adjusted by adjusting the butyl stearate to paraffin ratio. Wang et al. studied the effect of GO on the thermal properties of capric acid@UF microcapsules by adding various contents of graphene oxide (GO) [29]. It was found that the microcapsules with 0.6% GO had the highest enthalpy of 109.60 J/g and encapsulation ratio of 60.7%. The microcapsules with GO presented smoother surfaces and good thermal conductivity.

3.2 Inorganic shells

Microcapsules prepared by using organic polymeric shell materials are usually not suitable for application in some situations due to the low thermal conductivity, flammable nature and poor mechanical strength of the organic shell materials [30]. In recent years, inorganic shells due to their good thermal conductivity, high rigidity and high mechanical strength, have been progressively employed as an alternative shell material for microcapsule preparation [21]. The commonly used inorganic shell materials include Silica (SiO_2) [31], zinc oxide (ZnO) [32], titanium dioxide (TiO_2) [33] and calcium carbonate ($CaCO_3$) [34].

Silica because of its fire resistance nature, high thermal conductivity and ease of preparation are one the most commonly used shell materials for encapsulation of fatty acids [35], paraffin waxes [34] and inorganic hydrated salts [36]. Liang et al. prepared nanocapsules by encapsulating *n*-octadecane core material using silica as the shell material via interfacial hydrolysis and polycondensation of tetraethoxysilane (TEOS) in miniemulsion [37]. The thermal conductivity of nanocapsules so obtained was observed to be higher than 0.4 $Wm^{-1} K^{-1}$ with melting enthalpy and encapsulation ratio of 109.5 J/g and 51.5%, respectively. The enthalpy of the nanocapsules was not altered and no leakage was observed after 500 thermal cycles. However, the hydrolysis and polycondensation of TEOS, used as a silica precursor, could cause a reduction in the compactness of the silica shell and have a relatively weak mechanical strength. $CaCO_3$ shells have higher rigidity and better compactness compared to silica. Yu et al. employed $CaCO_3$ as shells material for encapsulating *n*-octadecane through the self-assembly method [12]. The microcapsules obtained were of spherical morphology with a uniform diameter (5 μm) and had good thermal stability, thermal conductivity, anti-osmosis properties and serving durability.

Metal oxides, like ZnO and TiO_2, owing to their multifunctional properties including photochemical, catalytic and antibacterial characteristics are frequently used as shell materials to obtain PCM microcapsules with some remarkable characteristics. Li et al. synthesized multifunctional microcapsules with latent heat storage and photocatalytic and antibacterial properties by using ZnO as the shell material and *n*-eicosane as the core material [38]. Similarly, Liu et al. utilized TiO_2 as shells material for encapsulating *n*-eicosane through interfacial polycondensation followed by ZnO impregnation [39]. The prepared microcapsules have both thermal storage and photocatalytic capacities with a melting temperature of 41.76°C and latent heat of 188.27 J/g.

3.3 Organic: inorganic hybrid shells

Organic–inorganic hybrid shells materials are used to overcome the shortcomings related to the individual organic or inorganic materials for encapsulating PCMs. In hybrid shells, organic materials offer structural flexibility while inorganic materials can improve thermal conductivity, thermal stability and mechanical rigidity [21]. Polymers (such as PMMA and PMF) based shells doped with SiO_2 or TiO_2 are extensively used to encapsulate PCMs [40]. Wang et al. prepared *n*-octadecane microcapsules using PMMA-silica hybrid shells via photocurable Pickering emulsion polymerization with good morphology and particles size of 5–15 μm [41]. The highest encapsulation efficiency (62.55%) was achieved with the weight ratio of MMA to *n*-octadecane of 1:1. Zhao et al. successfully synthesized bifunctional microcapsules by using PMMA doped with TiO_2 as the hybrid shell *and n*-octadecane as the core material [42]. TiO_2 was observed to improve microcapsules' thermal conductivity but reduce encapsulation efficiency and enthalpy. The initial degradation temperature of microcapsules with 6% TiO_2 reached 228.4°C, confirming good thermal stability of the microcapsules. Wang et al. prepared multifunctional microcapsules with regular-spherical morphology by using poly(melamine-formaldehyde)/silicon carbide (PMF/SiC) hybrid shells and *n*-octadecane as cores material [43]. The thermal conductivity of microcapsules with 7% SiC had improved by 60.34% compared to those microcapsules with no SiC, which is also accompanied by a significant increase in heat transfer rate.

4. Technologies for microencapsulation of PCMs

Microencapsulation techniques are of several types which are broadly classified into three major categories on the basis of fabrication mechanism: (1) physical methods (2) chemical methods and (3) physico-chemical methods. All these techniques involve the formation of a solid shell/coat around small liquid or solid particles of 1–100 μm diameter to accomplish the desired properties such as, protection competency, time-dependent release of material, provision of the substance to the particular target, minimize interaction with the environment, corrosion prevention, steadiness of function and to facilitate the use of toxic materials. The microencapsulation of PCMs is a special packaging methodology in which solid–liquid PCMs can be enclosed in some wall materials by using physical or chemical process to make small particles termed 'microcapsules' [44]. The PCM in a microcapsule is named as the core material while the outer shell which encloses the PCM from the surrounding environment is called the wall material. Microencapsulation as an emerging technology, commonly applied in many fields like thermal energy storage, medicine, food preservation, catalysis, dyes, textile, cosmetics, self-healing, coatings, engineering and defense [45]. A detailed classification of the microencapsulation methods is listed in **Figure 1**.

4.1 Physical methods

Physical methods involve involves physical processes, like drying, dehydration and adhesion in the formation of microcapsule shells. The most frequently used physical methods for PCMs encapsulation are spray-drying and solvent evaporation. The spray-drying process can be accomplished in the following steps: (1) preparation of oil–water emulsion comprising PCMs and shell materials, (2) spraying of the oil–water emulsion in a drying chamber via an atomizer, (3) drying of the sprayed droplets by using a stream of drying gas at a particular temperature, and (4) separating the solid particles by cyclone and filter [46]. Borreguero et al. employed a spray drying method for microencapsulation of paraffin Rubitherm®RT27 core using polyethylene EVA shell with and without carbon nanofibers (CNFs) [47]. The CNFs addition improved the thermal conductivity and mechanical strength of microcapsules, and the heat storage capability was retained. Also, the DSC analysis shown that even after the 3000- thermal charge/discharge cycles the microcapsules still had good thermal stability. Hawlader et al. synthesized spherical shape and uniform size microcapsule with paraffin core and gelatin and Arabic gum using spray-drying method [48]. The microcapsules prepared at the core-to-shell ratio of 2:1 have heat storage and release capacity reached 216.44 J/g and 221.217 J/g, respectively.

The solvent evaporation method includes: (1) preparation of polymer solution by dissolving shell materials in a volatile solvent; (2) addition of PCMs to the polymer solution to form O/W emulsion; (3) developing shells on the droplets by evaporating the solvent; (4) filtration and drying to obtaining the microcapsules. Lin et al. encapsulated myristic acid (MA) with ethyl cellulose (EC) using the solvent evaporation

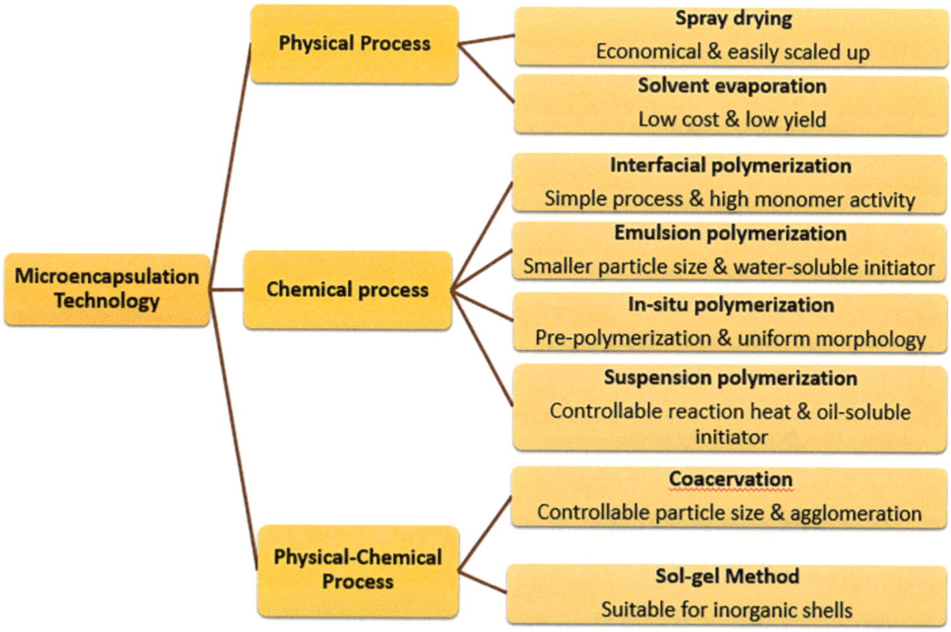

Figure 1.
Major physical and chemical microencapsulation methods for solid–liquid PCMs.

method [49]. The melting and solidifying temperatures were observed to be 53.32°C and 44.44°C, while the melting and solidifying enthalpies were found 122.61 J/g and 104.24 J/g, respectively. Wang et al. applied the solvent evaporation method to synthesize high-performance microcapsules by using sodium phosphate dodecahydrate (DSP) as the core and poly(methyl methacrylate) (PMMA) as the shell [50]. The optimal preparation temperature for the microencapsulation process was 80–90°C, reaction time 240 min, and stirring rate 900 rpm. The microcapsules obtained had an energy storage capacity of 142.9 J/g at the endothermic peak temperature of 51.5°C.

4.2 Chemical methods

In chemical methods, microencapsulation is done by the polymerization or condensation of monomers, oligomers, or prepolymers as raw materials to form shells at an oil–water interface. The chemical methods mostly involve in-situ polymerization, suspension polymerization, interfacial polymerization, and emulsion polymerization. The schematic diagrams of these four polymerization methods are shown in **Figure 2**. In situ polymerization method (**Figure 2(a)**), involves the formation of a shell on the surface of the droplet by polymerization of the prepolymers which can be accomplished in the following steps [52]: (1) preparation of the O/W emulsion by adding PCMs to surfactant aqueous solution; (2) preparation of a prepolymer solution; (3) addition of the prepolymer solution to the O/W emulsion, followed by adjusting the appropriate reaction conditions; and (4) microcapsule synthesis. Konuklu et al. successfully utilized in situ polymerization method for microencapsulation of decanoic acid using poly(urea-formaldehyde) (PUF), poly(melamine-formaldehyde) (PMF), and poly(melamine urea-formaldehyde) (PMUF) [53]. The microcapsules obtained by coating of PUF displayed higher heat storage capacity

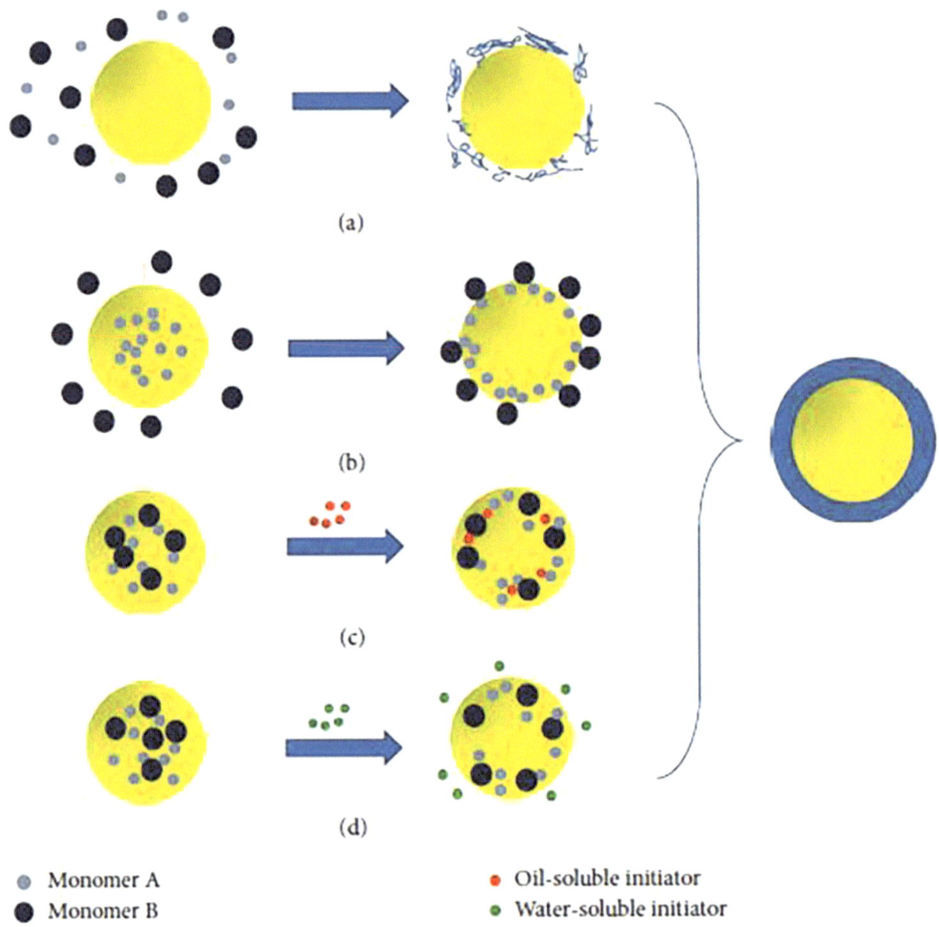

Figure 2.
Schematic diagrams of chemical methods for PCMs microencapsulation: (a) in situ polymerization, (b) interfacial polymerization, (c) suspension polymerization, and (d) emulsion polymerization [51].

but weaker mechanical strength and lower heat resistance, while the microcapsules coated with PMF shells had higher thermal stability but lower thermal energy storage capacity. However, the PMUF-encapsulated microcapsules possessed seamless thermal stability and no leakage was found at 95°C. Zhang et al. utilized in situ polycondensation method for synthesizing dual-functional microcapsules containing *n*-eicosane cores and ZrO_2 shells [54]. The microcapsules synthesized have a spherical shape with a size of 1.5–2 μm have good thermal energy storage and possessed better thermal stability, and thermal properties almost unchanged after 100 thermal cycles. Su et al. used methanol-modified melamine-formaldehyde (MMF) prepolymer as shell material for microencapsulation of dodecanol and paraffin via in situ polymerization [55]. They observed that the average diameter of dodecanol-based microcapsules sharply decreased and encapsulation efficiency increased with increasing stirring rates. The maximum encapsulation efficiency was found to be 97.4%.

Interfacial polymerization is used in the preparation of organic shell materials such as polyurea and polyurethane. In this method, two reactive monomers are separately dissolved in the oil phase and the aqueous phase, then in the presence of an initiator polymerization occurs at the oil–water interface as shown in **Figure 2(b)**. This method includes the following steps: (1) preparation of an O/W emulsion having hydrophobic monomer and PCMs; (2) addition of the hydrophilic monomer under proper conditions to initiate polymerization; (3) filtration, washing, and drying to get microcapsules. Ma et al. successfully used the interfacial polymerization method for microencapsulation of binary core materials like butyl stearate (BS) and paraffin with polyurea/polyurethane as the shell material [56]. The microcapsules phase change temperature was adjusted by changing the ratio of the two core materials. The microcapsules obtained possessed high thermal stability. Lu et al. encapsulated the butyl stearate core with a polyurethane-based cross-linked network shell via interfacial polymerization [57].

In the suspension polymerization method, the dispersed droplets of PCMs, monomers and initiators are suspended in a continuous aqueous phase by using surfactants and mechanical stirring. The oil-soluble initiator free radicals are then released into the emulsion system to initiate polymerization of the monomers at a suitable temperature and stirring rate [46], as presented in **Figure 2(c)**. Wang et al. successfully employed the suspension polymerization method to encapsulate *n*-octadecane with thermochromic pigment/PMMA shells at five different pigment/MMA ratios varying as 0, 1.4, 4.3, 7.1, and 14.3 wt.% [58]. It was observed that the microcapsules without pigment achieved the highest melting and crystallization enthalpies of 149.16 J/g and 152.55 J/g, respectively. Tang et al. prepared spherical shape microcapsules with an average diameter of about 1.60 μm using *n*-octadecane core material and *n*-octadecyl methacrylate (ODMA)-methacrylic acid (MAA) copolymer as shell material via the suspension polymerization method [59]. The microcapsules attained the highest phase change enthalpy of 93 J/g at monomers to the *n*-octadecane ratio of 2:1. Sanchez-Silva et al. microencapsulated Rubitherm®RT31 with polystyrene via suspension polymerization by using different suspension stabilizers [60]. The DSC investigations have shown that when PVP and gum Arabic were used as suspension stabilizers the microcapsules obtained presented the lowest thermal storage capacity of 75.7 J/g and highest of 135.3 J/g.

In emulsion polymerization (**Figure 2(d)**), first, the PCMs and monomers dispersed phase is suspended in a continuous phase in the presence of surfactants at constant stirring, followed by the addition of water-solution initiators to start the polymerization process [61]. This method is used to prepare microcapsule shells by polymerizing organic materials like PMMA and polystyrene. Şahan et al. encapsulated stearic acid (SA) with poly(methyl methacrylate) (PMMA) and four other PMMA-hybrid shell materials via emulsion polymerization technique [62]. The average diameter of microcapsules so obtained was found to be 110–360 μm, the thickness of 17–60 μm, heat storage capacity below 80 J/g and degradation temperature above 290°C. Sarı et al. successfully utilized the emulsion polymerization technique to microencapsulate paraffin eutectic mixtures (PEM) containing four different contents with PMMA shells [63]. The microcapsules obtained were spherical with a particle size of 1.16–6.42 μm, heat storage capacity of 169 J/g and melting temperature in the range of 20–36°C.

4.3 Physico-chemical methods

In the physical–chemical method, microencapsulation is accomplished by combining the physical processes like phase separation, heating and cooling, with chemical processes, like hydrolysis, cross-linking and condensation. Normally, the coacervation and sol–gel methods are the most frequently employed methods.

The coacervation method is of two types, one is single coacervation which requires only one type of shell material and the other is complex coacervation which requires two kinds of opposite-charged shell materials for microcapsules preparation. The microcapsules synthesized by the complex coacervation method usually have a more uniform size, better morphology and stability.

The complex coacervation processes involve the following key steps: (1) formation of emulsion by dispersing PCMs in polymer aqueous solution; (2) addition of a second aqueous polymer solution with opposite charges and deposition of shell material on droplet surface by electrostatic attraction and (3) Getting of microcapsules by cross-linking, desolation or thermal treatment. Hawlader et al. encapsulated paraffin cores with gelatin and acacia by using a complex coacervation process [48]. The melting and solidifying enthalpies of microcapsules obtained reached 239.78 J/g and 234.05 J/g, respectively, when the amount of cross-linking agent was 6–8 ml, homogenizing time was 10 min, and the ratio of core to the shell was 2:1. Onder et al. employed complex coacervation to microencapsulate *n*-hexadecane, *n*-octadecane and *n*-nonadecane core materials with natural and biodegradable polymers, like gum Arabic-gelatin mixture [64]. The microcapsules having *n*-hexadecane and *n*-octadecane cores showed good enthalpies of 144.7 J/g and 165.8 J/g, respectively, were obtained at the dispersed content of 80% in the emulsion and the microcapsules containing *n*-nonadecane prepared at the dispersed content of 60% in the emulsion presented enthalpy value of only 57.5 J/g.

The sol–gel method is a cheap and mild process for synthesizing PCMs microcapsules by inorganic shells, such as SiO_2 and TiO_2 shells. The major steps involved in the preparation of microcapsule by the sol–gel method are as follows: (1) preparation of colloidal solution by uniformly dispersing the reactive materials like PCMs, precursor, solvent and emulsifier in a continuous phase via hydrolysis reaction; (2) formation of a three-dimensional network structured gel system through condensation polymerization of monomers and (3) drying, sintering and curing processes to obtain microcapsules [65]. Cao et al. used the sol–gel process to microencapsulate paraffin core with TiO_2 shells. They found that the sample with a microencapsulation ratio of 85.5% had melting and solidifying latent heat of 161.1 kJ/kg (at the melting temperature of 58.8°C) and 144.6 kJ/kg (at the solidifying temperature of 56.5°C), respectively [66]. Latibari et al. successfully employed the sol–gel method to synthesize nanocapsules containing palmitic acid (PA) core with SiO_2 shell by controlling solution pH [67]. The nanocapsule obtained presented an average particle size of 183.7, 466.4 and 722.5 nm, at pH 11, 11.5 and 12, respectively, and the corresponding melting latent heats values of 168.16, 172.16 and 180.91 kJ/kg, respectively.

5. Design of microencapsulated PCMs for versatile application

5.1 PCMs microencapsulation with function inorganic shells

Microencapsulation with conventional polymeric, inorganic or composite shells can provide only protection for the PCM core, but at the same time, these inert wall materials cause a reduction in their latent heat-storage capacities which make them unsuitable for thermal energy storage and thermal management systems. In view of that various inorganic materials have a feature of functional diversity, it will be possible to synthesize bi-function PCMs-based microcapsules by encapsulating the PCM core with a functional inorganic shell. This idea was first used by Fei et al. [68] and successfully synthesized a novel multi-functional microcapsules based on an anatase

TiO_2 shell and *n*-octadecane/titania aerosol core via the hydrothermal method. The microcapsules obtained presented multi-functional properties with photocatalytic activity and UV-blocking effectiveness as well as a thermal energy-storage function. Chai et al. [69] introduced a new synthetic strategy by fabricating a well-defined core-shell structured PCM microcapsule based on a functional TiO_2 shell. The crystallization of amorphous TiO_2 was initiated by adding fluorine ions when the *in-situ* polycondensation of titanic precursors was performed in a nonaqueous O/W emulsion-templating system. The microcapsules so prepared have excellent thermal energy-storage capacity and show photocatalytic and antibacterial functions. Liu et al. [70] introduced a new technology by modifying the brookite TiO_2 shell of the *n*-eicosane core with graphene nanosheets. It was observed that graphene promotes the charge transfer and separation ability of microcapsule which leads to a significant increase in its photocatalytic activity. Liu et al. [39] also explored that modification of TiO_2 shell with ZnO boosts the latent heat-storage capacity and photocatalytic activity of the resultant microcapsules. A study on the utilization of microcapsules doped with ZnO presented good thermal regulation and thermal management properties when incorporated into the gypsum-matrix composites. These explorers make the modified microcapsules good candidates for direct solar energy utilization. Additionally, Liu et al. [71] introduced a morphology-controlled synthetic technology to fabricate PCM-based microcapsules with crystalline TiO_2 shells by using different structure-directing agents and effectively obtained the microcapsules in the tubular, octahedral and spherical shapes. They also studied the influence of structural morphology on the thermal energy-storage capacity of these microcapsules and observed the highest latent heat-storage efficiency with microcapsules of spherical morphology while the tubular ones displayed the fastest heat response rate. Li et al. [38] successfully encapsulated *n*-eicosane with ZnO shell via *in-situ* precipitation reaction of $Zn(CH_3COO)_2{\cdot}2H_2O$ and NaOH in an emulsion templating system. The microcapsule prepared exhibited good thermal energy-storage capability and high working reliability as well as high photocatalytic activities and antimicrobial effectiveness against *Staphylococcus aureus*. These microcapsules, therefore, have gained potential applications in medical care and surgical treatment. Gao et al. [72] designed multifunctional microcapsules by a microencapsulating *n*-eicosane core with a Cu_2O shell, through emulsion templated *in-situ* precipitation and reduction. The microcapsules obtained exhibited multifunctional properties of effective photothermal conversion, high latent-heat storage/release efficiency for solar photocatalysis and solar thermal energy storage, as well as demonstrated sensitivity to some toxic organics gases due to a *p*-type semiconductive feature of Cu_2O shell.

5.2 Advanced design of microencapsulated PCMs for versatile applications

In recent years due to the fast development in microencapsulation technology, a large number of innovative designs have been introduced for fabricating bi- or multifunctional PCMs-based microcapsules. Jiang et al. [73] designed magnetic PCM-based microcapsules as an applied energy microsystem for bio-applications as thermoregulatory enzyme carriers. They synthesized the magnetic microcapsules by encapsulating *n*-eicosane with a TiO_2/Fe_3O_4 hybrid shell by Pickering emulsion-templated interfacial polycondensation and then *Candida rugosa lipase* (CRL) was immobilized onto the microcapsules obtained by covalent bonds through a series of complicated surface modification and immobilization reactions. The microcapsules obtained were observed to have higher thermal stability, longer storage stability, higher biocatalytic activity and better reusability compared to traditional inert enzyme carriers. Likewise, Li et al. [74] also developed thermoregulatory enzyme carriers based on the magnetic microcapsules containing *n*-docosane core and SiO_2/Fe_3O_4 hybrid shell with α-amylase immobilized onto the microcapsule and examined the effect of ambient temperature on their biocatalytic activity. They found that the biocatalytic activity was increased considerably for the immobilized α-amylase on the developed enzyme carriers due to the thermoregulation microenvironment around the microcapsules. These innovative designs provide a novel approach for the preparation and applications of microencapsulated PCMs in areas of bioengineering and biotechnological.

Choi et al. [75] designed a novel, temperature-sensitive drug release system based on PCMs. They first prepared the gelatin nanoparticles containing fluorescein isothiocyanate-dextran as a drug via emulsification technique, and then 1-tetradecanol was used to synthesize the PCM-matrix microbeads containing these gelatin nanoparticles by using a simple fluidic device based on an O/W emulsion. Moreover, Wang et al. [58] designed and synthesized thermochromic microencapsulated PCM by encapsulating *n*-octadecane with PMMA shell with simultaneous dispersion of thermochromic pigments in core and shell by suspension-like polymerization. The microcapsules obtained showed a visible color change with change in temperature, confirmed the occurrence of thermal energy storage or release at the specific temperature.

Geng et al. [76] designed a three-component core consisting of 1-tetradecanol as a PCM, leuco dye and phenolic color developer as an electron donor and fabricated reversible thermochromic microcapsules for application in thermal protective clothing. They encapsulated the three-component core with a poly (methylated melamine-formaldehyde) (PMMF) shell via emulsion-templated copolymerization. The as synthesized microcapsules exhibited thermochromic reversibility with good energy storage/release capability and have a great potential for applications in thermal protective clothing of firefighters as well as intelligent textiles or fabrics, food and medicine package and so on. Wu et al. [77] synthesized reversible thermochromic microcapsules by encapsulating 1-hexadecanol with modified gelatin and gum Arabic via a complex coacervation process. The wall materials of this microcapsule system were fused with 2-phenylamino-3-methyl-6-di-*n*-butylamino-fluoran as a color former and 2,2-bis(4-hydroxyphenyl) propane as a color developer. The microcapsule prepared acts as an indicator for the states of energy saturation and consumption through color changes. In addition, Zhang et al. [78] introduced polysaccharide-assisted microencapsulation as an innovative methodology for encapsulation of volatile PCMs with a fluorescent retention indicator to determine the retention of microencapsulated volatile PCM in diverse working environments. They microencapsulated heptane core with polymeric shell by one-step *in-situ* polymerization path using Nile red as a fluorescent indicator which was incorporated into the heptane core during the synthetic process, and therefore the fluorophores in Nile red could give a clear indication for the core and shell structures of microcapsules.

6. Applications of microencapsulated PCMs

Microencapsulated PCMs due to their unique properties such as solid-to-liquid phase transition, chemical and thermal stability and higher amount of energetic changes, has received special attention for their applications in in our ordinary daily life and various industries. In recent years, PCMs have been designed and fabricated to meet the requirements around the world. The potential applications of PCM microcapsules are shown in **Figure 3**, and discussed as bellows:

6.1 Application in fibers and textiles

In textile industries, microencapsulated PCMs are embedded within the fibers or coated onto the surface of fabrics which are used in the preparations of outdoor dress such as snowsuits, trousers, gloves, ear warmers and boots etc. The microencapsulated PCMs enhance the thermal storage capacities of the fibers/fabrics (2.5–4.5 times) and thus protect from extremely cold weather [46]. Microencapsulation is a promising technology for applications in the textile industry such as agriculture textiles, medical textiles automotive textiles and sportswear/protective clothing. Scacchetti et al. explored the thermal and antimicrobial properties of cotton with silver zeolites functionalized via a chitosanzeolite composite and microcapsules of PCMs [79]. They suggested the use of chitosan zeolite for the production of textiles for superior antibacterial and thermoregulating properties. Microencapsulated PCMs increase the flame-retardant property thermal and comfort of the textiles, as these PCM microcapsules were scattered homogeneously onto textile substrates and were durable with repeated washings [80].

6.2 Application in slurry

PCM microcapsules with high latent heat are used in the slurry industry as an enhanced heat transfer fluid (HTF) and a thermal storage medium (TSM). Song et al. considered laminar heat transfer of PCM microcapsules slurry and proved that the heat transfer coefficient improved with increasing Reynolds number and volume concentration of microcapsules [81]. Roberts et al. compared the heat transfer capability of metal-coated and nonmetal-coated PCM microcapsules slurry and noticed an additional 10% increment in heat transfer coefficient and PCM microcapsules inducing pressure drop in slurry [82]. Zhang and Niu reported higher thermal storage capacity for PCM microcapsules slurry storage devices and stratified water storage tanks [83]. Xu et al. prepared PCM microcapsules with Cu-Cu_2O/CNTs shell and their dispersed slurry for direct absorption solar collectors [84]. They reported that the PCMs@Cu-Cu_2O/CNTs microcapsule slurry had high heat storage competency and outstanding photothermal conversion performance which made it as one of the most potential HTFs for direct absorption solar collector.

6.3 Application in energy-saving building

Another amazing application of PCM microcapsules is their utilization in building materials to overcome overheating problems in summer and provided a new effective solution for thermal management and energy saving in buildings. The PCM microcapsules in construction materials boost the thermal and acoustic insulation of walls. Usually, the PCM microcapsules are embedded into concrete mixtures, cement mortar, gypsum plaster, wallboards, sandwich, slabs, panels and to fulfill the energy demand of the building for heating, cooling, lighting, air conditioning, ventilation and domestic hot water systems [85]. Many researchers around the world worked on the application of PCMs microcapsules in the building industry. Cabeza et al. reported an innovative concrete material with high thermal properties by mixing it with PCM microcapsules [86]. It was found that the concrete wall with PCM

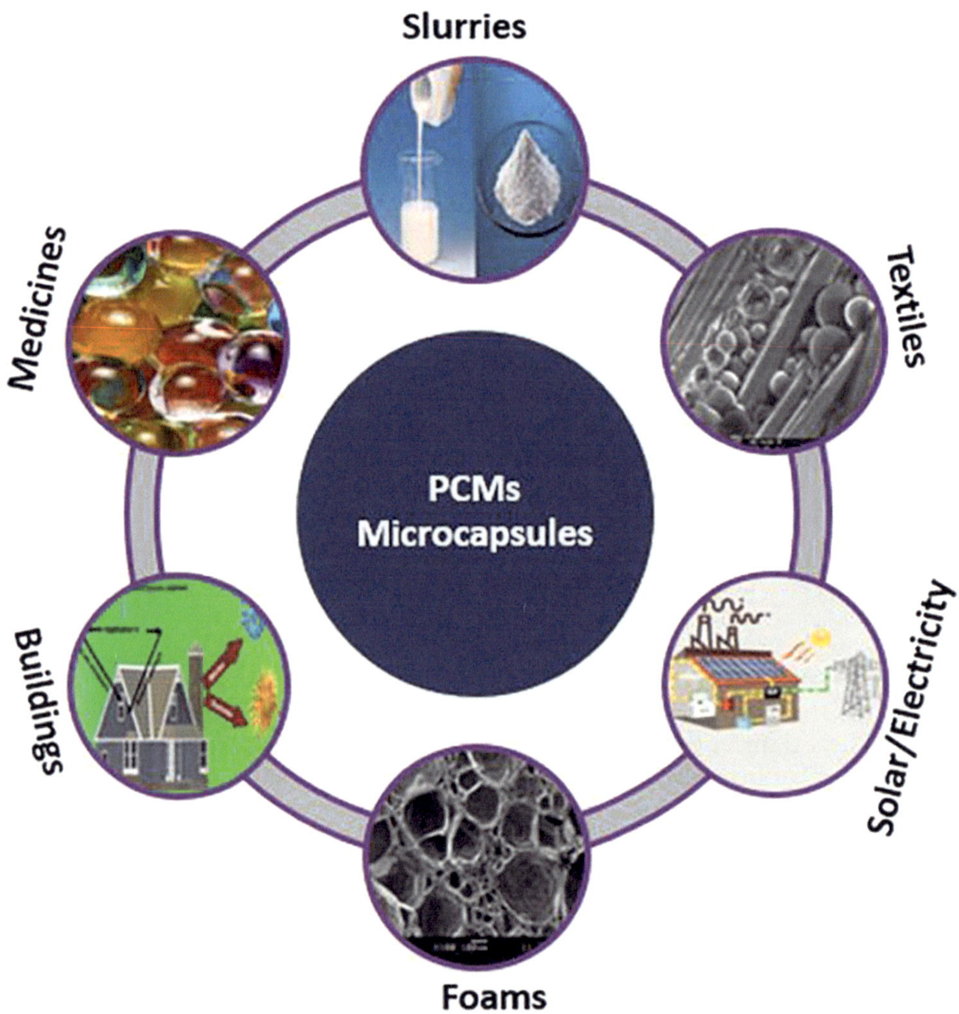

Figure 3.
Potential applications of PCMs microcapsules.

microcapsules increase its overall mechanical resistance and stiffness and causes even temperature fluctuations and thermal inertia, making it to be a promising technology to save energy for buildings [87]. Su et al. studied nano-silicon dioxide hydrosol as the surfactant for the preparation of PCM microcapsules for thermal energy storage in buildings [88]. Essid et al. investigated the compressive strength and hygric properties of microencapsulated PCMs concretes [89]. They reported that the use of concrete containing PCM microcapsules as structural material is sufficiently safe, though its compressive strength is lower and porosity is higher than the pure concrete. Schossig et al. [90] directly integrated formaldehyde-free micro-encapsulated paraffins in building materials and studied their effect for application in conventional construction materials. They observed that the utilization of these PCMs microcapsule could help to keep the indoor temperature up to 4°C lower than typical conditions and could reduce the number of hours that the indoor temperature was greater than 28°C.

6.4 Application in foams

The application of microencapsulated PCMs in foams can enhance their thermal insulating efficiency. Borreguero et al. reported that the thermal energy storage capacity of rigid polyurethane foams was improved when it was embedded with PCM microcapsules investigated rigid polyurethane foams containing and indicated that improved [91]. Li et al. introduced a new approach to enhancing the latent heat energy storage ability by embedding PCM microcapsules in metal foam [92]. They observed that compared to the surface temperature of virgin PCM modules, the surface temperature for the PCM microcapsule/foam and PCM/foam composite modules was reduced from about 90 to 55 and 45°C, respectively. PCM microcapsule/foam composites solved the problem of low thermal conductivity and leakage. Bonadies et al. synthesized poly(vinyl alcohol)- (PVA-) based foams containing PCM microcapsules and investigated their thermal storage and dimensional stability [93]. They observed that the formation of crystalline domain and amount of water uptake was influenced by microcapsules which in turn affected the number of intra- and intermolecular hydrogen bonds as many PVA –OH groups interact with microcapsule shells.

6.5 Others applications

There are many other potential applications of PCM microcapsules. These include biomedical applications, solar-to-thermal energy storage and electrical-to-thermal energy storage [65]. Zang et al. prepared multifunctional microencapsulated PCMs that can be used for sterilization [94]. They reported that these microcapsules have high antibacterial activity against *Escherichia coli*, *S. aureus*, and *Bacillus subtilis*, and the antibacterial efficiency of 2-hour contacting PCM microcapsules was inhibited up to 64.6%, 99.1%, and 95.9%, respectively. Zhang et al. also studied solar-driven PCM microcapsules with efficient Ti_4O_7 nanoconverter for latent heat storage [95]. The solar absorption capacity of the novel PCM microcapsules was found to be 88.28%, and the photothermal storage efficiency of the PCMs@SiO_2/Ti_4O_7 microcapsules was 85.36% compared with 24.14% for pure PCMs. Zheng et al. proposed a joule heating system to reduce the convective heat transferring from the electrothermal system of the surrounding by inserting the highly conductive and stable PCMs microcapsules [96]. They showed that the working temperature could be improved by 30% with the loading of 5% PCMs microcapsules even at lower voltage and ambient temperature.

7. Conclusion

Microencapsulation is a promising technology to fabricate PCM microcapsules for thermal energy storage and other versatile applications. Microencapsulation technology not only overcome the problem of leakage, volatilization and handling the difficulty of liquids PCMs but also improve the heat transferring ability of PCMs and thus makes them a favorable means for many broad range of applications. The thermal, physical, chemical and mechanical properties of PCM microcapsules are highly dependent on the type of the core materials, shell materials and synthesis processes. Encapsulation with inorganic shells can provide more advantages for microencapsulated PCMs and therefore will gain much more attention from fundamental research to commercial development in the future. The designing parameters, such as weight ratio of raw materials for core and shell, the selection of dispersion medium, reaction temperature, time, agitation speed, particles size and its distribution and other additives should be carefully addressed to obtain PCMs microcapsules with well-defined core-shell structured and good thermal energy-storage capability. Though PCM microcapsules are considered smart thermal energy storage materials, still much more new materials and synthetic techniques are to be explored to offer numerous possibilities for the design and fabrication of innovative bi- and multi-functional PCMs microcapsules with better properties and functions than the traditional ones.

Author details

Tahira Mahmood*, Rahmat Ali, Abdul Naeem and Murtaza Syed
National Centre of Excellence in Physical Chemistry, University of Peshawar, Peshawar, Pakistan

*Address all correspondence to: tahiramahmood@uop.edu.pk

References

[1] Pielichowska K, Pielichowski K. Phase change materials for thermal energy storage. Progress in Materials Science. 2014;**65**:67-123

[2] Khudhair AM, Farid MM. A review on energy conservation in building applications with thermal storage by latent heat using phase change materials. Energy Conversion and Management. 2004;**45**:263-275

[3] Kalnæs SE, Jelle BP. Phase change materials and products for building applications: A state-of-the-art review and future research opportunities. Energy and Buildings. 2015;**94**:150-176

[4] Sharma A, Tyagi VV, Chen C, Buddhi D. Review on thermal energy storage with phase change materials and applications. Renewable and Sustainable Energy Reviews. 2009;**13**:318-345

[5] Guney MS, Tepe Y. Classification and assessment of energy storage systems. Renewable and Sustainable Energy Reviews. 2017;**75**:1187-1197

[6] Su W, Darkwa J, Kokogiannakis G. Review of solid–liquid phase change materials and their encapsulation technologies. Renewable and Sustainable Energy Reviews. 2015;**48**:373-391

[7] Jamekhorshid A, Sadrameli S, Farid M. A review of microencapsulation methods of phase change materials (PCMs) as a thermal energy storage (TES) medium. Renewable and Sustainable Energy Reviews. 2014;**31**:531-542

[8] Delgado M, Lázaro A, Mazo J, Zalba B. Review on phase change material emulsions and microencapsulated phase change material slurries: Materials, heat transfer studies and applications. Renewable and Sustainable Energy Reviews. 2012;**16**:253-273

[9] Kant K, Shukla A, Sharma A. Advancement in phase change materials for thermal energy storage applications. Solar Energy Materials and Solar Cells. 2017;**172**:82-92

[10] Zhao C-Y, Zhang GH. Review on microencapsulated phase change materials (MEPCMs): Fabrication, characterization and applications. Renewable and Sustainable Energy Reviews. 2011;**15**:3813-3832

[11] He F, Wang X, Wu D. Phase-change characteristics and thermal performance of form-stable n-alkanes/silica composite phase change materials fabricated by sodium silicate precursor. Renewable Energy. 2015;**74**:689-698

[12] Yu S, Wang X, Wu D. Microencapsulation of n-octadecane phase change material with calcium carbonate shell for enhancement of thermal conductivity and serving durability: Synthesis, microstructure, and performance evaluation. Applied Energy. 2014;**114**:632-643

[13] Cao L, Tang F, Fang G. Preparation and characteristics of microencapsulated palmitic acid with TiO2 shell as shape-stabilized thermal energy storage materials. Solar Energy Materials and Solar Cells. 2014;**123**:183-188

[14] Khadiran T, Hussein MZ, Zainal Z, Rusli R. Advanced energy storage materials for building applications and their thermal performance characterization: A review. Renewable and Sustainable Energy Reviews. 2016;**57**:916-928

[15] Mondal S. Phase change materials for smart textiles–an overview. Applied Thermal Engineering. 2008;**28**:1536-1550

[16] Kenisarin M, Mahkamov K. Solar energy storage using phase change materials. Renewable and Sustainable Energy Reviews. 2007;**11**:1913-1965

[17] Zhang N, Yuan Y, Cao X, Du Y, Zhang Z, Gui Y. Latent heat thermal energy storage systems with solid–liquid phase change materials: A review. Advanced Engineering Materials. 2018;**20**:1700753

[18] Zeng J-L, Chen Y-H, Shu L, Yu L-P, Zhu L, Song L-B, et al. Preparation and thermal properties of exfoliated graphite/erythritol/mannitol eutectic composite as form-stable phase change material for thermal energy storage. Solar Energy Materials and Solar Cells. 2018;**178**:84-90

[19] Al-Shannaq R, Kurdi J, Al-Muhtaseb S, Farid M. Innovative method of metal coating of microcapsules containing phase change materials. Solar Energy. 2016;**129**:54-64

[20] Agyenim F, Hewitt N, Eames P, Smyth M. A review of materials, heat transfer and phase change problem formulation for latent heat thermal energy storage systems (LHTESS). Renewable and Sustainable Energy Reviews. 2010;**14**:615-628

[21] Umair MM, Zhang Y, Iqbal K, Zhang S, Tang B. Novel strategies and supporting materials applied to shape-stabilize organic phase change materials for thermal energy storage–a review. Applied Energy. 2019;**235**:846-873

[22] Brown EN, Kessler MR, Sottos NR, White SR. In situ poly (urea-formaldehyde) microencapsulation of dicyclopentadiene. Journal of Microencapsulation. 2003;**20**:719-730

[23] Huang R, Li W, Wang J, Zhang X. Effects of oil-soluble etherified melamine-formaldehyde prepolymers on in situ microencapsulation and macroencapsulation of n-dodecanol. New Journal of Chemistry. 2017;**41**:9424-9437

[24] Qiu X, Song G, Chu X, Li X, Tang G. Preparation, thermal properties and thermal reliabilities of microencapsulated n-octadecane with acrylic-based polymer shells for thermal energy storage. Thermochimica Acta. 2013;**551**:136-144

[25] Chen Z, Wang J, Yu F, Zhang Z, Gao X. Preparation and properties of graphene oxide-modified poly (melamine-formaldehyde) microcapsules containing phase change material n-dodecanol for thermal energy storage. Journal of Materials Chemistry A. 2015;**3**:11624-11630

[26] Mohaddes F, Islam S, Shanks R, Fergusson M, Wang L, Padhye R. Modification and evaluation of thermal properties of melamine-formaldehyde/n-eicosane microcapsules for thermo-regulation applications. Applied Thermal Engineering. 2014;**71**:11-15

[27] Alkan C, Sarı A, Karaipekli A. Preparation, thermal properties and thermal reliability of microencapsulated n-eicosane as novel phase change material for thermal energy storage. Energy Conversion and Management. 2011;**52**:687-692

[28] Ma Y, Chu X, Li W, Tang G. Preparation and characterization of poly (methyl methacrylate-co-divinylbenzene) microcapsules containing phase change temperature

adjustable binary core materials. Solar Energy. 2012;**86**:2056-2066

[29] Wang X, Chen Z, Xu W, Wang X. Capric acid phase change microcapsules modified with graphene oxide for energy storage. Journal of Materials Science. 2019;**54**:14834-14844

[30] Wu S, Yuan L, Gu A, Zhang Y, Liang G. Synthesis and characterization of novel epoxy resins-filled microcapsules with organic/inorganic hybrid shell for the self-healing of high performance resins. Polymers for Advanced Technologies. 2016;**27**:1544-1556

[31] Liu Z, Chen Z, Yu F. Preparation and characterization of microencapsulated phase change materials containing inorganic hydrated salt with silica shell for thermal energy storage. Solar Energy Materials and Solar Cells. 2019;**200**:110004

[32] Bao Y, Yan Y, Chen Y, Ma J, Zhang W, Liu C. Facile fabrication of BTA@ ZnO microcapsules and their corrosion protective application in waterborne polyacrylate coatings. Progress in Organic Coatings. 2019;**136**:105233

[33] Genc M, Genc ZK. Microencapsulated myristic acid–fly ash with TiO_2 shell as a novel phase change material for building application. Journal of Thermal Analysis and Calorimetry. 2018;**131**:2373-2380

[34] Jiang Z, Yang W, He F, Xie C, Fan J, Wu J, et al. Microencapsulated paraffin phase-change material with calcium carbonate shell for thermal energy storage and solar-thermal conversion. Langmuir. 2018;**34**:14254-14264

[35] Lin Y, Zhu C, Fang G. Synthesis and properties of microencapsulated stearic acid/silica composites with graphene oxide for improving thermal conductivity as novel solar thermal storage materials. Solar Energy Materials and Solar Cells. 2019;**189**:197-205

[36] Li M, Wang W, Zhang Z, He F, Yan S, Yan P-J, et al. Monodisperse $Na_2SO_4 \cdot 10H_2O$@ SiO_2 microparticles against supercooling and phase separation during phase change for efficient energy storage. Industrial & Engineering Chemistry Research. 2017;**56**:3297-3308

[37] Liang S, Li Q, Zhu Y, Chen K, Tian C, Wang J, et al. Nanoencapsulation of n-octadecane phase change material with silica shell through interfacial hydrolysis and polycondensation in miniemulsion. Energy. 2015;**93**:1684-1692

[38] Li F, Wang X, Wu D. Fabrication of multifunctional microcapsules containing n-eicosane core and zinc oxide shell for low-temperature energy storage, photocatalysis, and antibiosis. Energy Conversion and Management. 2015;**106**:873-885

[39] Liu H, Wang X, Wu D, Ji S. Fabrication and applications of dual-responsive microencapsulated phase change material with enhanced solar energy-storage and solar photocatalytic effectiveness. Solar Energy Materials and Solar Cells. 2019;**193**:184-197

[40] Li C, Yu H, Song Y, Liang H, Yan X. Preparation and characterization of PMMA/TiO2 hybrid shell microencapsulated PCMs for thermal energy storage. Energy. 2019;**167**:1031-1039

[41] Wang H, Zhao L, Chen L, Song G, Tang G. Facile and low energy consumption synthesis of microencapsulated phase change materials with hybrid shell for thermal

energy storage. Journal of Physics and Chemistry of Solids. 2017;**111**:207-213

[42] Zhao J, Yang Y, Li Y, Zhao L, Wang H, Song G, et al. Microencapsulated phase change materials with TiO2-doped PMMA shell for thermal energy storage and UV-shielding. Solar Energy Materials and Solar Cells. 2017;**168**:62-68

[43] Wang X, Li C, Zhao T. Fabrication and characterization of poly (melamine-formaldehyde)/silicon carbide hybrid microencapsulated phase change materials with enhanced thermal conductivity and light-heat performance. Solar Energy Materials and Solar Cells. 2018;**183**:82-91

[44] Dubey R. Microencapsulation technology and applications. Defence Science Journal. 2009;**59**:82

[45] Wazarkar K, Patil D, Rane A, Balgude D, Kathalewar M, Sabnis A. Microencapsulation: An emerging technique in the modern coating industry. RSC Advances. 2016;**6**:106964-106979

[46] Alva G, Lin Y, Liu L, Fang G. Synthesis, characterization and applications of microencapsulated phase change materials in thermal energy storage: A review. Energy and Buildings. 2017;**144**:276-294

[47] Borreguero A, Valverde J, Rodríguez J, Barber A, Cubillo J, Carmona M. Synthesis and characterization of microcapsules containing Rubitherm® RT27 obtained by spray drying. Chemical Engineering Journal. 2011;**166**:384-390

[48] Hawlader M, Uddin M, Khin MM. Microencapsulated PCM thermal-energy storage system. Applied Energy. 2003;**74**:195-202

[49] Lin Y, Zhu C, Alva G, Fang G. Microencapsulation and thermal properties of myristic acid with ethyl cellulose shell for thermal energy storage. Applied Energy. 2018;**231**:494-501

[50] Wang TY, Huang J. Synthesis and characterization of microencapsulated sodium phosphate dodecahydrate. Journal of Applied Polymer Science. 2013;**130**:1516-1523

[51] Peng G, Dou G, Hu Y, Sun Y, Chen Z. Phase change material (PCM) microcapsules for thermal energy storage. Advances in polymer technology. https://doi.org/10.1155/2020/9490873

[52] Zhang H, Wang X. Fabrication and performances of microencapsulated phase change materials based on n-octadecane core and resorcinol-modified melamine–formaldehyde shell. Colloids and Surfaces A: Physicochemical and Engineering Aspects. 2009;**332**:129-138

[53] Konuklu Y, Unal M, Paksoy HO. Microencapsulation of caprylic acid with different wall materials as phase change material for thermal energy storage. Solar Energy Materials and Solar Cells. 2014;**120**:536-542

[54] Zhang Y, Wang X, Wu D. Design and fabrication of dual-functional microcapsules containing phase change material core and zirconium oxide shell with fluorescent characteristics. Solar Energy Materials and Solar Cells. 2015;**133**:56-68

[55] Su J-F, Wang S-B, Zhou J-W, Huang Z, Zhao Y-H, Yuan X-Y, et al. Fabrication and interfacial morphologies of methanol–melamine–formaldehyde (MMF) shell microPCMs/epoxy composites. Colloid and Polymer Science. 2011;**289**:169-177

[56] Ma Y, Chu X, Tang G, Yao Y. The effect of different soft segments on the formation and properties of binary core microencapsulated phase change materials with polyurea/polyurethane double shell. Journal of Colloid and Interface Science. 2013;**392**:407-414

[57] Lu S, Shen T, Xing J, Song Q, Shao J, Zhang J, et al. Preparation and characterization of cross-linked polyurethane shell microencapsulated phase change materials by interfacial polymerization. Materials Letters. 2018;**211**:36-39

[58] Wang H, Luo J, Yang Y, Zhao L, Song G, Tang G. Fabrication and characterization of microcapsulated phase change materials with an additional function of thermochromic performance. Solar Energy. 2016;**139**:591-598

[59] Tang X, Li W, Zhang X, Shi H. Fabrication and characterization of microencapsulated phase change material with low supercooling for thermal energy storage. Energy. 2014;**68**:160-166

[60] Sánchez-Silva L, Rodríguez JF, Sánchez P. Influence of different suspension stabilizers on the preparation of Rubitherm RT31 microcapsules. Colloids and Surfaces A: Physicochemical and Engineering Aspects. 2011;**390**:62-66

[61] Huang X, Zhu C, Lin Y, Fang G. Thermal properties and applications of microencapsulated PCM for thermal energy storage: A review. Applied Thermal Engineering. 2019;**147**:841-855

[62] Şahan N, Nigon D, Mantell SC, Davidson JH, Paksoy H. Encapsulation of stearic acid with different PMMA-hybrid shell materials for thermotropic materials. Solar Energy. 2019;**184**:466-476

[63] Sarı A, Alkan C, Bilgin C. Micro/nano encapsulation of some paraffin eutectic mixtures with poly (methyl methacrylate) shell: Preparation, characterization and latent heat thermal energy storage properties. Applied energy. 2014;**136**:217-227

[64] Onder E, Sarier N, Cimen E. Encapsulation of phase change materials by complex coacervation to improve thermal performances of woven fabrics. Thermochimica Acta. 2008;**467**:63-72

[65] Arshad A, Jabbal M, Yan Y, Darkwa J. The micro−/nano-PCMs for thermal energy storage systems: A state of art review. International Journal of Energy Research. 2019;**43**:5572-5620

[66] Cao L, Tang F, Fang G. Synthesis and characterization of microencapsulated paraffin with titanium dioxide shell as shape-stabilized thermal energy storage materials in buildings. Energy and Buildings. 2014;**72**:31-37

[67] Latibari ST, Mehrali M, Mehrali M, Mahlia TMI, Metselaar HSC. Synthesis, characterization and thermal properties of nanoencapsulated phase change materials via sol–gel method. Energy. 2013;**61**:664-672

[68] Fei B, Lu H, Qi K, Shi H, Liu T, Li X, et al. Multi-functional microcapsules produced by aerosol reaction. Journal of Aerosol Science. 2008;**39**:1089-1098

[69] Chai L, Wang X, Wu D. Development of bifunctional microencapsulated phase change materials with crystalline titanium dioxide shell for latent-heat storage and photocatalytic effectiveness. Applied Energy. 2015;**138**:661-674

[70] Liu H, Wang X, Wu D. Fabrication of graphene/TiO_2/paraffin composite phase change materials for enhancement of solar energy efficiency in photocatalysis

and latent heat storage. ACS Sustainable Chemistry & Engineering. 2017;**5**:4906-4915

[71] Liu H, Wang X, Wu D, Ji S. Morphology-controlled synthesis of microencapsulated phase change materials with TiO2 shell for thermal energy harvesting and temperature regulation. Energy. 2019;**172**:599-617

[72] Gao F, Wang X, Wu D. Design and fabrication of bifunctional microcapsules for solar thermal energy storage and solar photocatalysis by encapsulating paraffin phase change material into cuprous oxide. Solar Energy Materials and Solar Cells. 2017;**168**:146-164

[73] Jiang B, Wang X, Wu D. Fabrication of microencapsulated phase change materials with TiO_2/Fe_3O_4 hybrid shell as thermoregulatory enzyme carriers: A novel design of applied energy microsystem for bioapplications. Applied Energy. 2017;**201**:20-33

[74] Li J, Liu H, Wang X, Wu D. Development of thermoregulatory enzyme carriers based on microencapsulated n-docosane phase change material for biocatalytic enhancement of amylases. ACS Sustainable Chemistry & Engineering. 2017;**5**:8396-8406

[75] Choi SW, Zhang Y, Xia Y. A temperature-sensitive drug release system based on phase-change materials. Angewandte Chemie International Edition. 2010;**49**:7904-7908

[76] Geng X, Li W, Wang Y, Lu J, Wang J, Wang N, et al. Reversible thermochromic microencapsulated phase change materials for thermal energy storage application in thermal protective clothing. Applied Energy. 2018;**217**:281-294

[77] Wu B, Shi L, Zhang Q, Wang W-J. Microencapsulation of 1-hexadecanol as a phase change material with reversible thermochromic properties. RSC Advances. 2017;**7**:42129-42137

[78] Zhang Y, Jiang Z, Zhang Z, Ding Y, Yu Q, Li Y. Polysaccharide assisted microencapsulation for volatile phase change materials with a fluorescent retention indicator. Chemical Engineering Journal. 2019;**359**:1234-1243

[79] Scacchetti FA, Pinto E, Soares GM. Thermal and antimicrobial evaluation of cotton functionalized with a chitosan–zeolite composite and microcapsules of phase-change materials. Journal of Applied Polymer Science. 2018;**135**:46135

[80] Alay Aksoy S, Alkan C, Tözüm MS, Demirbağ S, Altun Anayurt R, Ulcay Y. Preparation and textile application of poly (methyl methacrylate-co-methacrylic acid)/n-octadecane and n-eicosane microcapsules. The Journal of the Textile Institute. 2017;**108**:30-41

[81] Song S, Shen W, Wang J, Wang S, Xu J. Experimental study on laminar convective heat transfer of microencapsulated phase change material slurry using liquid metal with low melting point as carrying fluid. International Journal of Heat and Mass Transfer. 2014;**73**:21-28

[82] Roberts NS, Al-Shannaq R, Kurdi J, Al-Muhtaseb SA, Farid MM. Efficacy of using slurry of metal-coated microencapsulated PCM for cooling in a micro-channel heat exchanger. Applied Thermal Engineering. 2017;**122**:11-18

[83] Zhang S, Niu J. Two performance indices of TES apparatus: Comparison of MPCM slurry vs. stratified water storage tank. Energy and Buildings. 2016;**127**:512-520

[84] Xu B, Chen C, Zhou J, Ni Z, Ma X. Preparation of novel microencapsulated phase change material with Cu-Cu2O/CNTs as the shell and their dispersed slurry for direct absorption solar collectors. Solar Energy Materials and Solar Cells. 2019;**200**:109980

[85] Konuklu Y, Ostry M, Paksoy HO, Charvat P. Review on using microencapsulated phase change materials (PCM) in building applications. Energy and Buildings. 2015;**106**:134-155

[86] Cabeza LF, Castellon C, Nogues M, Medrano M, Leppers R, Zubillaga O. Use of microencapsulated PCM in concrete walls for energy savings. Energy and Buildings. 2007;**39**:113-119

[87] Giro-Paloma J, Al-Shannaq R, Fernández AI, Farid MM. Preparation and characterization of microencapsulated phase change materials for use in building applications. Materials. 2016;**9**:11

[88] Su W, Darkwa J, Kokogiannakis G. Nanosilicon dioxide hydrosol as surfactant for preparation of microencapsulated phase change materials for thermal energy storage in buildings. International Journal of Low-Carbon Technologies. 2018;**13**:301-310

[89] Essid N, Loulizi A, Neji J. Compressive strength and hygric properties of concretes incorporating microencapsulated phase change material. Construction and Building Materials. 2019;**222**:254-262

[90] Schossig P, Henning H-M, Gschwander S, Haussmann T. Micro-encapsulated phase-change materials integrated into construction materials. Solar Energy Materials and Solar Cells. 2005;**89**:297-306

[91] Borreguero AM, Rodríguez JF, Valverde JL, Peijs T, Carmona M. Characterization of rigid polyurethane foams containing microencapsulted phase change materials: Microcapsules type effect. Journal of Applied Polymer Science. 2013;**128**:582-590

[92] Li W, Hou R, Wan H, Liu P, He G, Qin F. A new strategy for enhanced latent heat energy storage with microencapsulated phase change material saturated in metal foam. Solar Energy Materials and Solar Cells. 2017;**171**:197-204

[93] Bonadies I, Izzo Renzi A, Cocca M, Avella M, Carfagna C, Persico P. Heat storage and dimensional stability of poly (vinyl alcohol) based foams containing microencapsulated phase change materials. Industrial & Engineering Chemistry Research. 2015;**54**:9342-9350

[94] Zhang X, Wang X, Wu D. Design and synthesis of multifunctional microencapsulated phase change materials with silver/silica double-layered shell for thermal energy storage, electrical conduction and antimicrobial effectiveness. Energy. 2016;**111**:498-512

[95] Zhang Y, Li X, Li J, Ma C, Guo L, Meng X. Solar-driven phase change microencapsulation with efficient Ti4O7 nanoconverter for latent heat storage. Nano Energy. 2018;**53**:579-586

[96] Zheng Z, Jin J, Xu G-K, Zou J, Wais U, Beckett A, et al. Highly stable and conductive microcapsules for enhancement of joule heating performance. ACS Nano. 2016;**10**:4695-4703